Language, from a Biological Point of View

Language, from a Biological Point of View: Current Issues in Biolinguistics

Edited by

Cedric Boeckx, María del Carmen Horno-Chéliz and José-Luis Mendívil-Giró

CAMBRIDGE SCHOLARS PUBLISHING

Language, from a Biological Point of View:
Current Issues in Biolinguistics,
Edited by Cedric Boeckx, María del Carmen Horno-Chéliz and José-Luis Mendívil-Giró

This book first published 2012

Cambridge Scholars Publishing

12 Back Chapman Street, Newcastle upon Tyne, NE6 2XX, UK

British Library Cataloguing in Publication Data
A catalogue record for this book is available from the British Library

ISBN (10): 1-4438-3781-4, ISBN (13): 978-1-4438-3781-1

TABLE OF CONTENTS

ACKNOWLEDGMENTS

This book originated as a series of lectures organized in 2009 by M.C. Horno-Chéliz and J.L. Mendívil-Giró, as part of the summer course offerings of the University of Zaragoza in its summer headquarters in Jaca.

The event, *Language, brain and genes: lectures in Biolinguistics*, brought together several authors of this volume (Benítez-Burraco, Boeckx, Cuetos, Laka, Lorenzo), as well as others, such as Andrea Moro, who unfortunately could not contribute to this volume. The success of the event among students led us to start thinking of a volume covering the topics of the lectures, as well as addressing issues for which the intensive, 3-day summer course had not left any room. The editors of the present volume invited Bridget Samuels to turn her poster presentation in Jaca into a chapter, and welcomed V. Longa, C. Baus, and M. Carreiras to the project.

The editors also asked S. Balari, T. Martín and J. Rosselló, founding members of the Biolinguistics Group in Barcelona, to write introductions for two of the parts of the book. The editors would like to thank all the authors for their collaboration, and their patience during the editing process.

Completion of this project was made possible by funds from the following sources, which are gratefully acknowledged: Marie Curie International Reintegration Grant from the European Union (PIRG-GA-2009-256413; recipient: CB), research funds from the Fundació Bosch i Gimpera, from the Gobierno de Aragón to the Sylex group (Universidad de Zaragoza), as well as grants from the Spanish Mininistry of Science and Innovation (FFI-2010-20634; PI: Boeckx), and from the Generalitat de Catalunya.

INTRODUCTION

INTEGRATING LINGUISTICS AND BIOLOGY

CEDRIC BOECKX,
MARÍA DEL CARMEN HORNO-CHÉLIZ
AND JOSÉ-LUIS MENDÍVIL-GIRÓ

The present volume offers a collection of essays covering a broad range of areas where currently a *rapprochement* between linguistics and biology is actively being sought. Following a certain tradition we call this attempt at a synthesis *biolinguistics*.

The goal of this introductory chapter is to examine, all too briefly, the nature of biolinguistics, a term that is encountered with increased frequency in linguistic, and more generally, cognitive science circles these days, and to offer an overview of the chapters to come.

The term *biolinguistics* is not new (its first appearance dates back to 1950), but recently, it has figured prominently in the titles of articles, books, book series, journals, courses, conferences, symposia, grant proposals, research interests on CVs, and research groups. In the following pages, we want to focus on two issues: (i) why this resurgent interest in biolinguistics? and (ii) the opportunities and challenges that this implies for linguists, which are treated in more detail in the chapters that follow.

1. Why Now?: Factors that led to the return of biolinguistic concerns

One can rarely, if ever, predict the course of events, in science or elsewhere, but with the benefit of hindsight, one can recognize several factors that contributed to the renaissance of biolinguistics. We say renaissance because there was a brief period during the 1970s where the term enjoyed a certain popularity. As mentioned above, the term *biolinguistics* appeared for the first time in a 1950 publication (Meader and Muysken 1950), but seemingly attracted no interest,[1] and was recoined by Massimo Piattelli-Palmarini (displacing its early competitor "bioanthropology") during events that led to the famous Chomsky-Piaget debate at Royaumont (Piattelli-Palmarini 1980). Alongside Piattelli-Palmarini, Salvador Luria was using the term in talks (advertising its promises), Lyle Jenkins was trying to launch a journal, and members of the MIT scientific community had formed a research group on biolinguistics (see Walker 1978).[2] All these activities were building on the research program initiated by Noam Chomsky, with its rejection of behaviorism and its embracement of ethology (Chomsky 1959), and the specific proposals made by Eric Lenneberg concerning the biological foundations of language (Lenneberg 1967). The climate at the time, well reflected at Royaumont, was supremely interdisciplinary.

All of this is strongly reminiscent of what is happening now in the context of "biolinguistic" activities.[3] The two periods are not identical, of course, but they share the same research focus, which is well captured in terms of Tinbergen's (1963) foundational questions for ethology, adapted to language (cf. Chomsky 1986, 1988):

I. What constitutes linguistic competence in humans?
II. How does this competence develop in the individual?
III. How is this competence put to use?
IV. How is this competence implemented in brain structures?
V. How did this competence evolve in the human species?

These are still the questions that dominate the current biolinguistic research panorama, but current biolinguistics benefits from recent advances in various areas of linguistics and biology that promise to overcome obstacles that may have proven too big to overcome the first time around (in the 1970s), leading to its temporary eclipse.

Let us, all too briefly, sketch some of these advances.

---

[1] Lenneberg mentions the term biolinguistics, and refers to this 1950 in the preface to his 1967 book, but does not use the term, given that what it referred to in 1950 was not the object of study Lenneberg (or Chomsky) had in mind.

[2] For quotes and other supporting documents, see Boeckx and Grohmann (2007), the introductory chapter in Di Sciullo and Boeckx (2011), and Jenkins (2000).

[3] Not by chance, many see aspects of the original Royaumont meeting being replayed at recent meetings; see Hornstein and Drummond's (to appear) review on the San Sebastián meeting (Piattelli-Palmarini, Uriagereka, and Salaburu 2009).

In mediatic terms, perhaps the major advance in biolinguistics has been the discovery of the *FOXP2* gene and, subsequently, its interactome.[4] It is clear to everyone that *FOXP2* is not "the" language gene, but it is also clear that *FOXP2* and the genes it interacts with provide a concrete example of the long-anticipated genetic basis of language (cf. Lenneberg 1967).[5] It thus allows for a connection with the intense genomic research line (the "omics") in biology. Moreover, the fact that *FOXP2* appears to be so well-conserved a gene allows for experiments with other species (mice, birds, bats, etc.)[6] that could not be possible with humans, if only for ethical reasons. Research on the bird variant of *FOXP2* (*foxp2*) has led to a renewed appreciation of the many parallelisms between human speech (the externalization aspect of human language) and birdsong at various levels (genetic, neuronal, functional), parallelisms that many now believe will vindicate the productivity of Darwin's remarks concerning human speech and birdsong in *The Descent of Man* (1871). In addition, the *FOXP2* discovery promises to shed light on the nature of various linguistic disorders and deficits, which for linguists and biologists alike, have always been phenomena of choice to shed light on the nature of normal biological functions (cf. the logic of monsters, as Alberch 1989 called it). It is to be hoped that such improved understanding will not only advance basic research but also lead to improved therapeutic strategies in a medical context.

Equally central to the reemergence of biolinguistic concerns has been the shift of perspective in comparative psychology, extremely well captured in the following passage from de Waal and Ferrari (2010):

> Over the last few decades, comparative cognitive research has focused on the pinnacles of mental evolution, asking all-or-nothing questions such as

---

[4] Fisher *et al.* (1998), Lai *et al.* (2001), MacDermot *et al.* (2005), Vernes *et al.* (2007, 2008), Spiteri *et al.* (2007), and Konopka *et al.* (2009). For a comprehensive treatment, see Benítez-Burraco (2009), and for a linguistics-oriented overview, see Piattelli-Palmarini and Uriagereka (2011); for other overviews, see Fisher, Lai, and Monaco (2003), Marcus and Fisher (2003), Vargha-Khadem *et al.* (2005), Fisher (2006), Ramus (2006), and Fisher and Scharff (2009).

[5] On Lenneberg's position concerning this genetic basis, see Boeckx and Longa in press.

[6] Enard *et al.* (2009), Shu *et al.* (2005), Haesler *et al.* (2004, 2007), Teramitsu *et al.* (2004), Webb and Zhang (2005), Scharff and Haesler (2009), Teramitsu and White (2006), Li *et al.* (2007), and Bolhuis, Okanoya and Scharff (2010). For studies concentrating on FOXP2 in the *homo* lineage, see Krause *et al.* (2007). On the evolution of FOXP2, see Enard *et al.* (2002), and Berwick (2011).

> which animals (if any) possess a theory of mind, culture, linguistic abilities, future planning, and so on. Research programs adopting this top-down perspective have often pitted one taxon against another, resulting in sharp dividing lines. Insight into the underlying mechanisms has lagged behind. A dramatic change in focus now seems to be under way, however, with increased appreciation that the basic building blocks of cognition might be shared across a wide range of species. We argue that this bottom-up perspective, which focuses on the constituent capacities underlying larger cognitive phenomena, is more in line with both neuroscience and evolutionary biology.

It is of great significance for current biolinguistics that an influential article advocating this shift of perspective, this new call for meaningful, descent-oriented, properly Darwinian comparative psychology, was co-authored by Noam Chomsky. We are here referring to Hauser, Chomsky, and Fitch (2002), where it is proposed that investigations into the nature of human language may prove more fruitful if one distinguishes between a faculty of the language in the narrow sense (FLN), that which is unique to language, and a faculty of language in the broad sense (FLB), that which is not. Although much attention was devoted to what Hauser, Chomsky and Fitch suggest may be in FLN—suggesting a residue of the old top-down approach discussed by de Waal and Ferrari,[7] it is perhaps the research on FLB that has so far proven far more productive.[8]

Also of great importance has been the emergence of (calls for) a new, expanded synthesis in biology (Pigliucci and Mueller 2010; Gould 2002), one that emphasizes the severe limitations of naive adaptationism and genocentrism (Gould and Lewontin 1978), and the need for a more pluralist, internalist, structuralist, generative (Webster and Goodwin 1996) biology.[9] This internalism resonates strongly with the anti-behaviorist,

---

[7] Witness the debate between Pinker and Jackendoff (2005), Fitch, Hauser and Chomsky (2005), and Jackendoff and Pinker (2005); and the attention to the negative results of Fitch and Hauser (2004) regarding the inability of certain primates to learn recursive structures, and to the positive results of Gentner *et al.* 2006 regarding the ability of certain birds to do so. For sobering notes regarding these results see Liberman (2006), Hochmann, Azadpour and Mehler (2008), and van Heijningen *et al.* (2009).

[8] Endress *et al.* (2007, 2009, 2010), Endress and Hauser (2010, 2011), Endress and Mehler (2009, 2010), Endress, Nespor and Mehler (2009), Gervain and Mehler (2010), Berwick, Okanoya, Beckers and Bolhuis (2011), Abe and Watanabe (2011), and Bloomfield, Gentner and Margoliash (2011).

[9] The term "Evo-devo" is perhaps the most popular term to characterize this new synthesis in biology, but I prefer the term "expanded thesis", as the best-known

innatist stance in Chomskyan linguistics (see Boeckx 2011c, Fodor and Piattelli-Palmarini 2010), and at once allows linguists to move beyond the meaningless "Chomsky vs. Darwin" discussions that plagued biolinguistics when evolutionary psychology advertised itself as the only road towards biocognition (see Dennett 1995, Pinker 1994, Pinker and Bloom 1990).

A fourth factor that facilitated the return of biolinguistics is the recognition of a central problem of interdisciplinarity in the context of language, particularly salient in the context of neurolinguistics (linking mind and brain). The problem has been most clearly articulated by David Poeppel (Poeppel 2005; see also Poeppel and Embick 2005), who notes that quite apart from the seemingly inherent locationist/phrenologist tendencies in cognitive neuroscience, interactions between neuroscientists and linguists have led to impasses (Poeppel calls it "interdisciplinary cross-sterilization") because of a granularity mismatch: currently, there is a lack of a common level of representation at which the two disciplines investigate processes and the fundamental elements used. This consequently prevents the formulation of theoretically motivated, biologically grounded and computationally explicit descriptions of language processes in the brain.

To better relate to neuroscience, Poeppel says that

> Linguists and psycholinguists owe a decomposition (or fractionation) of the particular linguistic domain in question (e.g. syntax) into formal operations that are, ideally [from the perspective of neuroscience], elemental and generic. The types of computations one might entertain, for example, include concatenation, comparison, or recursion. Generic formal operations at this level of abstraction can form the basis for more complex linguistic representation and computation.

Fitch reinforces this point when he says in his "Prolegomena to a science of biolinguistics" (2009b) that

> [w]e need to distill what we know from linguistic theory into a set of computational primitives, and try to link them with models and specific principles of neural computation.

Thus we need linguistic models that are explicit about the computational primitives (structures and operations) they require, and that attempt to define linguistic problems at a fine enough grain that one can discuss

---

strand of evo-devo remains strongly geno-centric (see, e.g., Carroll 2005). For an excellent discussion of the many meanings of evo-devo, and what this implies for linguistics, see Benítez-Burraco and Longa (2010).

algorithmic and implementational approaches to their solution. We need a list of computations that linguistic theorists deem indispensable to solve their particular problem (e.g., in phonology, syntax, or semantics).

We say that the recognition of this central interdisciplinary problem is a key factor in the return of biolinguistics because, as is well known, recognizing the problem and articulating it in detail is often half its solution.

In addition, the idea that "the linking hypotheses between language and brain are most likely to bear fruit if they make use of computational analyses that appeal to generic computational subroutines" (D. Poeppel) directly relates to the fifth, and most specifically linguistics-internal, factor that led to the renaissance of biolinguistics: the formulation of a minimalist program in theoretical linguistics.

This is not the place to provide a comprehensive overview of the minimalist program for linguistic theory first formulated in Chomsky 1993.[10] Suffice it to say that it consists in approaching the content of Universal Grammar (the grammatical basis for the human language capacity that is hypothesized to be part of our biological endowment) from below. In the words of Chomsky (2007):

> At the time [pretty much throughout the history of generative grammar], it seemed that FL must be rich, highly structured, and substantially unique. [...] Throughout the modern history of generative grammar, the problem of determining the character of FL has been approached "from top down": How much must be attributed to UG to account for language acquisition? The M[inimalist] P[rogram] seeks to approach the problem "from bottom up": How little can be attributed to UG while still accounting for the variety of I-languages attained.

As several linguists have pointed out (see Boeckx 2006: chap 4, Hornstein 2009, among others),[11] it is clear that, although independently formulated, the minimalist program, rigorously pursued, addresses Poeppel's concerns regarding the granularity mismatch (a granularity mismatch which, by the way, need not stop at the neuronal level, as it also applies in

---

[10] See Boeckx (2006, 2010a, 2011b) for overviews.

[11] Jackendoff (2011) takes issue with the minimalist program in the context of biolinguistics, but we think that his conception of what minimalism is is much narrower than what the literature warrants. Moreover, a careful reading of his paper (for which we don't have room in this introduction, but see Boeckx 2011c) suggests that his vision of biolinguistics is much closer to that of proponents of minimalism than he acknowledges.

full force at the level of genetics, and the relation between genes and cognitive functions). It also resonates strongly with themes like optimization, specificity, laws of form that are at the heart of the new biology (see the material and references of part I of Fodor and Piattelli-Palmarini 2010, especially chapter 5); like the new comparative psychology, it adopts a bottom up approach to cognitive faculties, welcoming the evidence of powerful computational resources in other species/domains of cognition, since it allows for a more deflated characterization of Universal Grammar, and more plausible descent scenarios (Longa, Lorenzo and Uriagereka 2011, Chomsky 2007, 2010b). Furthermore, one of the running themes in minimalism in recent years is that there is a basic asymmetry between the way syntax subserves the sound and the meaning components it interfaces with (Chomsky 2007, 2008; Berwick and Chomsky 2011), being more optimized for the latter. This fits rather well with the research in birdsong that indicates that the basic apparatus for externalization (speech/sound) can be found in the absence of meaning (Berwick, Okanoya, Beckers, and Bolhuis 2011; Samuels 2009a, 2011; Samuels, Hauser and Boeckx, To appear; Bloomfield, Gentner and Margoliash 2011).

As Eric Reuland (2011) put it, the linguistic principles formulated before the advent of the minimalist program were too good to be false (they accounted for a surprisingly wide range of data, and formed the basis of the success of modern theoretical linguistics), they were nevertheless too "language" specific to be true: this is the granularity mismatch problem again. The pre-minimalist linguistic principles, not being decomposed into their elementary components, looked like nothing else in cognition and biology: as such (i.e., undecomposed), they were not biologically plausible (only the existence of some biological components responsible for language development was). The highly-specific, strongly modular nature of these principles required an evolutionary miracle for their emergence, which is why no alternative to the adaptationist scenario could reasonably be offered until the advent of minimalism (witness Piattelli-Palmarini 1989).

In sum, the minimalist program formulated within the confines of linguistic theory converges with the other forces that favored the return of biolinguistic concerns. Minimalist ideas, if pursued rigorously, can help put an end to the still dominant isolationism in theoretical linguistics, where the emphasis is on how language is distinct from other cognitive domains, and in so doing, minimalism promises to stop the "alienation" of linguistics from other branches of cognitive (and biological) sciences diagnosed by many at the turn of the century (see, e.g., Jackendoff 2002).

But the key (or the catch) here is in the phrase "(minimalism,) *if rigorously pursued*". It should be clear that linguists have a central role to play in biolinguistics: they are the ones to provide the elements that researchers from other fields must look for at the neural and genetic levels, and whose evolutionary origins must be traced back. But this is only feasible to the extent that linguists are willing to engage in this interdisciplinary dialogue; that is, it's only feasibly if linguists are willing to become genuine biolinguists. What the opportunities and challenges of this change are is the topic of this book.

2. A wonderful window of opportunity, *if linguists are really willing to take advantage of it*

The linguists who read thus far are likely to point out that biological concerns have been present in theoretical linguistic studies for over 50 years, and that, as far as they are concerned, they have always been biolinguists. After all, already back in the 1970s, Chomsky was saying that linguistics is biology at a suitable level of abstraction.

All of this is true, but the emphasis in theoretical linguistics has all too often been (and continues to be) on the modification "at a suitable level of abstraction". The safe distance of abstraction, we fear, has become a license to posit entities and processes whose biological (neural, genetic, etc.) foundations are, at best, unclear. When this is pointed out to them, linguists often say that we know so little about the brain, about how genes relate to cognition, that it is best not to speculate. Again, this is not incorrect; we indeed have a lot more to learn about mind and brain, genes and cognition, but the fact that we know little in these areas does not mean we know nothing, and it does not mean that we should not engage with the relevant literature in adjacent fields. Too often theoretical linguists adopt an imperialist attitude, assuming that it is the task of others to find biological correlates of theoretical entities, not questioning whether these entities have the right "format" suitable for biological integration.

Put differently, linguists all too often ignore Marr's 1982 vision that a truly productive cognitive science must relate description at the computational level to description at the levels of mechanisms (what Marr calls the algorithmic level) and at the level of brain implementation (what Marr calls the implementational level).

It is with this stance in mind that we decided to assemble this volume. We asked (mostly, theoretically-trained, mostly 'Chomskyan') linguists to show how a genuine engagement with the existing literature in biology could redirect linguistic inquiry.

The volume is organized into three parts (I. *Language and Cognition*, II. *Language and the Brain*, III. *Language and the Species*), each with their own introduction.

The first part (*Language and Cognition*) includes three chapters that deal with issues traditionally associated with linguistic theory, such as the architecture of the language faculty (*The I-language Mosaic*, by C Boeckx), the process of language development and acquisition (*Theoretical linguistics meets development*, by V. Longa and G. Lorenzo), and the problem of the unity and diversity of languages and the nature of linguistic change (*The Myth of Language diversity*, by J.L. Mendívil-Giró).

The second part (*Language and the Brain*) consists of three chapters that introduce the reader to what is known and is currently being investigated in the area of brain processes relating to language. The first one of these chapters (*The role of aphasic disorders in the study of the brain-language relationship*, by F. Cuetos) presents the analysis of different aphasias as a way of addressing the study of language-brain relations; the second one (*Word Processing* by C. Baus and M. Carreiras) introduces the study of language use in real time, focusing on comprehension and production of words, both in oral languages as in sign languages. The third chapter (*More than one language in the brain,* by I. Laka), focuses on multilingualism and its brain basis.

The third part of the book (*Language and the Species*) addresses issues ranging from the molecular and genetic basis of language (*The "Language Genes"* by A. Benitez-Burraco), to the problem of unravelling the evolution of the human faculty of language (*The evolution of the Faculty of Language*, by G. Lorenzo), and to the comparison of the capabilities required for human language with those found in other species (*Animal minds and the roots of Human language*, by B. Samuels).

Taken together, these nine chapters offer a comprehensive overview of issues at the forefront of biolinguistic research. Each contribution highlights exciting prospects for the field, but they also point to significant obstacles along the way.

Perhaps the biggest challenge for linguists will be that of open-mindedness. The field of language acquisition, and the field of biology at large, offers strong reasons to adopt a stance of theoretical pluralism. Instead of viewing different theoretical proposals as competing with one another (the traditional stance in linguistics, where Chomskyan and non-Chomskyan, formalist and functionalist approaches are often contrasted), it is becoming increasingly clear that ingredients from each one of these theoretical perspectives will help shed light on the complexity of the human language faculty. Adopting a pluralist stance should not only enable us to make faster progress, it would also give the field a more united look and feel, a welcome feature at the time of recruiting students. But this, of course, will have to come at a price: each one of us will have to recognize that we've been wrong. Much like there is likely to be some truth to each one of our theoretical proposals, there is also likely to be quite a few things that are wrong, not in detail, but in terms of the big picture. The age of theoretical exclusivity (necessarily leading to unfortunate exclusion), much like the age of theoretical specificity (leading to isolation), will have to end if interdisciplinarity is to reign and if biolinguistics is to flourish.

For students and teachers alike, the prospects of interdisciplinarity will require a shift of emphasis with respect to the material being covered. Instead of focusing on the descriptive tools that today constitute the main (exclusive?) focus of courses, more fundamental and foundational aspects of our field will have to be taught (aspects that too many students and teachers today only master imperfectly). For example, Gallistel and King (2009) make a strong case for the relevance of foundational results of the earliest days of generative grammar and mathematical linguistics in guiding computational neuroscience (Fitch 2010b makes a similar point in the context of comparative cognition; witness O'Donnell, Hauser, and Fitch 2005, and Berwick, Okanoya, Beckers and Bolhuis 2011). Paradoxially, such foundational results are only covered in detail in one introductory syntax textbook (Lasnik 2000), and (admittedly only briefly) in one general text about linguistics as cognitive science (Boeckx 2009b).

Students in the language sciences should also expect their next jobs to be outside of the humanities, if biolinguistics is to become productive. This gradual displacement already took place with psychology, and there is no reason to think that linguistics will be an exception. True, this will require other departments to open their door to linguists, but we are confident that if linguists succeed in demonstrating the relevance of their results to experts in adjacent fields, this should be possible, particularly as deans lay emphasis on interdisciplinary enterprises. Importantly, linguists should expect to work on things that prima facie look nothing like human language, such as bird song or bat echolocation, ant navigation or chimpanzee tool use, baboon social structure or vervet monkey call system, but this fact was already predicted by Darwin when he wrote at the end of *Origin of Species* (Darwin 1859) that if he was right about evolution (and who now doubts this these days?), psychology would be based on a new foundation. Following Darwin, biologists have come to learn about essential components of higher-order organisms like humans by experimenting with model organisms that seemingly display none of these essential components.[12] Once decomposed, though, these apparently irreducibly complex, atomic, and unique properties yield to the logic of descent with modification and tinkering.

[12] Consider the fact that it has become standard to explore mental retardation in children by studying specific mutants in drosophila, to look for the roots of autism in rats, and so on (see the bottom-up research program outlined in Manolio *et al.* (2009).

PART I

LANGUAGE AND COGNITION

INTRODUCTION TO PART I

FOR AN APPROACH TO LANGUAGE WITH BIOCOGNITIVE IMPORT: THE DISMISSAL OF SPECIFICITY AND OTHER ISSUES

JOANA ROSSELLÓ AND TXUSS MARTÍN

Under current circumstances, any field of research can easily be split up into two kinds of people and/or activities: those that are representative of the normal practice and those that are one step beyond, usually concerned with possibly unsettled questions and potentially conflicting conceptual issues in the domain of study. Sometimes when clear progress is made in the more conceptual side, conflicts arise with the everyday endeavors in the field. When such thing occurs, not a bad thing in itself, it is normally time to rationally assess what the appropriate next step is and then act accordingly. Otherwise, the risk of enlarging the gap between both perspectives can jeopardize the whole domain.

In the field of Chomskian generative linguistics (CGL), we are now likely in such a crossroad. Time is ripe to dedicate efforts to study the origin and evolution of language, but a plausible approach in this area is at odds with what is still proposed for language acquisition. Conflicts also come up when comparing the standard apparatus used to explain the content of an I-language and what is seen as feasible at the level of its neural implementation. Thus, it seems to us that it is more than appropriate to confront the challenge. CGL will keep being considered relevant within the multidisciplinary ally that focus on the human mind only if it addresses the faculty of language in a sound biological sense. That means that we have to enhance the biocognitive import in our approach to language even in normal practice as theoretical linguists. Language, the most distinctly human hallmark in our otherwise almost unremarkable primate mind, will then reveal its secrets to us.

The three following chapters show us that there are linguists with proposals that aim at bridging the gap between common practice, remotely imbued with biology at most–biolinguistics in the weak sense–, and an elaborated and strongly biological view coming from sources like evolutionary developmental biology (evo-devo) which are now penetrating the otherwise also stagnant standard genocentric biology–biolinguistics in the strong sense (Boeckx & Grohmann 2007).

Among the topics running through the three chapters in this section, the issue of specificity stands out because of its being in the middle of the current crossroad. Wherever you look at, the question arises on whether it is or not correct to assume that X is a specifically linguistic property. And the pronouncements lead us to affirm that the lack of specificity is in order in all those aspects with a prominent biological dimension. Put it otherwise, to find linguistic specificity you have to recluse in history, in the sociocultural dimension of language, much wider than usually acknowledged (see Koster 2009). Another possibility, but not a very interesting one, is to attribute specificity not to basic properties but to the global outcome of some basic not specifically linguistic traits. In a sense, emergence will be at stake in such a case. But before dealing with emergent properties, let us see how specificity lost terrain, with some detail.

As Boeckx reminds us in his chapter, minimalism is since its birth an invitation for the generative enterprise to go in search of minimal design specifications for language. It also invites to go "beyond explanatory adequacy" (Chomsky 2001a) and recommends that Universal Grammar (UG) be "approached from below", i.e. minimally (Chomsky 2007). That being so, we could confidently conclude that minimalism goes beyond Occam's razor as a methodological precept: the minimalist program is not only that, which could hardly count as a novelty in CGL, but rather a strong hint in favor of a minimal design for language in the narrow sense (Hauser, Chomsky & Fitch 2002) and, therefore, a strong hint in favor of unspecific mechanisms in general.

Boeckx suggests that this overall change happened when in the nineties "Chomsky got the impression that the overall approach was well-established and that it was time to take the next step". For us, this sounds too optimistic. We think that at the root of the change there were the conflicting views emerging from the common practice in the different arenas (investigation of steady, normal I-languages, and their acquisition and abnormal deficits) and the combination of a rigorous conceptual analysis with a serious consideration of the evolutionary dimension of language. How can we otherwise understand the incompatibility between the high degree of specificity that is posited for language acquisition by the majority of its practitioners and what is required for a plausible proposal in the evolutionary side, namely an as much unspecific apparatus as possible?

If the allegedly Chomskian optimism was true, the field of language acquisition wouldn't need to be rethought as we, with Longa and Lorenzo (L&L), think it does. Moreover, it is not only because of the evolutionary demands that the innate contribution to the steady linguistic state of knowledge (I-language) has to be recast. As L&L show in chapter 2, it's also because of neural implementation demands that the necessity of an integral redesign of language acquisition becomes indisputable. In other words, as we see it, the responsibility for the changes we are witnessing lies in the pressure brought to bear on explanatory adequacy (language acquisition) not only from evolutionary adequacy, since especially the programmatic paper by Hauser, Chomsky and Fitch (2002), but also from neural adequacy—if we can use this, to our knowledge, new term. Equivalently, one could say that the imperative approach to Darwin and Broca's problems have paved the way for a new approach to the older Plato's problem.

A strong biolinguistic view of language requires, indeed, a convergent solution to the different problems at stake. What if in a given moment the solutions are divergent? How to know the right direction? For concreteness, couldn't it be that the lack of specificity was the wrong hint in the case of language? Our answer is negative for empirical, methodological and conceptual reasons. Let us illustrate each of these three sides.

On the empirical facet, recent findings in paleogenetics–think of *FOXP2*, for instance, and see footnote 4 in Boeckx's chapter–and the archeological record suggest that the language faculty is only 200,000 years old at most. Language is, then, a very recent achievement in evolutionary terms. This circumstance makes implausible any attempt to naturalize all the complexities that, for this reason, could not have philogenetically evolved, and should rather be the outcome of cultural processes.

The methodological reasons that favor unspecific rather than specific mechanisms have to do with the "granularity mismatch" (Poeppel and Embick 2005) easily at hand in neurolinguistics (Broca's problem). Independently of the ultimate nature of the linguistic operations in the brain, it is clear that the level of abstractness and specificity at which the principles of grammar are currently formulated makes them unmanageable at the brain level. The interaction, then, becomes, in Poeppel's words, "cross-sterilization". To avoid this annoying situation, in fact the current one, it would be better to recast linguistic formulations in more basic and unspecific terms. That would not preclude that at the same time one had to keep installed at a more abstract level in order to meaningfully capture the linguistic workings of the system.

Finally, there is a point to be made on a rather conceptual issue. It is, simply, that it is easier to add than subtract. Take for instance Universal Grammar (UG) under the orthodox view of the Principles and Parameters framework. What could we subtract from this artifact? Of course, any part you took away from such an intertwined and specific top-down system would take it to pieces. All in all, the only possible real alternative would be starting it anew from scratch, building a UG from below. According to this strategy, we could even expect there to be no mechanism specific to the faculty of language and, then, no UG. However, Chomsky, the very proponent of it, would insist on maintaining UG even in the case that recursive Merge, which is present in other systems (arithmetic, music, etc,), was its only ingredient. One way to obviate such an apparent blatant inconsistency would be to vindicate the linguistic primacy of Merge, which would appear only derivatively in arithmetic, music and the like. This is, in fact, Mukherji's (2010) approach. However, Chomsky would posit UG even in the case of a truly unspecific Merge by saying that "there still must be a genetic instruction to use Merge to form structured linguistic expressions satisfying the interface conditions" (Chomsky 2007: 7). This doesn't seem convincing to L&L, who overtly criticize this move, but it could be sustained by Mendívil-Giró who in his chapter assumes a more lax view which equates UG with the initial state of the language faculty. Be that as it may, with or without (a minimal) UG, the demise of specificity is nowadays beyond dispute.

All this, of course, doesn't invalidate either the tautologic claim that he faculty of language as a whole (FLB, in the sense of Hauser, Chomsky & Fitch 2002) is a specifically linguistic outcome or the obvious statement that the faculty of language is specifically human in the sense that it is unique to our species. As said before, this is almost uninformative by itself. However it suffices to combine the first claim with our previous stand in favor of unspecific properties to see that what ensues from it is the conception of the faculty of language as an emergent property and this is, by contrast, a very interesting issue.

This concept of emergent property is one of the fundamental blocks of the multidisciplinary research paradigm known under the umbrella term of Complex Systems Theory. This research paradigm, inspired by figures like D'Arcy Thompson or Alan Turing, has produced some of the most exciting ideas of current science with people like Ilya Prigogine, Brian Goodwin, Stuart Kauffman, or Ricard Solé, among many others. In many of these works, the center of the explanatory stage is occupied by concepts such as emergence, chaos, self-organization, and dynamics. The main issue there explores the conditions under which novel and coherent structures, patterns and properties emerge during the process of self-organization in complex systems out of their parts, parts where you don't find those properties. To put it in the words of Kauffman quoted in Mendívil-Giró's chapter, life is not to be found as the property of any single molecule of a living being, but rather as the collective emergent property of the whole those molecules create: "The collective system is alive. Its parts are just chemicals". A more systematic definition of this concept is offered by Jeffrey Goldstein in the inaugural issue of the journal Emergence (Goldstein 1999). This author considers the minimal common characteristics of emergent properties to be the following: (1) radical novelty (features not previously observed in the system); (2) coherence or correlation, meaning integrated wholes that maintain themselves over some period of time; (3) a global or macro "level", i.e. there is some property of "wholeness"; (4) it is the product of a dynamical process, that it to say it evolves; and (5) it is "ostensive"—it can be perceived (Goldstein 1999, apud Corning 2002).

If we stick to Goldstein's proposal, language can be considered an emergent property, as it fits with each one of the five properties Goldstein 1999 proposes for this kind of system: (1) radical novelty: none of the component parts of FL in a broad sense (FLB, in the terminology of Hauser, Chomsky, & Fitch 2002) is in itself linguistic; (2) coherence or correlation: language can be seen as a integrated system, something that sometimes turns out to be equivocal, as language should be seen as a complex system formed by many different parts; this is indeed the rationale for (3) seeing language as a global or macro "level" where we find some property of "wholeness"; (4) language is also the product of a dynamical process, that it to say it evolved in the Homo genus and appeared, probably abruptly 200,000 years ago with the emergence of Homo sapiens; and finally (5) it is an "ostensive" system that can be perceived.

This view is fully coherent with the one advocated for by L&L in their chapter, where they want to reduce the role of genocentrism, the pre-specified endowment, in favor of epigenetism, where from a rather underspecified initial state, a cascade of epigenetic biophysical phenomena make language appear in our species. Such a process of emergence, the authors claim (see also Lorenzo & Longa 2003), should be the same at both the ontogenetic and the phylogenetic level. That is, FL emerged in the species for the same reason it emerges once and again in every normal human individual: as an effect of the interaction of a minimum linguistic initial state of the system with two independently evolved performance systems, via general laws of biophysical efficiency (epigenesis).

In any case, and contrary to what Mendívil-Giró calls the functional-cognitivist paradigm (FCP), that is the approaches to language that attribute the properties of language to its communicative and social functions, CGL remains committed to an internal and biological view of language, to the primarily individual and internal nature of the human faculty of language. In this perspective, language is a natural, rather than a cultural object. That is to say, in CGL, as opposed to the FCP, the object of study is not external languages, English, or Swahili, the scientific status of which is at least fuzzy, The object rather are the restrictions on possible human languages imposed by the computational system of language. That is to say, CGL focuses, on the internal computational system, the FL, which is common for the whole of the species. That emphasis on what all human languages share is opposed to the endeavors of the FCP, which focuses on how different particular languages are.

Summary of Part I Chapters

The I-Language mosaic (Cedric Boeckx).

Boeckx presents a programmatic contribution to the biolinguistic enterprise understood in its strong sense. In order to measure it against the still predominant view in the generative enterprise, which is of weaker biologic concern, he first summarizes Lasnik's authorized state of the art, 'Grammars, Levels and Biology' representative of the standard practice. Next he substantiates the drastic change of perspective and attributes a great deal of responsibility in it to Chomsky's ideas as expressed in his last linguistic essays. Following that, Boeckx goes on to give a specific articulation of the faculty of language that, in contrast with that advocated in Lasnik's essay, can make true sense biologically, namely ontogenetically and phylogenetically.

With this goal in mind, he effectuates a decomposition—hence "mosaic" in the nice title—where the relevant "levels" cease to be determined by cross-linguistic differences but by more psychologically sound considerations instead. Thus, syntax (narrow syntax), seen as the only purely biological level of language where recursion lies, consists of two alternating and, hence, symmetry breaking operations: one, "densely recursive", is merge (external and internal), and the other, "sparsely recursive", is a process of partial transfer. Syntax, reduced in this way, becomes the only level of the faculty of language that is completely language-free. This reduction paves the way for an easy enough (antiadaptive) evolutionary scenario in which syntax is the biophysical outcome of a self-organization process once a neural complexity threshold was attained. To complete the very inside of an I-language, two more levels have to be added, a (post-syntactic) lexicon and a grammar, which stabilizes the syntactic machine through a grammaticalization process encoding external properties and developing in a great part inductively. Finally, with respect to the syntax-meaning interface, Boeckx, in line with Pietroski, assumes a linguistic semantics relatively empoverished and embraces the view that syntax and semantics are isomorphic. Regarding the syntax-sound (or gesture) interface instead, a much more kludged adjustment is posited.

Implicit in the previous account of Boeckx's proposal is the demise of specificity and the related mandate to approach language from below or, in other words, to proceed according to truly minimalist premisses.

Theoretical linguistics meets development. Explaining FL from an epigeneticist point of view (Víctor M. Longa and Guillermo Lorenzo).

Longa and Lorenzo argue that genocentrism is preformationist and preformationism is wrong because it neglects the role of development. After a thoroughly criticism of rampant genocentrism and the ensuing defense of epigenetism, Longa and Lorenzo make a comprehensive review of chomskian nativist stance. They go on to demonstrate that it is mainly genocentric although Chomsky himself didn't explicitely embrace this particular form of innateness until the eighties. Grammatical geneticism with its preformationist view of Universal Grammar (UG), however, is not the only way to sort out the poverty of linguistic stimulus. An alternative nativist and developmentalist (epigenetic) account is in order also for language as they neatly show shedding a renewed light on the structure-dependent character of linguistic rules.

Surprisingly enough however, their case study on the structure dependence can hardly be more classical since they deal with the fronting of the (main) auxiliar verb in polar questions in English, surely the oldest and most discussed case in the hot debate around linguistic nativism, since Chomsky attracted attention to it.

The novelty of the two steps approach they undertake resides in two characteristics. The first one, deployed thoroughly, is the overt resort to a "data-driven" learning usually banned in orthodox nativist accounts. In the first approximation, this data-driven learning is combined with and based on an "innate-primed" and specifically linguistic procedure that refrains the child to entertain any rule not being structure-dependent. The second new characteristic, which appears in the second and final proposal, is the denial of any linguistic specificity in the structural principle at work. They substantiate their claim by appealing to the cortical growth that underlies an enhanced work memory, which in turn enables the structural processing of a sequential stimulus. The dismissal of any specificity in the structural basis of linguistic rules in favor of a view where merge/structure-dependence is seen as a mere by-product of cortical growth amounts to neglect the necessity of UG, namely of a specifically linguistic factor causally entering the state of knowledge finally attained. Were this approach generalized, UG would only be the limit space for linguistic variation. In this way, the divorce between the theory of linguistic variation and the theory of the acquisition of language will be ratified. By contrast, we will be able to aptly envisage a convergent solution for the origin of language in the individual and in the species.

The myth of language diversity (José-Luis Mendívil-Giró).

Mendívil-Giró explores the myth of language diversity by discussing "The Myth of Language Universals", a target paper by Nicholas Evans and Stephen Levinson (Evans & Levinson 2009; E&L henceforth). E&L represents to a good extent the general standpoint of the FCP, and defends the idea that the diversity of languages must be the essential topic of the study of language from a biological point of view. For those authors, the idea of a Universal Grammar is nothing but a myth invented by Chomskyan linguists and constitutes in fact an obstacle to the progress of cognitive science in general. This view, rooted in neo-Darwinism, entails an externalist, functionalist, and adaptationist conception of languages as historical objects created by its development through successive adaptive changes. However, Mendívil-Giró argues that such a view can easily lead to linguistic relativism, which he vigorously rejects. If languages are external objects, and languages and brains coevolve, then different types of languages might have ended up yielding different types of brains, and as a result a particular language would drastically determine the cognition of its speakers. As a corollary, there might be languages (and therefore cognitions) in different stages of evolution, i.e. more primitives than others. Against that view, Mendívil subscribes to the so-called *Language Uniformity Hypothesis*, according to which all languages have the same degree of development or evolution, and hence there are no primitive languages. As a result, the study of the individual faculty of language, uniform for all humans, and probably the most defining trait of the biological species *Homo sapiens*, is a crucial part of the study of the biology of language. This is indeed opposed to limiting that biology of language to the study of language diversity, the claim by E& L.

CHAPTER ONE

THE I-LANGUAGE MOSAIC*

CEDRIC BOECKX

1. A point of departure: "Grammar, Levels, and Biology"

The range of issues I want to address in this essay—the evolving conception of the human language faculty in the context of the biolinguistics—corresponds fairly closely to the topics addressed in a chapter in the *Cambridge Companion to Noam Chomsky* written by Howard Lasnik, entitled "Grammar, Levels, and Biology" (Lasnik, 2005)[1]. I decided to structure the present work around the points made by Lasnik, to see what has changed in the intervening years. Although published in 2005, Lasnik's chapter was written around the year 2000, the year when Lyle Jenkins's book *Biolinguistics* came out (Jenkins, 2000). I learned the very word *Biolinguistics* from Lyle's book, when my graduate career was drawing to a close. I think this is quite telling, for this is only a decade ago, and I was educated in a department where (at the time) the Chomskyan approach to language was taken Very Seriously (something I have tried to convey in Boeckx (2009b)). And yet, biolinguistics was somehow not part of our vocabulary. It could have been, for the origin of

* Although he may not agree with everything I say here, Massimo Piattelli-Palmarini played a central role in shaping my thinking on Biolinguistics, as did Marc Hauser, Dick Lewontin, Juan Uriagereka, and Noam Chomsky. But perhaps the most central role was (and continues to be) played by my students—Dennis Ott, Bridget Samuels, Hiroki Narita, to whom I am indebted. The present work is supported by a Marie Curie International Reintegration Grant from the European Union (PIRG-GA-2009-256413), research funds from the Universitat de Barcelona, as well as a grant from the Spanish Mininistry of Science and Innovation (FFI-2010-20634; PI: Boeckx).

[1] I was Lasnik's Research Assistant when he started working on his chapter, and I remember taking part in its elaboration. I therefore feel in a privileged position to make use of it.

the term (in its current meaning) goes back to the early 1970s (the "Debate on Bio-linguistics" organized by Massimo Piattelli-Palmarini in 1974, the prequel to the famous Chomsky-Piaget debate; Piattelli-Palmarini (1980)), and all the relevant foundational works (from Chomsky's review of Skinner's *Verbal Behavior to Aspects and Language and Mind*) were high on our lists of required readings[2]. But the general impression (never fully articulated or stated explicitly) was (and for many, I'm sure, still is) that theoretical linguistics is biology "at a suitable level of abstraction" (Boeckx and Piattelli-Palmarini (2005, 462)); oftentimes so abstract a level that any attempt at a rapprochement, at a more concretely biological investigation seemed too difficult, or premature, and therefore non-advisable. This is arguably why the word 'biology' does not occur often in Las- nik's chapter, despite the fact that it is part of the title. By my count, the word occurs exactly three times, in passages that I reproduce here:

> For a half century, Noam Chomsky has been developing a theory of language that deals with these two questions [What is knowledge of language? and How is that knowledge acquired?], by positing explicit [i.e., generative] formulations of human language capacity in terms of a productive 'computational' system, most of whose properties are present in advance of experience, 'wired in' in the structure of the human brain. Thus Chomsky conceives of his enterprise as part of psychology, ultimately biology. (p. 60)

> Explanatory adequacy in Chomsky's sense concerns language acquisition. Theories that seek to attain explanatory adequacy must posit some innate structure in the mind. This is surely indisputable; while a human being can learn language, a rock, or a gerbil, cannot. The research question, ultimately a question of biology, concerns just what this innate structure is. (p. 62)

> The levels [of representation] are part of 'universal grammar', a wired in part of the 'language acquisition device' that constitutes part of a human being's genetic endowment. Of course, the linguist does not know in advance of research what the levels are. Determining them is a scientific question, one of psychology, the theoretical biology of the mind. (p. 64)

Reading these passages, one is compelled to endorse the statement made by Akeel Bilgrami and Carol Rovane later on in the very book that contains Lasnik's chapter:

---

[2] Interestingly, I don't recall any explicit mention of Lenneberg (1967), a must-read for anyone interested in the biological foundations of language.

> Though eventual intergration with biology is the goal, it is a distant goal. In the interim, scientists work with the data and theoretical resources available to them, at a level of description and explanation which it allows them.They have the scientific goals of describing and explaining the language faculty which is present in the entire species as a biological endowment, but at a level of description and expanation which in the interim is bound to be a cognitive and computational level, with the properties of internality, universality, innateness, domain-specificity, among others, all of which Chomsky's own successive theories of grammar over the last few decades have exemplified. (p. 182)

This is why Lasnik's chapter focuses on the notion of "level of representation", the second term in the chapter's title. (The remaining term, "Grammar", is there presumably to remind the reader that "the central concept throughout is 'grammar', not 'language'. The latter is derivative [...]" (Chomsky 1981: 4)). The term "level of representation" figures prominently in Chomsky's *magnum opus*, *The Logical Structure of Linguistic Theory* (1955). The role of "level" is made very clear in Lasnik's chapter:

> On first examination, human languages appear to be almost overwhelmingly complex systems, and the problems, for the linguist, of successfully analyzing them, and for the learner, of correctly acquiring them, virtually intractable. But if the system is broken down into smaller parts, the problem might likewise be decomposed into manageable components. (p. 62)

This divide-and-conquer (modular) approach is the rationale behind identifying levels of representation (i.e., structure/analysis). In the words of Chomsky (1955: 63):

> A language is an enormously complex system. Linguistic theory attempts to reduce this immense complexity to manageable proportions by the construction of a system of linguistic levels, each of which makes a certain descriptive apparatus available for the characterization of linguistic structure. A grammar reconstructs the total complexity of a language stepwise, separating out the contributions of each linguistic level.

A level, according to Chomsky (1955, 5), "consists of elementary units (primes), an operation of concatenation by which strings of primes can be constructed, and various relations defined on primes, strings of primes, and sets and sequences of these strings."

My objective in this paper is not to review the various levels and their properties that have been posited during the generative enterprise from its inception to the present. Lasnik does an excellent job at this, and I recommend his paper to anyone interested in these issues. Rather, I would like to rethink the notions of "level" and "grammar" when the role of "biology" (the three terms in the title of Lasnik's chapter) assumes a more prominent role[3], as is the case in the context of what I like to call the "revival of biolinguistic concerns", a revival responsible for the greater visibility that the term "biolinguistics" is currently enjoying.

2. Biological concerns: New directions for Biolinguistics

Since Lasnik drafted his chapter much has happened in the realm of biolinguistics, so much so that students today often gets the impression that biolinguistics is a completely new field. Let me be clear again: biolinguistics is only comparatively new. Ignoring its now 60 years of history (as well as its conceptual lineage, which goes much further back; Boeckx 2009a) would be a serious mistake. However, it is true that the field is enjoying a new lease of life, and (in my view) promises to go much beyond what it could achieve in its first incarnation typified by Lenneberg (1967). As I discussed in Di Sciullo and Boeckx (2011), part of this revival is down to sheer luck, the sort of luck that characterizes scientific discoveries: there was no way to predict the success in 2001 at identifying the exact locus and nature of the genetic mutation affecting the KE family that initiated what Piattelli-Palmarini and Uriagereka (2011) call "the *FOXP2* saga". Since then *FOXP2* has been a key area of research at the intersection of genetics, neurobiology, and linguistics (see Benítez-Burraco (2009) for what is to my mind the best illustration of this synthesis). There has been a steady stream of papers focusing on the gene, its evolutionary history, and more recently its interactome. Although it is as yet far from clear how all the results in this domain fit together and what this entails for the nature of the language faculty, there is little doubt that the *FOXP2* discovery will continue to offer a unique opportunity for the experimental branch of biologuistics in the years to come. Lenneberg would have been thrilled.

But the new genetic data is only part of what made biolingistics rise again. On the more theoretical side, our conception of the language faculty has undergone a dramatic shift in perspective in the last ten years, and it is this shift that I think will play an even more influential role in the

[3] This prominent role would correspond to what Boeckx and Grohmann (2007) call the "strong" sense of biolinguistics, to be contrasted with the "weak" sense that the term assumes in Lasnik's chapter and in most publications in theoretical linguistics.

development of the field. The shift I have in mind is well captured by Chomsky when he writes in Chomsky (2007, 2; 4),

> Throughout the modern history of generative grammar, the problem of determining the character of FL has been approached "from top down": How much must be attributed to UG to account for language acquisition? The M[inimalist] P[rogram] seeks to approach the problem "from bottom up": How little can be attributed to UG while still accounting for the variety of I-languages attained?

Readers familiar with Lasnik's chapter quoted above will know that the top-down approach is assumed throughout that chapter. Lasnik starts with the logical problem of language acquisition, and goes on to motivate the notion of level in terms of simplification of the acquisition task. All the literature on levels reviewed by Lasnik assumed (and many linguists today continue to assume) that the language faculty "must be rich, highly structured, and substantially unique" (Chomsky (2007: 2)). The bottom-up approach is deeply suspicious of this richness. It does not, of course, dispute the richness of the knowledge attained, but it suspects that much of that complexity is epiphenomenal, and emergent; attainable on the basis of very simple properties and processes.

Interestingly, a substantial portion of the motivation for this bottom-up approach is downright biological: it takes seriously the idea that a theory of Universal Grammar is, ultimately, a statement of human biology, and, as Yang (2010: 1160) correctly points out, "one needs to be mindful of the limited structural modification that would have been plausible under the extremely brief history of Homo sapiens evolution"[4]. The recognition of

---

[4] Yang's statement assumes that the language faculty is indeed of very recent vintage. One could, of course, dispute this, but I tend to think that the evidence is now even stronger than what many suspect. It's now generally agreed upon, based on converging evidence, that the guess we are talking about is as good as any: the emergence of new tools, cultural artifacts, signs of trade, cave paintings, and so on, that we find in the archeological record, first in Africa and then in Europe (McBrearty and Brooks 2000, Mellars *et al.* 2007) point to a significant evolutionary transition. I tend to agree with Diamond (1992), Tattersall (1998), and many others that it is hard to imagine the emergence of these artifacts and signs of modern human behavior in the absence of the language faculty. But, as I pointed out in Boeckx (2011b), we can now make an even stronger case for the suddenness of the emergence of the language faculty. Recent genetic data suggest that Homo Sapiens split into two sub-populations around 150,000 years ago, which remained separated for about 100,000 years (Behar *et al.* 2008). If this interpretation of the data is correct, it suggests that the language faculty was already in place 150,000

this fact is not unique to the generative enterprise (witness Tomasello (1999)), but in a generative context, which relied on a rich UG for much of its history, due to its focus on the richness of the knowledge attained, biological realism, as one may call it, requires a drastic rethinking. Fortunately, much generative research was already going in that direction under the umbrella of the minimalist program.

At the beginning of the nineties, Chomsky got the impression that the overall approach was well-established, and that it was time to take the next step on the research agenda of the generative enterprise. The next step amounts to an attempt to go beyond explanatory adequacy. As Lasnik (2005) reviews, Chomsky (1965) distinguishes between two major kinds of adequacies: descriptive, and explanatory, and, not surprisingly, puts a premium on explanatory adequacy. The aim of (generative) linguistics was first and foremost to account for the amazing feat of human language acquisition in all its subtleties. Once it was felt that the model was sufficiently well-established, it became natural to ask how one could make sense of the properties of the language faculty that the model posits—how much sense can we make of this architecture of language? Put differently, why does the language faculty have this sort of architecture? Quite reasonably, Chomsky formulated this quest beyond explanatory adequacy in the most ambitious form (what is known as the strong minimalist thesis), in the form of a challenge to the linguistic community: Can it be shown that the computational system at the core of the language faculty is optimally or perfectly designed to meet the demands on the systems of the mind/brain it interacts with? By optimal or perfect design Chomsky meant to explore the idea that all properties of the computational system of language can be made to follow from minimal design specifications, a.k.a. 'bare output conditions'—the sort of properties that the system would have to have to be usable at all (e.g., all expressions generated by the computational system should be legible, i.e., formatted in a way that the external systems can handle/work with). Put yet another way, the computational system of language, minimalistically construed, would consist solely of the most efficient algorithm to interface with the other components of the mind, the simplest procedure to compute (generate) its

---

years ago. Now, if we combine this with the *FOXP2* evidence that suggests something significant happened, at the latest, 200,000 years ago (Piattelli-Palmarini and Uriagereka 2005), the time of appearance of homo sapiens, and you get a window of time of just about 50,000 years. So all in all, I think we should really try to keep the evolutionary add-ons for language to a bare minimum, with virtually no internal modularization to speak of.

outputs (expressions) and communicate them to the organs of the mind that will interpret them and allow them to enter into thought and action. If the strong minimalist thesis were true, FL would be an "ideal" linguistic system. But it should be stressed that the point of the minimalist program is not to prove the validity of this extreme thesis, but to see how far the thesis can take us, how productive this mode of investigation can be. The strong minimalist thesis amounts to asking whether we can make perfect sense of the properties ascribed to the language faculty. Asking this question is the best way to find out how much sense we can make out of them. The points where the minimalist program fails will mark the limits of our understanding. If one cannot make perfect sense of some property P (i.e., if P can be given a minimalist rationale in terms of computational efficiency towards interface demands), then P is just something one must live with, some accident of history, a quirk of evolution, some aspect of language that one must recognize in some brute force fashion[5]

To me the most interesting, innovative aspect of minimalism is the fact that, strategically speaking, the program forces linguists to reformulate previous findings in terms of elementary units, operations, and interface conditions. Many of these, one may anticipate, will have such a generic flavour to them ("combine", "map onto a linear sequence", etc.) that they are plausibly not specific, or unique to the language faculty. This should be very good news to researchers in other areas, as the concepts articulated by minimalists may find an equivalent in their own field, or be more readily testable using familiar techniques. These generic operations may

---

[5] I would be remiss if I didn't point out that isolating those quirks of evolution is not devoid of biological research value. As Marcus (2008) points out, much of what is being studied in biology is the result of evolutionary inertia (what Coyne 2009 calls "vestiges"). It's one of the strongest arguments offered against intelligent design. (Dawkins 2009 aptly calls it "unintelligent design").

Jackendoff (1997) in fact uses the well-established profusion of biological quirks to argue against the premises of the minimalist program (see also Kinsella 2009, Kinsella and Marcus 2009): "It is characteristic of evolution to invent or discover 'gadgets'. (...) The result is not 'perfection'". Jackendoff goes on to say that he would "expect the design of language to involve a lot of Good Tricks (...) that make language more or less good enough. (...) But nonredundant perfection? I doubt it".

Having said this, Jackendoff offers nothing in terms of mechanisms that may have produced the sort of highly specific Universal Grammar he continues to assume (witness Pinker and Jackendoff 2005, Jackendoff and Pinker 2005). As a result, his approach to language, contrary to his rhetoric, offers remarkably little in terms of genuine integration with biology.

indeed make it more plausible to entertain "descent with modification" scenarios concerning the evolution of language. At the same time, these generic operation may be more palatable to researchers in neurolinguistics, who have so far been confronted with an overwhelming feeling of what David Poeppel has called the granularity mismatch: a persistent incommensurability between the units and processes assumed by linguists and those assumed by neurologists (for discussion, see Poeppel and Embick 2005, Poeppel 2005).

In sum, linguistic minimalism contributes to the end of I call linguistic isolationism—an inevitable period of internal over-modularization in generative grammar during which the language faculty as a whole was studied sui generis, as an autonomous system bearing little or no resemblance to other cognitive modules in humans, or other species. To return to the terms in Lasnik's paper, the introduction of a new level of adequacy ("beyond explanatory adequacy") brings properties of universal grammar closer to the familiar concerns and assumption in (neuro)biology. No wonder this new level of adequacy has been called "phylogenetic adequacy" (Richards 2009), "natural adequacy" (Boeckx and Uriagereka 2007), "evolutionary adequacy" (Fujita 2009, Longobardi 2005) and "biological adequacy" (Narita 2010).

By a remarkable coincidence (or is it the *Zeitgeist*?), a converging bottom-up approach emerged in comparative psychology, especially the part of it touching on aspects long assumed to be specific and unique to *Homo sapiens*. Whereas early works in animal studies seemed to point to the unbridgeable gap between human language and other communication systems, more recent work stresses that a move away from communication, and a more decompositionalist approach is far more constructive. Here is how De Waal and Ferrari (2010) characterize this shift:

> Over the last few decades, comparative cognitive research has focused on the pinnacles of mental evolution, asking all-or-nothing questions such as which animals (if any) possess a theory of mind, culture, linguistic abilities, future planning, and so on. Research programs adopting this top-down perspective have often pitted one taxon against another, resulting in sharp dividing lines. Insight into the underlying mechanisms has lagged behind. A dramatic change in focus now seems to be under way, however, with increased appreciation that the basic building blocks of cognition might be shared across a wide range of species. We argue that this bottom-up perspective, which focuses on the constituent capacities underlying larger cognitive phenomena, is more in line with both neuroscience and evolutionary biology.

To linguists, the most well-known example of this trend in what is now called 'biocognition' or 'cognitive biology' is of course Hauser, Chomsky, and Fitch (2002), who, more than anything, drew attention to the richness of the Faculty of Language in the Broad Sense, and, in so doing, reasserted the plausibility of a minimal amount of linguistic specificity (the Faculty of Language in the Narrow Sense) to capture the biological character of the language faculty. This is the point where (as Jackendoff and Pinker correctly pointed out) the Hauser-Chomsky-and-Fitch vision meets minimalism[6].

Both strands of research take a deflationist, bottom-up stance on Universal Grammar. Hauser, Chomsky, and Fitch (2002) also make clear that this thing we call human language (even in the I-sense) is not a monolithic object: at some explanatory level, it is more appropriate to treat the language faculty as a mosaic of traits. This is what the title of this essay alludes to, and this is something I will return to below, for it is a rather different sense of modularity than the one Lasnik (2005) had in mind when he discussed the nature of levels.

This remarkable convergence, or consilience, lends a greater sense of unity to the modern biolinguistic agenda. By reducing the gap between sapiens and other species at the computational level, it affords researchers to entertain evolutionary scenarios that may not always be gradualist, but at least fully in line with standard Darwinian continuous assumptions (for a recent survey, see Fitch 2010). The fact that many aspects of FOXP2 appear to have been deeply conserved (Enard *et al.* 2002) also facilitates the elaboration of evolutionary scenarios that, because of the shared bottom-up perspective, have a higher degree of plausibility in the eyes of the linguists. This in turn invites a more active involvement on the part of the linguists in this kind of research, leading to a greater sense of genuine interdisciplinary dialogue. Finally, apart from evolutionary questions, the development of a commensurable perspective across fields offers opportunities at other levels of biological research, such as development and brain mechanisms.

All in all, recent developments in the interfield of biolinguistics have led to an increased research focus on questions that did not figure in Lasnik's chapter, questions that Chomsky (1986, 1988) (following

---

6 Perhaps I should soften this assertion and say something like "may meet minimalism", for it is fair to say that much (most?) "minimalist" work continues to adopt a top-down approach, as I review in Boeckx (2010/In progress). This may explain why both Marc Hauser and Tecumseh Fitch have been reluctant to jump on the minimalist bandwagon.

Tinbergen 1963) saw as fundamental to the generative enterprise, but about which little could be said in terms of concrete hypotheses. It seems that linguistics is becoming biology at a less remote level of abstraction.

3. A specific articulation

In this section I would like to sketch a way in which the new trends in biolinguistics discussed above could help articulate a model of grammar that is quite distinct from the one we are accustomed to.

3.1. Preliminary remarks

I stress that what follows is but one of the possible implementations of what I understand to be natural biological guidelines, and is intended as illustration only.

Let me begin by reiterating Chomsky's and Lasnik's assertion that languages are overwhelmingly complex systems. And although languages is the end point of the acquisition process that is so central to the generative enterprise, it is not clear to me that languages, even in the 'I-' sense of Chomsky (1986) (internal, individual, intensional), should be the focus of our investigation. As Chomsky (1995, Introduction) remarks,

> Thus, what we call "English" or "French" or "Spanish" and so on, even under idealizations to idiolects in homogeneous speech communities, reflect the Norman conquest, proximity to Germanic areas, a Basque substratum, and other factors that cannot be regarded as properties of the language faculty. Pursuing the obvious reasoning, it is hard to imagine that the properties of the language faculty—a real object of the natural world—are instantiated in any observed system. Similar assumptions are taken for granted in the study of organisms generally.

Even Joe's (I-)English or Pierre's (I-)French may not devoid of properties that go beyond the narrow confines of their language faculty. So I think the right focus, when it comes to biolinguistics, should be on the initial stage of the language faculty, and on the factors that determine its development. Even at this level, the complexity can be overwhelming. Consider the fact that at a minimum one must recognize as part of Universal Grammar a lexicon, a syntactic component, an interface with thought systems and an interface with externalization systems. None of these levels are simple. Take the lexicon: is this to be thought of as a pre-syntactic resource component? A post-syntactic morphological component? An encyclopedia? A repository of idiosyncracies? And, if the latter, Of which kind? At Word/morpheme level, or at the Construction/idiom/phrase level? Consider also the interface with thought systems. How many systems are we dealing with here? Do they have the same interface requirements? And when it comes to externalization, is it legitimate to collapse, as we typically do, sound systems and signs system? Do these really impose the same interface requirements? Finally, let me note that what we mean by syntactic component may be quite different, depending on whether we emphasize the interface with meaning or the interface with sounds/signs. It is becoming increasingly clear that the situation is far from symmetric: whereas the nature of the interface between syntax and meaning appears to be very transparent (so much so that some entertain the radical thesis that the two systems are one and the same; cf. Hinzen (2006, 2007)), the interface between syntax and sounds/sings is far more opaque. On the so-called PF-side of the grammar, few are under the illusion that syntax and phonology are the same (cf. Bromberger and Halle 1989). That is to say, whereas syntax is fairly naturally construable as a program for meaning, it is more like data on the externalization side. (On the difference between program and data in a biological context, see Atlan and Koppel 1990.)

Given the richness of the life sciences ("biology"), it is not at all clear that all these layers of structure ought to receive the same explanation. It is perhaps useful to recall that behind the term biolinguistics there is an intended symmetry between two disciplines (linguistics and biology). It is true that one's view of language will determine the range of hypotheses one is willing to entertain when it comes to the biological bases of language, but the latter also depend on one's view of biology. It is noteworthy that both disciplines are currently undergoing a conceptual expansion, with theoretical linguistics recognizing a greater role to non-specifically-linguistic processes ("Faculty of Language in the Broad Sense") and theoretical biology adding to the conceptual panoply of the modern synthesis, with a more prominent role given to non-genetic processes (cf. Pigliucci and Müller 2010).[7] The growing pluralism in both

[7] In this context it is worth recalling Ernst Mayr's statement that "Darwinism is not a simple theory that is either true or false, but rather is a highly complex research program that is being continuously modified and improved", a statement that parallels the frequent characterization of minimalism as "a program, not a theory." In related work (see Boeckx 2010/In progress) I have argued that the parallelism runs deeper, in fact. Both programs actually exist in a "hardened" form, found in textbooks and mainstream journals, a hardened form that for too long blackboxed developement, and entertained single-level theories (geno-centrism and lexico-

disciplines increases the complexity of the hypothesis space for the biolinguist.

To cut through all these layers of complexity, I suggest we distinguish between two notions. The first one could be called "emergent complexity" (the result of interactions among simple processes), a kind of complexity for which a certain minimalism is not only appropriate, but in fact de rigeur. (It is interesting that a minimalist outlook goes virtually unquestioned—except, curiously enough, in linguistics—when it comes to accounting for the emergence of new forms or "major transitions" such as the origin of life or the origin of the immune system.). The second could be called "cumulative or irreducible complexity" (the result of "complexification", of historical additions, of evolutionary/developmental intertia, of "one damned thing after another"). For the latter, a more pluralist, multi-factorial, explanatorily less ambitious, approach seems more appropriate.[8]

3.2. (Narrow) Syntax, Lexicon, and Grammar

If the emergent/cumulative complexity distinction is adopted, it now becomes important to figure out which aspect (which level) of the language faculty out to fall into which kind of complexity (and which kind of explanation).

Consider the lexicon. Are lexical entries complex? Yes, under anyone's theory. But can some of that complexity be seen as emergent? This is the line of inquiry that has its roots in Hale and Keyser's (1993, 2002) theory of argument structure, and that grew into distributed or exoskeletal models of the lexicon (Halle and Marantz 1993, Borer 2005). In such models, aspects of lexical entries that used to be stipulated for

---

centrism for biology and linguistics, respectively) instead of a more pluralist, interactionist (neither overwhelmingly internalist nor externalist), constructivist alternative that now seems more adequate. Concretely, for generative linguists, this means that explanations should rely less on unanalyzed, underived lexical features to explain both form and variation.

[8] My use of the term emergence here and below in the context of syntax is not to be confused with the way it is used in O'Grady (2008). Space prevents me from contrasting the two notions of emergence here. Suffice it to say that my use of the term corresponds to what Deacon (2003, 2006) calls second-order emergence, whereas O'Grady's corresponds to Deacon's third-order. The two notions of emergence also map on distinct dimensions of evolution discussed in Jablonka and Lamb (2005).

familiar "top-down" reasons are now constructed bottom-up, emerging via the simple, universal rules of syntax. Pursuing this line of inquiry myself, I have suggested (Boeckx 2009d, 2010c,b,f, 2011b,d, 2010/In progress) that the pre-syntactic lexicon can be (in fact, should be) reduced to a pool of formally identical conceptual addresses (Akin to roots in Distributed Morphology) whose sole computationally-relevant property ("feature", if one wishes to use this term) is their potential to merge.

Once this minimalist vision of the pre-syntactic lexicon is adopted, narrow syntax itself can be drastically simplified. It boils down to a machine that never ceases to Merge (externally and internally). All it does is construct sets (about the simplest thing one can ask of syntax). To be usable at all by the external systems, such a Merge α machine must nevertheless be regulated. Otherwise, all that would be delivered to the external systems would be a completely uninformative homogeneous soup of conjoined conceptual atoms. Fortunately for us, the problem is familiar to specialists in complex systems, who, since Turing (1952), standardly appeal to self-organization and periodic spontaneous symmetry breaking processes to generate patterns out of initial homogeneity. I have argued in the works referred to in this section that what linguists have called Cyclic/Phasal Transfer or Multiple Spell-Out (Chomsky 2000a, Uriagereka 1999) is but a specific manifestation of this generic and robust method of pattern formation. That is to say, the coupling of (unconstrained, iterated) Merge and Periodic Transfer is akin to a reaction-diffusion mechanism (technically, a Clock-Wavefront model) whose dynamics is—I claim—sufficient to shape elementary syntactic structures. I say "elementary" for the syntactic structures thus built are radically underspecified when it comes to all sorts of properties that are traditionally seen as central to linguistic analysis, such as category, "parameter", selectional restrictions, and the like. All of these will have to be seen as either emergent properties at the interfaces (more accurately, they will have to be regarded as emergent interpretations imposed by the external systems on the basis of their own resources, which are presumably part of the Faculty of Language in the Broad sense—not specific to language, and perhaps not even specific to humans),[9] or else they will have to seen as the outcome of various factors, heavily dependent on specific developmental pathways, as I have argued is the case with the many patterns of variation that many linguists still try to capture 'neatly' in terms of single-level,

[9] Readers will know that this is exactly how Hale and Keyser (1993, 2002) think of theta-roles.

parametric theories. (See Boeckx 2009c, 2008b, 2011a, 2010e, as well as Yang 2004a, 2010 for relevant discussion.)

Because this is a chapter that does not intend to focus on evolutionary or neurological details, I will refrain from going into the "wet-ware" basis of the very minimal syntax model I am advocating. Suffice it to say that the critical mass of conceptual addresses that I claim form the pre-syntactic lexicon is a very plausible outcome (spandrel) of the well-documented brain growth pattern in the homo lineage. As for reaction-diffusion models, there are so general processes at both the organic and inorganic levels that it is safe to assume that they could be implemented as brain processes.

The present approach finds a new, biologically-motivated use for an old concept like the syntactic cycle, and takes the (contours of the) asymmetric character of syntax to emerge dynamically (as anticipated by Moro 2000), the result of processes that Alan Turing took to underlie morphogenesis more generally. As for recursion, that key property of the syntactic component, its specifically-linguistic character is to be decomposed (much like I argued in Boeckx 2009e should be done in the case of Merge) and conceived of as the result of a two-step process: an iterative merge process, which is "densely recursive" (every merge product is directly embedded in a new merge relation), and a periodic, partial transfer process, which is "sparsely recursive":[10] not every merge product is immediately transferred; periods of transfer and non-transfer alternate in the course of the derivation, generating a structural rhythm that serves as the mode of instruction to the external systems and that receives a very plausible neurological interpretation in terms of brain oscillations (see Boeckx 2010/In progress).

The message for theoretical linguists is that much of the syntactic order we tend to think requires heavy (complex) lexical instructions to be generated can come about for free, if we are willing to make use of the

[10] Because they focus on macro-properties like morpho-syntactic categories, which it is true do not typically self-embed immediately (e.g., a TP immediately dominating another TP), Arsenijevic and Hinzen (2010), Hinzen (2011b) argue that true, immediate or "direct" recursion (what I have called "dense recursion") is not a primitive of the syntactic component. The discussion in the text shows why this is incorrect. Moreover, their position forces Hinzen and Arsenijevic to downplay the role of such a desirably generic operation as Merge, and to rely on underived (unexplained) templates not unlike those taken as irreducible constructions/idioms in Construction Grammar.

conceptual and technical resources provided by studies in Complex Systems, Chaos Theory, and the like.[11]

Please note that I say "much" of the syntactic order. I do not say all of it, nor do I say that all linguistic or grammatical properties come in such a free fashion. In this I follow Kauffman (1993), and to some extent Gould (2002), who recognize that the creative aspect of life is not solely (and perhaps not even primarily) to be explained in terms of (natural selection). As Jacques Monod would have it, next to chance, one must also recognize the role of necessity. I follow these scientists in recognizing a filtering, or selecting function after the primary structure role of self-organization. For me, this role is played by morpho-phonology, which freezes the derivational dynamics of syntax and stabilizes it (much like genes do; cf. West-Eberhard 2003, Pigliucci and Müller 2010) into distinct phenotypes we standardly call idiolects/(I-)languages—a collection of more-or-less flexible constructions/idioms that we must code in largely arbitrary features, morphemes, morpho-syntactic categories, and the like. Note that the position I am advocating here amounts to saying that as the child acquires a language, she moves from a more to a less flexible system, where the dynamics of syntax is taken over by the static character of morpho-phonological rules. This increasing amount of what I will call "grammaticalization" in the text below is as radical as the shift from a caterpillar to a butterfly. One can think of this shift as a move from a Generative Semanticist to a Construction Grammarian. The first stage is almost pure bio-physics ("third factor" in Chomsky's 2005 sense), the other encodes many more properties of the environment, as a result of learning and induction.[12]

The stance I am making here has important implications for a recent debate regarding the distinction between syntax and morphology, or between what Hale and Keyser (1993) called "s-syntax" and "l-syntax". Some syntacticians (the names of Mark Baker, Richard Kayne, Rita Manzini, Michal Starke come to mind) have suggested that morphology really is no different from syntax. Hale and Keyser were criticized for maintaining a distinction that they describe, following their critics, as "probably an onerous distinction, perhaps nothing more than a temporary terminological convenience". Nevertheless, Hale and Keyser stuck to their guns, and I think they were exactly right in doing so. As they say,

---

[11] I may be wrong, but this is the message I read into Uriagereka (1998).

[12] For what I take to be a similar intuition, see Uriagereka (2008a).

> We must nevertheless assume that there is something lexical about any verbal/lexical entry. [...] What is it that is lexical about the entry corresponding to *shelve*? Clearly, it is a lexical fact that *shelve* exists as a simple transitive verb in English. [...] In reality all verbs are to some extent phrasal idioms, that is, syntactic structures that must be learned as the conventional "names" for various dynamic events.

In effect, Hale and Keyser are pointing out that their "l-syntax" is a syntax in the representational sense (a post-syntax, a morphology, in my terminology), whereas "s-syntax" is a syntax in the dynamic, derivational sense (narrow syntax, for me). Confusing the two would be like confusing genetics and epigenetics. One is static, the other one dynamic. One relies on the other, but they are not to be collapsed. One is completely generic and invariant (narrow syntax), the other is necessarily language-(equivalently, morpheme-/feature-)specific. In writing this, I am reminded of Lasnik's (2005) discussion about levels and their motivation, especially when he wrote that Chomsky, from LSLT onwards, was never shy of positing an extra level of structure if by doing so, each level was simplified. By positing a morphological component in addition to a narrowly syntactic component, I am not introducing any new device into the theory: every linguist will have to have language-specific, morpheme-specific, feature-specific conditions somewhere. But by sharply separating these from syntax per se, I am allowing for the later to be formulated completely (i.e., more naturally/simply) in terms of self-organization. In a biolinguistic context, this should be seen as an attempt to carve the language faculty at its (bio-physical) joints. It is true that by separating syntax from morphology I am making it impossible to seriously entertain concepts like comparative syntax (at least in so far as the comparisons involve different human languages; comparison with other species, or across cognitive domains, still make a room for something like comparative syntax), or for thesis titles like *French Syntax* or *Italian Syntax*. If most of the things that get most linguists excited fall outside of the purview of (my sense of) syntax, so be it. I tend to think that this is significant biolinguistic progress. Biolinguistics is not (and should not be) philology by other means. It's a very different conception of language, one that leads to a notion of syntax that looks nothing like what traditional grammarians would recognize. Please note that this discussion is not a trivial choice of terminology. It is about completely different levels of linguistic analysis, with a level being argued to be completely language-free.

The present conception of grammar partially (I insist, partially) agrees with the idea articulated over the years by Ray Jackendoff that the lexicon and the grammar are not distinct entities (see Jackendoff 1997, 2002, 2005, 2010).[13] Where Jackendoff and I part company is at the level of syntax. Jackendoff's conception of syntax (well articulated in Culicover and Jackendoff 2005) is completely language-specific, and Construction-based. It is much like what I call morphology, or what Hale and Keyser would call l-syntax. It's a more or less articulated catalogue of idioms, ranging in productivity from the very rigid (think *kick the bucket*) to the very flexible (think X-bar schema), which I suspect will ultimately be motivated by memory-efficiency considerations (along the lines of Yang 2005). I view the Grammar (and the lexicon) with exoskeletal lenses, from the bottom-up: as something that must be constructed. Yes, there is a level of linguistic analysis that one may call Grammar, but, if the attempt to go beyond explanatory adequacy is to go anywhere, one must also recognize a distinct level of syntax that constructs the grammar, that grammaticalizes.[14]

Let me conclude this discussion by pointing out that in many ways the model of grammar I am advocating here radicalizes intuitions that, to the best of my knowledge, were first formulated by Carlos Otero.[15] Otero (1976, see also Otero 1983, 1996) clearly saw the need to recognize two distinct grammatical components, which he called "syntagmatic" (my narrow syntax) and "paradigmatic" (my morphology). Like many current models of the lexicon, Otero advocated a process of very late lexical insertion, and understood well that "[a] syntagmatic grammar is essentially

---

[13] Jackendoff here sides with Construction Grammarians. I too recognize the existence of Constructions, but, unlike Jackendoff and the Construction Grammarians, I recognize a distinct level of (narrow) syntax that allows me to construct the constructions and the templates, which they simply take for granted. Theirs is still a top-down approach, far away from the bottom-up perspective that progress in biolinguistics seems to demand. This is all to say that in linguistics, pre-formationism still dominates. If syntax is to be of any use, it should adopt the role of cognitive embryology, where epigenesis and growth dominate. We still have a lot to learn from Piaget (as I discuss at length in Boeckx 2010d).

[14] Arguably, my sense of "grammaticalization" is more meaningful, from a biolinguistic perspective than the standard definition, due to A. Meillet: "the attribution of grammatical character to an erstwhile autonomous word." It is now clear that words are pieces of grammar (think of Marantz's "Cat is a phrasal idiom"), so Meillet's definition is not describing a process from non-grammar into grammar, but merely a process of re-grammaticalization: a relocation of the furniture, more than the building of a new house.

[15] Much to my regret, Otero (1976) remains unpublished. For an exhaustive discussion of Otero's proposals, with extensive quotes, see Boeckx (2010/In progress).

universal (biologically given in essence), [whereas] a paradigmatic grammar is, to a considerable extent, a historically evolving subsystem, burdened with the weight of the past, like other cultural systems. Only a paradigmatic grammar can be fossiliferous. As Otero wrote, it is only "the syntagmatic grammar [that] can be assumed to be a fairly direct reflection of the language faculty of the mind/brain", hence (in my view) the need to focus away from paradigmatic aspects in a biolinguistic context.[16]

Let me stress once again that I think that the discussion revolves around far more than merely academic/terminological distinctions. As Koster (2009) makes clear, if "Syntactic structures are not generated by lexicon-independent rules (like phrase structure rules or Merge) but as the spelling out of the contextual properties of lexical items ("valency")"—as virtually all models of grammar do, including Koster's—then "the capacity for language, even in its narrowest sense, is [bound to be] seen [not] as a matter of biology but as applied biology, i.e., a technology belonging not primarily to individuals but to their shared culture". The syntax/lexicon distinction, at bottom, is about the well-foundedness of the biolinguistic enterprise.

---

[16] My urge to set variation aside may shock some linguists, as it goes against statements like "Comparative syntax has become an indispensable, if not privileged, part of our attempt to understand the (syntactic component of the) human [language] faculty" (Kayne, 2005, 55). My position may also surprise many (non-)linguistis. After all, isn't biology all about variation? Variation is, in fact, "a key concept in biology (cf. Hallgrímsson and Hall 2005). As Ernst Mayr once wrote, "Natural selection would be meaningless without variation". The two great books by Darwin (1859, 1871) in fact start with the very issue of variation.
The point is well taken. In fact, some biologists are reluctant to view universal processes of self-organization, which I take to be so central to the biolinguistic enterprise, as genuine biology. The role of variation touches on central issues in the philosophy of biology, such as the essence of biology and how it differs from physics, and on the very possibility of a theoretical biology. I cannot address this topic here, and refer the reader to Boeckx (2006, chap. 4) and Boeckx (2009a). Suffice it to say that I honestly believe that not making variation so central will help bridge the gaps between linguistics and neurobiology, and uncover the biological foundations of language. By emphasis deep homology/conservation and the like, Evo-devo is in fact moving in the same direction I am advocating. Having said this, biolinguistics, like biology, is destined to be a very large, pluralist field, which will ultimately have to take population-level aspects of language into account, as biology routinely does. For a discussion of such population-level aspects of language in a (broad) biological context, see Mufwene (2001), Lorenzo (2010), as well as Givon (2002).

3.3. Interface levels

Having talked about the lexicon and narrow syntax, it is time to turn to the interface levels. It is here that we find the most important contribution of evolutionary history to the constitution of the language faculty. As recent works have made clear (see, in particular, Hurford 2007, Carruthers 2006 on the "meaning" side, and Fitch 2010, Samuels 2009b, 2011, Samuels, Hauser and Boeckx To appear, on the "externalization" side), much of what is needed to adequately account for how syntactic form gets interpreted and externalized is already in place in some form. To be sure, some modification of what was already available was needed. Everything indeed points to the fact that the emergence of narrow syntax was quite transformative: interface levels had to "grow" along with syntax, but evolutionary history had given the owners of the language organ a fair head start, with only adjustments, as opposed to *de novo* mechanisms and structures, to be developed.

Consider the meaning side. Although most semantic theories posit a richness equal to, or reaching beyond what GB syntacticians were willing to posit as part of the biological endowment, recent "minimalist" inquiries into the nature of semantic knowledge, most especially work by Paul Pietroski (Pietroski 2005, 2011, To appear), point to the sufficiency of a minimal code that is as underspecified as the syntactic code discussed in the previous subsection. As Pietroski argues, knowledge of meaning should not be equated with Meaning as a whole, which surely goes beyond the narrow confines of the language faculty to involve the mind at large (Jackendoff's "Conceptual Structures"), and should be thought of as *Begriffspläne* (not Fregean *Begriffsschriften*); instructional sketches[17] that guide (structure, constrain) meaning, but do not completely determine it. Under this more adequately modest reconception of what semantics is, Pietroski shows that minimal Neo-Davidsonian representations (decomposed event structures), like (1), are sufficient to capture the character of logical forms.

(1) $\exists e\,[\ldots\&\ \theta_{ext}(x, e)\ \&\ \ldots\&\ \theta_{int}(y, e)\ \&\ \ldots\&\ ad(w, e)]$

I have argued (Boeckx 2009f, 2010/In progress) that such Neo-Davidsonian representations are isomorphic to the elementary structures

[17] Much like the various sketches in Marr (1982).

generated by Merge coupled with Cyclic Transfer:[18] Merge is the syntactic correlate of the semantic operation of Conjunction for Pietroski, and points of Cyclic Transfer correspond to special interpretive instructions like "∃" or "θ", or "ad". Although these instructions are captured by means of distinct markers, they are in fact much more similar in nature than they look: all boil down to locative functions, anchoring participants in events, adjuncts onto main events, and events in speech acts, which strengthens the idea that syntax need not make use of categorial features to generate logical forms.

Of course, all of this offers but the roughest of meaning sketches. But once enriched (post-syntactically) with the interpretive resources of the thought systems (the concepts behind the conceptual addresses/lexical items), they can go a long way into capture our core semantic competence, as Pietroski has shown.[19]

The key, of course, here is to recognize the semantic richness of pre-linguistic systems, where extraordinary conceptual richness resides (again, see especially, Hurford 2007, Carruthers 2006). What the interface with syntax adds is an ability to decompose events into connected subevents (something with Hurford 2007 argues lies out of reach of other animals), and combine any two concepts together, given that the rules of composition (like compose with like) take place at the level of lexical items (conceptual addresses, which are all on a par), and not at the levels of concepts, whose wide range of adicities (valencies) dramatically reduces compounding options. With syntax, semantic composition is truly algebraic, and transcends the boundaries of modular thought, as well as the bounds of perception (hence the stimulus-free, non-strictly-referential character of semantics emphasized by Chomsky over the years; see

---

[18] In a different way, Hinzen (2006, 2007) has argued for as perfect a syntax-semantics isomorphism as I am claiming here, but Hinzen goes too far in concluding that semantics loses its independent status. After all, the two sides of the same coin still could as distinct! Moreover, because he appears not to recognize the available semantic richness outside of language, he is forced to posit a much richer syntactic ontology to derive the semantic fact that he deems essential. For a clear expression of his departure from a minimal(ist) syntax, see Hinzen (2009, 2011a).

[19] To give but one example, which is slowly becoming the consensus position, a pre-linguistic social cognition of the type found in baboons Cheney and Seyfarth (2007), supplemented with the recursive template constructed by narrow syntax, may be just what is needed to account for our species-specific theory of mind. On the still-much-underappreciated richness of non-human, pre-linguistic social cognition, see Tomasello (2009a), Fitch *et al.* (2010).

Chomsky (2000b)). The end result is a sort of conceptual *lingua franca*, a truly universal language of thought. (This kind of cross-modular thought is what I argue in Boeckx 2009b, 2011c, 2010/In progress made us sapiens; for reflections along similar lines, see Mithen 1996, Carruthers 2006, Spelke 2003, Pietroski 2007, 2008, Ott 2009, Hauser 2009b,a).

On the externalization side, things are, I am afraid, much messier (in the sense of "less isomorphic"), but in a way that will look familiar to the evolutionary biologist, who expects nothing else from the shrewd tinkerer that is nature. Given the "big facts" about this interface (it deletes copies, collapses the hierarchies, proceeds from left to right, uses repair and readjustment rules, etc.), it looks like syntax was never meant to be externalized. If they had known more about the language organ, Dawkins (2009) and Coyne (2009) could have included the syntax–Phon interface in their survey of "unintelligent design" and evolutionary vestiges. If Marcus (2008) had been careful about fragmenting the language faculty as a whole, he could have been right about some aspects of our language faculty (FLB!) being klugy. The design of the syntax–Phon interface, being so unlike the syntax–Sem interface (where copies are interpreted, hierarchies retained, no repair/readjustment rules are needed, and where the bottom-up order is retained as the basis for compositionality), could indeed be used as an argument in favour of the idea that the externalization of linguistic structures was an evolutionary afterthought.[20] This is what Chomsky when he writes in Chomsky (2008) that "[i]t may be that there is a basic asymmetry in the contribution to language design of the two interface systems: the primary contribution to the structure of [the] F[aculty of] L[anguage] may be optimization of the C-I [sense] interface".

None of this should be surprising: long-silenced genetic pathways for vocal learning and for finely-tuned motor control had to be re-opened (re-

---

[20] By using the term "afterthought", I do not want to imply that externalization systems evolved after the syntax-thought interface was in place. There is in fact reasonably good evidence that Neanderthals were capable of vocal learning and flexible vocalization, but little evidence that they matched Sapiens's conceptual complexity. This suggests that much of what now constitutes PF, including our sense of rhythm and concomitant body entrainment (Patel *et al.* 2009, Schachner *et al.* 2009, Fitch 2009a)—which may well have given rise to sounds and signs simultaneously—was recruited in our lineage after having re-evolved in previous stages of homo evolution. For the exploration of a scenario that goes in this direction, see Mithen (2005). Darwin's (1871) idea concerning the role of music in the context of speech may indeed be on the right track. See Fitch (2010) for a recent reappraisal.

teaching old genes long-forgotten tricks),[21] and had to adjust to the modified hardware, now including this new piece called narrow syntax. Moreover, since the maintenance of an externalization system only makes sense in the context of inter-individual communication, such a system must be sensitive to learning, which rests on unfailingly imperfect imitation, and is often invariably used to mark (epigenetic) "group" (and thus, anti-group) "cultural" boundaries.[22]

All of this has the effect of promoting the sort of irreducible complexity I discussed above. If, as I have argued syntax is invariant, morpho-phonology is necessarily full of exceptions,[23] with tendencies (more or less general rules), instead of laws. Modifying a well-known statement by Edward Sapir, one could say that syntax never leaks, but morpho-phonology always does. As Jespersen had it, one can dream of a universal syntax, but no one would ever dream of a universal morpho-phonology.[24] Bromberger and Halle (1989) are right: phonology is different. In this respect, morpho-phonology is much closer in nature to the kind of systems studied by evolutionary biologists, often described as the domain of the "higgledy-piggledy". This may well explain the success of a pluralist, weakly/minimally nativist, history-heavy approach to phonology of the sort explored by Blevins (2004) (for a much-improved technical articulation, see also Samuels (2009b, 2011, this volume).

As an illustration of the contrast between narrow syntax and morpho-phonlogy, consider how the two systems relate to one another (the "PF-interface"), and contrast it with the way syntax relates to thought systems

---

[21] I would not be surprised if in the next few years, as we learn more about vocal learning in songbirds and other species, this will come to be seen as a textbook case of deep homology.

[22] For illuminating discussion of the role of "culture" in the context of I-phonology, see Samuels (To appear). For "live" evidence of gradual stabilization of the phonological system (phenotypic accommodation), see the extraordinary evidence from Al-Sayyid Bedouin Sign Language discussed in Aronoff *et al.* (2008). Of relevance as well is the neuronal recycling hypothesis first explored by Dehaene (2005) in the context of reading and arithmetic, and now extended to music (Patel 2010) and word meaning (Koster 2010).

[23] Morris Halle once told me that SPE provides a theory of exceptions. On the importance of exceptions in phonology, see also Halle and Nevins (2004).

[24] I boldly claimed above that there cannot be any comparative syntax, nor any language-specific syntax. Just the opposite holds for phonology. In syntax you can write a book called Syntactic Structures, period (no language name needed). In phonology, you cannot write a book called "Sound Patterns". It has to be Sound Patterns of English or Sound Patterns of Russian.

(as per our discussion above). On first approximation, the connection with narrow syntax was as transformative on the sound/sign side as it was on the meaning side. As Boeckx and Samuels (2009) argue, just like a recursive syntax allows for new conceptual compounds to be formed, it forced the emergence of morpho-phonlogical units, viz. underspecified forms that require repair/readjustments rules for which morpho-phonology is now well-known. This is easy to understand: the formation of underspecified forms is nothing more than an attempt to optimally learn an unbounded input in terms of necessarily abstract/space-saving schemas. Syntax also provided natural boundaries on the sound/sign side (the equivalent of the Neo-Davidsonian markers discussed above): these are the prosodic markers that current research suggests is much more aligned with phasal boundaries than is standardly assumed (contrast the confident assertion of syntax-phonology mismatches in Jackendoff 2002 with the robust findings of perfect match in Wagner 2010, Samuels 2009b, 2010, and references therein).[25]

However, the semantics/phonology parallelism quickly breaks down, for, if elementary syntactic structures are virtually ready for immediate semantic use, they are only the first step of a longer, more complex (and partially language-specific) process leading to externalization. In the current literature, this complexity is often hidden under the simple-sounding rubric of "linearization" or "transfer/spell-out" (perhaps under the influence of Kayne 1994, who enriched narrow syntax considerably to suggest that linearization was a simple mapping). But terms like linearization or Spell-Out/Transfer are to be understood as "operation complexes" (as recognized by Idsardi and Raimy In press). That is to say, the PF-interface level consists of many sub-levels. In Boeckx (2009g, 2010/In progress), I argue that at least three such (sub)levels must be recognized if the conception of narrow syntax advocated in this essay is adopted.

First, various asymmetries must be anchored or grafted onto cyclic transfer (dubbed "the mother of all asymmetries" in Boeckx 2009d, as it makes a difference between elements transferred and elements remaining syntactically active, part of the workspace): Categorization (featurization), Agreement-valuation/Case-assignment, copy deletion, labeling/ establishment of head-complement relation (X0-YP polarization), Prosodic phrasing, etc.

---

[25] This sort of perfect alignment is of great help to the child in the process of language acquisition, as the literature on 'prosodic bootstrapping' attests (see Millotte and Christophe 2009 for an overview). Paraphrasing Anne Christophe, intonation is the syntactic skeleton made (loud and) clear.

Let me note that at this level, a lot of filtering[26] must take place to cut down the amount of symmetry produced by Merge in narrow syntax.[27] A term like "Demerge" (Fukui and Takano 1998) aptly captures what must be taking place.

Next, a level of actual Vocabulary insertion must be posited. Here too, a lot of selection/filtering will have to take place in the form of morpheme competition mechanisms (principles like the Elsewhere Condition, The Subset Principle, or the Superset Principle, regulating Late Insertion in frameworks like Distributed Morphology and Nano-Syntax).

Finally, there will have to be a third level, where actual linearization, complete flattening of the morpho-syntactic structure will take place. The whole structure will have to be re-scanned, and all loops created by specific properties of the inserted vocabulary items (those giving rise to reduplication, e.g.) will have to be removed (along the lines explored in Raimy 2000, Samuels 2009b). It's also at this stage that item-specific readjustment rules are called for (repair of haplology, re-activation of copies to meet the stranded affix filter, etc.).[28]

The preceding discussion strongly suggests that in a certain sense elementary structures were never "meant" to be externalized. They constitute too raw material for communication, and this is why so much extrinsic information has to be added (violating the minimalist guideline of Inclusiveness as Chomsky 2001b recognized early on) along the way to make externalization possible. It may not be too surprising that it is at this level of fine-tuning, where cross-level coordination is essential, that most "linguistic" disorders are found.[29] Given the complexity of the process, break-downs are to be expected.

---

[26] Very local, "nano"-parameter settings can be reconstructed as rules at this level.

[27] Note that because so many conditions in language depend on features, they will have to wait until these are inserted. In other words, these conditions, often assumed to be syntactic, will have to be reconstructed as representational, interface conditions ("filters"). Such will be the fate of Relativized Minimality, Anti-Identity Conditions, Person Case Constraint, Islands, and the like. See Boeckx (2008a,c, Submitted,I) for supporting material.

[28] Some of the filters applying at the first level of asymmetrization will have to reapply, although this time they will apply to tokens (specific morphemes/words), as opposed to types (categories). For evidence of recycling of PF-principles, see Walter (2007) on haplology/identity avoidance, and of Rezac (2011) on **me–lui* and other clitic-combination restrictions.

[29] It is tempting to speculate that the various subtypes of SLI (Specific Language Impairment) distinguished in, e.g., Van der Lely (2005) arise as progressive deficits, encompassing more PF-sublevels (getting closer and closer to, but

If the emergence of narrow syntax sketched above looks eerily similar to that of a spandrel—an inevitable, lawful appearance given certain initial conditions (such as the joining of arches of the joining of mental modules), the emergence of morpho-phonology looks like the ways spandrels acquire their decorations. Though such decorations are limited by the space defining the spandrel, such geometrical constraints are very weak indeed, and much of the decorative process is determined by the idiosyncratic, contingent inspiration of the artist. As Gould and Lewontin (1979) noted, it is often the elaborate decorations that attract the art student or the tourist, much like it is often the idiosyncratic details of languages that attract the linguist or the language amateur, but it would surely be a mistake to ignore the laws of growth making such decorations possible, as it would surely be a mistake to reduce narrow syntax to its morphological effects.

Let me conclude this section by addressing a common misperception in the minimalist/biolinguistic literature. One often hears at professional meetings, or reads in publications such as Culicover and Jackendoff (2005) or Hinzen (2009) that minimalists arrive at a very simple model of syntax but only at the cost of burdening the external systems. The model sketched in this section may suffer from the same criticism, but I think that it would be unfair, for I do not think that I have added extra operations to the external systems or the interface levels. I have merely made the fine-grained nature of such levels and systems explicit (witness my treatment of "linearization"), and have consolidated processes (such as lexical insertion which is, I claim, uniformly "late"/post-syntactic) to a single component of the grammar, instead of distributing them in ways that invariably lead to unanalyzed, underived and therefore unexplained constructs (e.g., pre-syntax lexical bundles or templates). So, I contend that the simplification of syntax advocated here is real, and leads to an overall more biologically plausible model of the language faculty, albeit one that is almost unrecognizably far away from traditional notions of grammar. But, as I stressed in the opening section, this is the path that biolinguistics must take if it is to become integrated into the natural sciences.

---

crucially for me not affecting narrow syntax). For a partially converging view, from a different, connectiionist perspective, Joanisse (2004), Joanisse, Manis, Keating, and Seidenberg (2000).

4. Concluding remarks

As Lasnik made clear in his (2005) essay, which I have used as a foil for this chapter, "the research question [for biolinguistics], ultimately a question of biology, concerns just what this innate structure is" (p. 62). For most of the past half century, generative contributions to biolinguistics have sought to demonstrate that there must be some innate structure for language (the range of "poverty of stimulus" evidence for a "language instinct" or for a "universal grammar"; see Jackendoff 2002, chap. 4, Boeckx 2009b, Parts I-II), but have been rather vague as to which kind of innateness that was. To be sure, they did highlight many specific aspects of knowledge of grammar that had to be innate, but they did not really ask what aspects of biology would make these aspects true. References to our "genetic endowment" can often be found, but nothing more precise. The biology part of the generative discourse remained at a suitably very high level of abstraction.

Asking which innateness, which kind of biology is appropriate for language is in my view the central question on today's biolinguistic agenda. It forces linguists to engage with current debates in biology about the role of genes: are they leaders or followers?; is the dominant geno-centric, adaptationist stance in evolutionary biology adequate?; how do we get from genes, which (as Marcus 2004 correctly points out) build neural structures—not behaviour, to what these neural structures actually do (the mind as "what the brain does")?; what are the relative contributions of the three factors involved in all aspects of biological organization? (Seilacher 1970, Lewontin 2000a, Gould 2002, Chomsky 2005). Asking these (hard) questions not only force linguists to be more than aware of the controversies in biology, it could in fact enable them to contribute to these very controversies (well captured in Pigliucci and Müller 2010, much like the early generative engagement with the ethology literature (Chomsky 1959a, Lenneberg 1967), made significant contributions to cognitive science and ultimately bioanthropology (the term that was originally chosen for what become biolinguistics in the circles of the Royaumont Center for a Science of Man in the 1970s).

Biolinguistics today is both ripe with challenges that could be the causes of its (temporary) demise (perhaps the issues are just too hard) but also with opportunities that could bring linguistics back to the high table of biocognition after, it is safe to say, a few decades of (perhaps unavoidable) isolation (the effects of modularity, which Piattelli-Palmarini 2006, 2010 aptly calls the "age of specificity").[30]

In this essay I have tried to show that the notions of "levels" and of "grammar" in Lasnik's chapter are still of utmost importance, but they acquire meanings and roles when the third term in his title, "biology", is taken at a less abstract level of analysis and involvement. The demands for simplicity, and for decomposition into levels, remain the same,[31] but they are pressed into new service in a biolinguistic context, where the question of language acquisition (ontogeny) has to be treated on a par with the questions of language evolution (phylogeny) and brain implementation (physiology). The primes (elementary units) and rules of combinations posited to account for language development must also be made compatible with everything else we know about what the brain can do, and what evolution can do. From the logical problem of language acquisition, we are moving to the biological problem of language acquisition. The decomposition of the language faculty into its component parts is no longer solely motivated on the basis of purely (cross-)linguistic arguments, but instead by considerations of comparative psychology (Hauser, Chomsky, and Fitch's FLB vs. FLN). The modules posited are no longer moving into the direction of ever-increasing (linguistic) specificity (from the levels of LSLT to the modules of GB to the features of minimalism), but in the direction of modified descent (cf. Marcus 2006, Karmiloff-Smith 1992, Boeckx 2010/In progress). Properties of the language faculty are no longer approaches from the top, but from below, and although the

[30] Effects which I hasten to add Piattelli-Palmarini does not see in as negative a light as Jackendoff (2002), who speaks of the "alienation of linguistics".

[31] Witness the demands of decomposition in discussions of what biolinguistics needs: "Linguists and psycholinguists owe a decomposition (or fractionation) of the particular linguistic domain in question (e.g. syntax) into formal operations that are, ideally, elemental and generic. The types of computations one might entertain, for example, include concatenation, comparison, or recursion. Generic formal operations at this level of abstraction can form the basis for more complex linguistic representation and computation" (Poeppel 2005). "We need to distill what we know from linguistic theory into a set of computational primitives, and try to link them with models and specific principles of neural computation... Thus we need linguistic models that are explicit about the computational primitives (structures and operations) they require, and that attempt to define linguistic problems at a fine enough grain that one can discuss algorithmic and implementational approaches to their solution. We need a list of computations that linguistic theorists deem indispensable to solve their particular problem (e.g., in phonology, syntax, or semantics)" (Fitch 2009b: 298).

results of both perspectives should ultimately converge, it is I think increasingly clear that the resulting picture of the language organ will be more complex, more multi-factorial, more generic than most (generative) linguists would have been willing to entertain even twenty years ago. Aspects of positions that were once considered wrong (e.g., Piaget's insistence of development and epigenesis, or the connectionist's bottom-up approach that, much like my view of syntax, lets structure emerge; cf. McClelland *et al.* 2010) will be reconsidered in a new light, and proven right.[32] One should perhaps begin to speak of an extended or expanded generative (or biolinguistic) enterprise, much like they have begun to do in biology (Pigliucci and Müller 2010). Yes, the so far typical generativist's nativist stance (with its high degree of specificity) will have to make way to a softer, less specific, less genetic nativism, but such nativism will emerge much stronger, much more robust, much better grounded in biology. The aim of the minimalist program of using only notions that are "virtually conceptually necessary" will lead to a kind of elementarity that will be so simple as to be unobjectionable (and no longer just simple enough as to be learnable). The level of biological abstraction will not be so far removed from the concrete concerns of biologists, but the level of linguistic abstraction will be much further away from the concrete concerns of the philologists. The abstraction will be, to use a phrase from Gallistel (2009), "foundational". Piattelli-Palmarini (2006, 45) noted that "things are changing" in biology, and likewise (in Piattelli-Palmarini 2001: 3):

> Something is happening in, and to, cognitive science. We are being progressively dislodged from the neat, consistent, and exciting scheme [called "classical cognitive science"] in which we had settled for a long time [...] [T]he scene is changing, possibly irreversibly.

The emerging I-language mosaic, to use the title of this essay, will make us reconsider Plato's Problem (language acquisition) in light of Darwin's problem (language evolution) and Broca's problem (brain implementation). Both brain studies and evolutionary studies point to decentralized, multi-causal, complex explanations that will help us move

---

[32] I note in passing that quite a few names whose approaches to biology are highly appreciated in generative circles, such as C.H. Waddington or S. Kauffman, have strong ties to psychological approaches that are at variance with tenets of generativism; Piaget was strongly influenced by Waddington and Kauffman was McCullock's student.

beyond the impasse[33] that Massimo Piattelli-Palmarini must surely have felt when he used (Piattelli-Palmarini 1989) all the modularity evidence from theoretical linguistics to argue against facile adaptationist scenarios (of the Pinker and Bloom 1990 kind), knowing full well that the kind of specific modularity he was using could only come about by the sort of gradual processes that evolutionary psychologists promote and rely on to construct the very same facile adaptationist scenarios that Massimo rejected.[34] This impasse can be avoided if we carefully distinguish between the biological foundations of language and the final product we call I-language (equivalently, between language and a language),[35] and take seriously the role of development leading from one to the other. This is all in good, current evo-devo style. It's also in good Marr (1982) style, who would have been pleased by the prospects of smooth integration among the various levels of explanation afforded by the mosaic vision of the language faculty.

---

[33] The same impasse, which Poeppel (2005) calls "cross-sterilization", is being felt in neurolinguistics, where modular, locationist, neo-phrenological assumptions continue to dominate.

[34] Jerry Fodor saw this very clearly, as the following passage from Fodor (1998) reveals: "If the mind is mostly a collection of innate modules, then pretty clearly it must have evolved gradually, under selection pressure. That's because [...] modules contain lots of specialized information about problem-domains that they compute in. And it really would be a miracle if all those details got into brains via a relatively small, fortuitous alteration of the neurology. To put it the other way around, if adaptationism isn't true in psychology, it must be that what makes our minds so clever is something pretty general".

[35] I once heard Leila Gleitman say that it's easy to learn language (it's easy because you don't learn it, it's innate), but it's hard to learn a language (so many factors, linguistic and non-linguistic come into play, from the level of words (see Bloom 2000) to the level of grammatical constructions (see Yang 2010, Boeckx 2010e). The same could be said about evolution (another parallelism between Plato's Problem and Darwin's Problem, between devo and evo, between ontogeny and phylogeny; cf. Boeckx (2009e): it's easy to evolve language (it's easy because you really don't evolve it gradually, most of it emerges spontaneously, epigenetically), but it's hard to evolve a language (the grammaticalization process is a long and slow). One level is a crystal, the other is a kluge.

CHAPTER TWO

THEORETICAL LINGUISTICS MEETS DEVELOPMENT: EXPLAINING FL FROM AN EPIGENETICIST POINT OF VIEW*

VÍCTOR M. LONGA
AND GUILLERMO LORENZO

According to Crain and Pietroski (2002: 172), "One can't avoid nativist conclusions by refusing to do linguistics". We think they are right, but we urge to add that one can't avoid reflecting on the ultimate meaning of nativist conclusions by refusing to do some biology. So in this chapter we aim to reappraise the question of linguistic nativism from both a biological and a linguistic point of view. In so doing, we depart from the strong geneticist stance of biology and (generative) linguistics of the second half of the 20th century and embrace an increasingly influential developmentalist standpoint which is presently gaining popularity in the study of organic systems, including cognitive faculties.

One of the reasons for conceptualizing the Faculty of Language (FL) as an organic system is that it follows a characteristic developmental schedule (Lenneberg 1967). The completion of such a schedule depends on environmental stimulus in rather obvious ways, but it is customarily assumed that its unfolding exhibits program-like properties and, as a consequence, one can safely conclude that it obeys a plan somehow encoded within the human genotype. An extreme version of this view is that genes encode not just a chart of the developmental process leading to

* This work has been carried out through the project *Biolingüística: evolución, desarrollo y fósiles del lenguaje* (FFI2010-14955/FILO) of the Spanish Ministerio de Ciencia e Innovación and partially co-funded by FEDER funds.

fullly-fledged mature language, but also a blueprint comprising the essentials of particular linguistic systems. Nothing thus far said comes as a surprise, as it represents a predictable adhesion to the preformationist stance that pervaded biological thinking for the most part of the past century by linguists accepting the organic status of FL.

However, a growing body of literature convincingly shows that the genocentric stance of contemporary biology is extremely problematic for several reasons. Firstly, it entrusts the genes with tasks they obviously cannot carry out—such as containing timetables or drafts. Genes are not representational devices endowed with intentional powers—apart from a metaphorical, but explanatorily empty interpretation of what they do. Secondly, it is also at this metaphorical level that the contention is made that genes comprise programs, the running of which relies on data provided by the environment. This kind of parlance can be useful as a first approximation to recalcitrant problems—as developmental issues indeed are, but cannot be sustained for long, at the risk of being taken as true explanation. And finally, by exclusively focusing on genes this view ignores the putative causal role in development of many other non-genetic factors, in the absence of which genes are "among the most impotent and useless materials imaginable" (West-Eberhard 2003: 93).

This chapter relies on the conviction that the biolinguistic approach is mature enough to accept the challenge of articulating a theoretical model which frames the explanation of FL as an outcome of complex, but still decipherable developmental processes, instead of presupposing and ultimately ignoring the fact of development. So the first part of the chapter is devoted to a critical scrutiny of current geneticism in biology at large and to present the advantages of taking an alternative epigeneticist view on development, theoretically updating ideas of long and venerable tradition. This done, the second part of the chapter focuses on how the study of language can benefit from this non-geneticist, non-deterministic, and pluralistic view of development, and establishes the bases for applying it to the explanation of a particularly significant property of linguistic systems.[1] Our main theoretical conclusion is that the view on linguistic

[1] As we are aware that the structure and the argumentative strategy of the chapter mimics that of Fodor and Piattelli-Palmarini's (2010) critique of Darwinian thinking, we have decided to (partially) name the main parts of our chapter after the titles of their book's first and second parts. Besides, there exist obvious conceptual connections between both enterprises. See also Boeckx (2010e), a paper similarly inspired in Fodor and Piattelli-Palmarini's book and congenial with our chapter.

development defended in this chapter is congenial with the Minimalist Program—the latest outcome of generative linguistics (Chomsky 1995, and subsequent works), but just under a particular interpretation in which this theoretical trend is seen as a relative rupture from previous incarnations of generativism, such as Principles and Parameters Theory (Chomsky 1986).

1. The biological argument: Two (not equally plausible) conceptions about the generation of form

It seems evident that the primacy of the gene as the core explanatory concept of biological structure and function is more a feature of the twentieth century than it will be of the twenty-first.
(Keller 2000: 9)

One of the more important and complex issues in biology is how it is possible that the egg cell develops into an embryo and ultimately into an adult being. Such an issue raises many questions: How is biological form generated? Where does it come from, that is, what are its sources? How does it develop? Where does the information required for development reside? Questions like these have been at the very heart of biology and a strong controversy has traditionally surrounded them. A pretty clear example, very relevant for our purposes, took place in the 17th and 18th centuries, when a heated debate arose about how generation and development of biological form should be regarded. Disagreement gave place to two competing positions, "preformationism" and "epigenecism" (see Müller and Olsson 2003 and Maienschein 2005; for an in-depth analysis, Pinto-Correia 1997 and Roe 1981). Preformationists considered that the fetus preexisted as a homunculus (a fully formed being in miniature), located either in the ovulum ("ovists") or in the sperm ("animalculists"). According to epigenesists, the fetus did not preexist at all, but developed in a gradual and successive manner.

More than two centuries later, updated versions of the two aforementioned positions (with no appeal to the homunculus, in the case of preformationism, or vitalism, in the case of epigenecism) continue to be in conflict. Modern preformationism, represented by Neo-Darwinian genocentrism and based on the notion of genetic program, has dominated the biology of the 20th century. However epigenecism, represented by what has been called the "developmentalist challenge" (Griffiths and Knight 1998, Weber and Depew 2001; see section 1.2.) has increasingly gained credibility in the last few decades. The next two sections are devoted to a brief presentation of both positions.

1.1. Modern preformationism

As biology historically advanced, it became clear that preformationism lacked empirical support, for the supposed homunculus that the theory relied on did not exist at all. However, two centuries later the Neo-Darwinian movement that grew up from Evolutionary Synthesis resurrected the preformationist position, assuming a stance according to which "embryonic development only consists of the elaboration of the preexistent form" (Müller and Olsson 2003: 115).[2]

This point requires further clarification. Neo-Darwinism did obviously not accept the existence of the homunculus, but nevertheless resurrected preformationism on the base of the notion of "information", a quality exclusively attributed to the genetic level of analysis. Accordingly, genes were given priority over other developmental levels and resources, in such a way that they were considered the only possessors of the essential information guiding the growth and maturation of organic structures. This is the idea underlying "genocentrism", a key Neo-Darwinian stance (also shared by strong geneticist versions of Evo-devo) that entails endowing genes with a "special directive, formative, or informative power" (Oyama 2001: 178). The following quotes illustrate the point:

> The instructions for assembling every organism on the planet—slugs and sequoias, peacocks and parasites, whales and wasps—are all specified in DNA sequences […] (Lander and Weinberg 2000: 1777)

> […] each egg contains, in its genes, a set of instructions for making the appropriate adult. […] it is the information contained in the genes that specifies the adult form. (Maynard-Smith and Szathmáry 1999: 2)

---

[2] For justification of the preformationist nature of Neo-Darwinism, see Bateson (2001), Gottlieb (2001), Griffiths and Stotz (2000), Lewontin (2000), Longa (2008), Maienschein (2005), Moore (2001), Oyama (2000a) and Weber and Depew (2001), among others. Those scholars use several labels in addition to 'preformationism': 'neo-preformationism' (Griffiths and Stotz 2000: 34), 'weak preformationism' (Weber and Depew 2001: 241) or 'new predeterminism' (Maienschein 2005). All these labels make the same point: "like the old preformation theory of the embryo it [modern preformationism] denies that the order manifested in the developed organism actually originates during development" (Griffiths and Stoz 2000: 34).

In the entire complement of DNA of a species (the genome) there exists the information for building that animal. The instructions for making five fingers, or two eyespots, or six legs, or black and white stripes are somehow encoded in the genomes of the species that bear those traits. (Carroll 2005: 35)

These quotes also reveal a central tenet of molecular Neo-Darwinism: while classical Neo-Darwinism defended that the information required for development lies in genes, its molecular version adds the core assumption that such information is coordinated by means of a 'genetic program'. This notion arose from the discovery of the difference between regulatory and structural genes in Jacob and Monod's research on enzyme induction in *Escherichia coli*:

The discovery of regulator and operator genes [...] reveals that the genome contains not only a series of blueprints, but a coordinated program of protein synthesis and the means of controlling its execution. (Jacob and Monod 1961: 354)

Thus, genes are equated with a self-contained and self-sufficient program, which includes both information concerning patterns of structural organization and guidelines for their unfolding in space and time: "The development of form depends upon the turning on and off of genes at different times and places in the course of development" (Carroll 2005: 11). According to this view, the sources of development are rooted in the information contained within nuclear DNA and development entails spreading out something already contained in the genes: hence the preformationist position.[3] So Lewontin's (2000b: xii) question does not come as a surprise: "What important difference is there except in mechanical details between a preformed individual and all the information necessary to specify that individual?" In both cases, "adult organisms are merely expanded versions of the fertilized egg" (Bateson 2001: 156).

Consequently, Neo-Darwinism has assumed the idea that the genome is the true source of form and that the answer to the question of how form arises is explained through a prior instance of that very form, which is the central tenet of the notion of genetic program (Oyama 2000a: 1). Genes do have a central role: they are the main (or unique) causal agents, whereas the remaining developmental factors are supposed to have a merely supportive role, lacking information in any relevant sense. The result is a strong asymmetry that generates a hierarchy of causes (Oyama 2000a: 17).

---

[3] As Mayr (1982: 106) self-acknowledges, development "is controlled by something preformed now recognized as the genetic program".

Genetic causes are essential and the other causes are thought to be mere details inserted in a preformed structure. According to Robert (2004: 39), a "modern consensus on development" exists that unites all this into three basic tenets:

(1) genetic informationism: genes are the only resources that possess developmental information;
(2) genetic animism: that information consists of a genetic program in the zygotic DNA which controls development; and
(3) genetic primacy: genes are the only units of heredity, the main promoters of development, and the elements offering both the materials and organization for development to proceed).

It is thus clear that, according to this position, development is simply bringing something preexistent to light, as aptly explained by Monod (1970: 87):

> No preformed and complete structure preexisted anywhere; but the architectural plan for it was present in its very constituents. It can therefore come into being spontaneously and autonomously, without outside help and without the injection of additional information. The necessary information was present, but unexpressed, in the constituents. The epigenetic building of a structure is not a creation; it is a revelation.

1.2. Modern epigenecism

Genocentrism is currently being contested by a number of theories, like developmental psychobiology, developmental systems theory or Jablonka and Lamb's theory of the parallel inheritance channels. Although these views have different assumptions and conceptions, they share the aim of "taking development seriously", to quote from Robert's (2004) title. That perspective, which can be dubbed as "developmentalism",[4] is fostering "a new synthesis which challenges the gene-centered version of Neo-Darwinism that has dominated biological thought for the last fifty years" (Jablonka and Lamb 2005: 1).

---

[4] For a presentation of its tenets, see Blumberg (2005), Gottlieb (2001), Johnston (1987), Johnston and Edwards (2002), Lickliter (2008), Moore (2001), Oyama (2000a), Oyama *et al.* eds. (2001), Robert (2004), Smith and Thelen (2003), or Spencer *et al.* (2009).

The defining feature of developmentalism is the rejection of any vestige of preformation, as the view is committed to showing that any attempt to unravel development should be built around the notion of epigenesis, which states that:

> An organism develops through a process in which the proximate cause of each step is the total set of interactions at the immediately preceding state, starting with the procreative act, moving through the differentiation and articulation of physical and psychological traits, and ending with the initiation of another cycle of life. (Weber and Depew 2001: 241)

According to developmentalism, therefore, traits (physiological or cognitive) cannot be specified in advance nor contained within the genome. This amounts to saying that the notion of genetic program as the only source of information for development distorts the understanding of developmental processes, for it ignores the contribution of many factors and resources located between genes and traits, without which development would not take place. This move has important consequences, as it breaks the conflation between form and genetic codification, causes the rejection of the genome as the source of organic form, and promotes the denial of the preexistence of forms. Instead, it defends that forms progressively emerge during development itself, which entails the rejection that genes are the main or unique causal agents in organic growth, as well as the associated theses of asymmetry and hierarchy of causes explained above.

The next section is devoted to show why development cannot simply be reduced to the action of genes.

1.3. On the notion of genetic program—and why it is untenable

This section offers a synthesis of the reasons highlighted by developmentalism in order to cast doubt on the primacy of genes, as envisioned by preformationists. As we will show, the key tenets underlying the notion of genetic program are highly problematic.

1.3.1. Genes are central because they have causal primacy

The Neo-Darwinian assumption that genes have a central role can hardly be justified. An example from cellular biology illustrates the point. To develop, the cell egg must contain many structures in advance without which DNA is completely useless: basal bodies, cytoplasmic chemical gradients which are crucial for the differentiation of the embryo, patterns of DNA methylation, cell membranes or organelles, in addition to DNA itself (Moss 2003: Ch. 3). For example, the cell must have an organization based on membranous partitions, not derived from genetic dictates, which is responsible for its three-dimensional structure (Moss 2003: 80). The source of this organization "is passed on from one generation to the next by way of the maternal egg cell" (Moss 2003: 95): in other words, it is epigenetically inherited.[5]

To further appreciate the point, some other facts can also be pinpointed. For example, within the cell membrane (a membrane which constitutes the very boundary of the cell), membrane-based cellular structures exist which extend from the nucleus to the cell membrane. These structures are central for DNA to be interpreted, so the relationship between the membrane system and DNA "is that of codependence and causal reciprocity" (Moss 2003: 77). Besides, all proteins except those whose target destination is the nucleus of the cell enter the membrane system at the endoplasmic reticulum (the outer surface of the nucleus). The proteins that pass through the endoplasmic reticulum and reach the membrane system possess a signal sequence at the end of the polypeptide chain that is encoded in DNA but, according to Moss (2003: 82), "its function can be realized only in the context of a receptor complex already present in the ER [endoplasmic reticulum]". Finally, the movement of proteins whose targets are the different levels of the membrane system is constrained by several types of filamentous structures, which are known as cytoskeleton. The conclusion is that protein transport is crucially based on non-DNA activity (see Moss 2003: 87-90).

To sum up, non-DNA cellular structures must pre-exist in order for the genes to be interpreted. Otherwise they would be biologically pointless:[6]

> Biochemically distinct membranous bodies constitute the necessary and irreplaceable templates of their own production and reproduction, are passed along from one generation to the next, and provide the unavoidable context in which DNA can be adequately interpreted, that is, in which *genes can be genes*. (Moss 2003: xviii)

In the light of the example, the causal primacy of genes seems a doubtful contention. Instead, a deep causal equality among different developmental interactants seems a much more reasonable position, which

---

[5] Therefore, "If cellular membrane organization is ever lost, neither 'all the king's horses and the king's men' nor any amount of DNA could put it back together again" (Moss 2003: 95).

[6] The discussion also nicely illustrates that "a gene initiates a sequence of events only if one chooses to begin analysis at that point" (Oyama 2000a: 40).

substantiates in Oyama's (2000a: 201) "parity of reasoning" (Oyama 2000b for discussion) and Griffiths and Knight's (1998: 254) "parity thesis". According to these proposals, the importance of genes in development can hardly be denied, but if the cell can interpret DNA sequences at all it is because it possesses a complex non-DNA machinery formed by elements and resources as necessary as DNA itself. Therefore, "It makes no more and no less sense to say that the other resources 'read off' what is 'written' in the genes than that the genes read off what is written in the other resources" (Griffiths and Gray 1994: 284).

So according to the developmentalist position, development does not originate from any essential cause, but emerges from interactions among many genetic and non-genetic resources. Genes cannot be endowed with a special formative power, as they require a context that is as important as genes themselves.

1.3.2. Genes contain blueprints or master plans of organisms

Genes cannot pre-specify the plans of organisms—or its component parts. The reason is that the only information that can safely be attributed to genes is the sequence of amino acids that forms the primary structure of proteins (Bateson 2001, Godfrey-Smith 2007, Griffiths 2001, Johnston 1987, Moore 2001, among many others). This makes the idea that genes hoard phenotypes in any relevant sense unfeasible. As Bateson (2001: 157) puts it, "Genes store information coding for the amino acid sequences of proteins; that is all. They do not code for parts of the nervous system and they certainly do not code for particular behavior patterns".

1.3.3. Genetic programs are self-sufficient and self-contained

As pointed out by Gottlieb (2001: 47), the divide between essential or fundamental causes and accessory or merely supportive causes leads to assuming the encapsulation of the genetic material, isolated from any supragenetic influences. This view reinforces the idea of the genome as a blueprint of the organism. However, genes are highly dynamic entities and highly sensitive to contextual contingencies (Oyama *et al.* 2001: 3, Robert 2004: xiii). The following phenomena are illustrative of this point:

(a.) Alternative splicing causes the same DNA sequence to give rise to several mRNAs (messenger RNA) and, consequently, to alternative proteins,[7] depending on the specific context of its production. Accordingly, the function of genes does not solely depend on their own sequential composition, but also on their specific genetic, epigenetic, cytoplasmic, or metabolic context. Thus, as explained by Keller (2000: 71-72), the concept of 'gene' actually refers to at least two different entities: a 'structural' gene, or DNA sequence, and a 'functional' gene, which emerges from complex interactions, the structural gene being only one of the interactants.

(b.) Genes specify the amino acid sequences of proteins, but crucial aspects like the three-dimensional structure of protein molecules, on which functional specificity crucially depends, does not derive from genetic information at all. As a matter of fact, this structure depends on certain parameters of the internal (i.e. intracellular) environment, like temperature or pH (Johnston 1987: 160).

(c.) The biological function of proteins does not only depend on their accurate synthesis, but also on pre and post-translational modifications and their target destination within the organism (see Mattick *et al.* 2009, and Moss 2003: Ch. 3). So in order to become functional, many proteins must be modified after their synthesis and, in addition, many of them require integration within multiprotein complexes. Besides, they must act in the right cellular or extracellular locations which requires complex intra and extracellular transport mechanisms. As stated by Carroll (2005: 208), "everything about a toll kit protein's action depends on context".

(d.) The process of cell differentiation (which in human beings gives rise to about 350 different types of cells from the egg cell) depends on conditions arising in other cells of the internal environment (Moore 2001: Ch. 5). As many experiments with embryos demonstrate, the final form of a cell "is not determined solely by information contained within the cell", but "differentiation results from interactions between neighboring cells" (Moore 2001: 87)—a fact that can be particularly appreciated in the first stages of embryonic development, where the key factor is the position occupied by the cells in the gastrula.

Examples like these suggest a very different picture from the alleged self-contained and self-sufficient character of genetic programs, an idea nicely captured by Gottlieb's (1997, 2001) concept of 'probabilistic epigenesis', according to which development is guided and constrained by

[7] As explained in Jablonka and Lamb (2005: 66-67), the "cSlo" gene, which expresses in the hair cells of the chicken inner ear, has the astonishing number of 576 alternative splice variants.

complex and multidirectional interactions holding at many levels (genetic, epigenetic, neural, behavioral, environmental, an so on).

1.3.4. The genetic program is the only source of developmental information

The genetic program is customarily seen as the only entity endowed with relevant developmental information, all remaining factors just providing raw material and support for development to take place. This intimate link between genes and information is the main reason to degrade the developmental role of factors other than the genetic one (Maynard-Smith 2000; for discussions on this point, see Godfrey-Smith 2007, Griffiths 2001, and Jablonka 2002).

However, the only information that can be safely attributed to genes is that which codes the amino acid sequences of proteins, as already discussed in 1.3.2. Besides, the reason for which the genome is supposed to contain information can be equally applied to other developmental factors, as the following examples illustrate:

(a.) "Genomic imprinting" is a mechanism related to "epigenetic inheritance" (non-genetic cellular inheritance; see Jablonka and Lamb 2005), which "can cause one member of a pair of genes to be silenced while the other is expressed" (Blumberg 2005: 61). This fact is rooted in the progenitor from which the relevant gene derives and due to the different marks possessed by both alleles—for example, DNA methylation, where a methyl group silences a DNA sequence. Thus "some genes become activate only if they have a paternal 'imprint', others only if they have a 'maternal' imprint" (Jablonka and Lamb 2005: 254). The phenomenon affects how genes respond to cellular signals during development, in such a way that the differential activation of the alleles may have phenotypic consequences, obviously not due to the chain of nucleic acid. Consequently, it seems clear that epigenetic inheritance phenomena offer relevant developmental information.

(b.) "Behavioral inheritance" is a similar case. We can adduce an example, taken from Jablonka and Lamb (2005: 162 and ss.), which refers to substances influencing early food preferences in rabbits. Although the mother does not provide pups with information about food sources—she does not instruct them about what kinds of food they should eat or avoid, they know what is good and safe to eat. Actually, the pups know that because their mothers give them that information in the form of chemical cues in the uterus through the amniotic liquid and placenta.

Both examples show that many developmental resources should be given the same informational status as the genome. They also illustrate why, according to developmentalism, traits cannot be said to preexist, as information for development is actually distributed among many different sources. As Griffiths and Knight (1998: 258) conclude: "Development is an orchestra with no conductor".

1.3.5. Genes are the only currency of inheritance

According to the premises of Neo-Darwinism, inheritance is a purely genetic matter, an idea that fits in well with the notion of genetic program and associated assumptions: "the development of complex organisms depends on the existence of genetic information, which can be copied by template reproduction" (Maynard-Smith and Szathmáry 1999: 2; see also Lander and Weinberg 2000). Complementarily, "nongenetic inheritance is either something that does not exist or something without evolutionary significance" (Mameli 2004: 36).

However, even if one accepts the standard mechanism of natural selection—survival and differential reproduction that expresses fitness differences, the theory only requires the existence of some type (or types) of inheritance, without imposing an exclusively genetic mechanism. As a matter of fact, several developmentalist theories now defend that "there is more to heredity than genes" (Jablonka and Lamb 2005: 1) and, accordingly, they are replacing the traditional notion of inheritance with an "extended" version which incorporates many extra-genomic components (see Blumberg 2005, Griffiths and Gray 2001, Jablonka 2001, Jablonka and Lamb 2005, Longa 2009, Mameli 2004, 2005, or Sapp 2003). Furthermore, an important consequence of non-genetic inheritance is that it breaks down the traditional Neo-Darwinian conflation between "genetic" and "innate"—see 1.3.8 below.

1.3.6. Genes code for phenotypical traits

The notion of "genetic program" implies that traits are directly codified in the genes by means of a one-to-one relationship, a received idea that derives from classical (Mendelian-Morganian) genetics (see Moore 2001: Ch. 7, for a general discussion, and Johnston 1987: 160, for the particular case of behavioral traits). This conception, which assumes that a gene is both a necessary and sufficient condition for a trait to develop, is highly problematic. Mutated versions of genes can actually induce characteristic phenotypical effects—as in the case of *FOXP2*, mutated versions of which can induce certain linguistic impairments, but it cannot be inferred from this fact that the non-mutated versions of the corresponding genes are the ultimate originators of the relevant 'normal' traits. This would be equivalent to claiming that a unique gene is responsible for the development of a particular trait, when in fact the route from genotype to phenotype is much more complicated. As Jablonka and Lamb (2005: 6) comment, this simplistic assumption underlies "the idea that there is a gene for adventurousness, heart disease, obesity, religiosity, homosexuality, shyness, stupidity, or any other aspect of mind or body has no place on the platform of genetic discourse", which pervades (and distort) explanations of development at the behavioral level.

1.3.7. Genotype and phenotype are linked by a simple function

A central idea of the 'modern consensus on development' is that development derives from an asymmetrical interaction between genes and environment: genes are central, while environmental inputs simply trigger or complement their activity. This view, referred to as 'popular' (Oyama 2000a) or 'conventional' interactionism (Oyama 2000c), entails ignoring the vast range of biological factors and levels which mediate between genes and environment, turning them into useful developmental units. Traits only come into existence due to interactions among many developmental resources, concurrently working at the right time and place. These resources include: DNA sequences, the membrane-based cellular structure, mRNA, the extracellular matrix, cytoplasm, hormones, enzymes, chemical cues, and so on; but also external factors like behavior, habitat, temperature, sunlight, social structure, nutrition, or niche construction.

So genes are simply a piece of the whole developmental puzzle, sometimes acting very early, sometimes being recruited along the tortuous path leading to mature traits. Their effects are very indirect, because many other developmental resources intervene in the causation of phenotypes beyond them. Furthermore, development is not a linear concatenation of stages, nor a unique or direct function of the genotype, but a continuous transformation where "biological products are built up, deformed, broken down, distributed or deformed" (Oyama 2000a: 133). As Robert (2003: 96) puts it:

Constructive causal interactions in development involve inducing, facilitating, maintaining, and participating in time-sensitive feedback loops at multiple levels within and beyond the developing system—only some of which might be characterized as gene activation. The interactions comprising organismal development are complex, and their effects are not simply additive.

To sum up, causal interactions underlying developmental episodes do not simply hold between genes and environment, but also affect the whole range of developmental factors (Johnston and Edwards 2002) and imply complex 'cascading events' (Moore 2001: Ch. 4). Thus, a sophisticated notion of epigenesis is required, according to which "development emerges via cascades of interactions across multiple levels of causation" (Spencer *et al.* 2009: 79), far from the idea that organic forms are predetermined in a single and somewhat central level of analysis.

1.3.8. If it is not in the environment, then it must be in the genes

Ignoring all the intermediate factors between genes and environment is a strategy that directly connects with which Stich (1975) calls the 'input-output' model: if a mismatch exists between poor environmental conditions—or experience—and the rich structural pattern of organization —or state of knowledge—ultimately attained, these results must be conceptualized as 'innate'. This non-problematic conclusion, however, is routinely translated into the problematic idea that if they are innate, they must be somewhat present in the genes.

Let us consider again the example about food preferences in rabbits—1.3.4 above. Knowledge about food sources in young rabbits does not derive from experience (they are not instructed by their mothers), so the input-output model would automatically attribute this knowledge to the genes. A wrong conclusion—as we saw before, in that it ignores that information can also come from other organism-internal sources. As Wimsatt (1999: 160) concludes: "Equating 'innate' with 'genetic' is a kind of functional localization fallacy", which wrongly implies "assuming that the function of a larger system or subsystem is realized completely in a part of that system".

As an alternative, one can embrace a phenotypic conception of traits, based on which Stich (1975) refers to as a 'dispositional model' (see also Maclaurin 2002). Within this model, traits are explained as the result of multiple informative factors provided by highly heterogeneous sets of developmental resources, and innate traits are simply those that reliably appear at certain points of a species-typical path of development (see Longa and Lorenzo 2008: 547). As Maclaurin (2002: 126) puts it, all that is required in order to deem a certain trait innate is "that there exists within the population some mechanism or process that maintains the developmental resources which very reliably produce the trait in question".

1.4. Does the idea that genes contain traits really explain development?

The reasons thus far discussed convincingly show that:

> Despite the widespread talk of genetic blueprints and programs in contemporary biology, there is no scientifically defensible sense in which a subset of developmental resources contains a program or a set of instructions for development. (Oyama *et al.* 2001: 5)

Furthermore, assuming the genetic program metaphor transforms development into a theoretically irrelevant notion, as already pointed out by Lehrman (1953), with the consequence of barring from explanations a plethora of factors in the absence of which the task of understanding how organisms—and their component parts—come into existence becomes hopeless (Ariew 1999, Oyama 2000a, Johnston 1987, Moore 2001, and Spencer *et al.* 2009), a lesson that we should promptly project to the particular case of the language faculty.

2. The conceptual situation: On orthodox and unorthodox forms of being a linguistic nativist

Attributing behavior to the genes does not solve the problem of its development, it simply refuses to address the problem.
(Johnston 1987: 153)

We argued in the previous section that preformationism leaves the fact of development in any organic realm unexplained. Cognition—by which we mean a significant part of the nervous system's activity—is not an exception to this remark. Preformationism, however, is the idea underlying most versions of nativism as applied to cognitive capabilities conceptualized as "instincts" or "faculties", a blueprint of which is thought to exist written in the symbolic code of genes. This is the "consensus view" in the case of linguistic theories with a biological orientation (Hornstein *et al.* 2005), a case worth considering in special detail as the concept of "language-as-a-faculty" (Chomsky 1965)—or "language-as-an-instinct" (Pinker 1994)—has been a main source of inspiration and has permeated cognitive science at large, starting with the ground-breaking ideas put forward in Chomsky (1959a). So, it comes as a surprise to discover that Chomskyan linguistic nativism—an unmistakable form of preformationism—does not necessarily follow from the argument from which it is said to be the consequent.

This fact does not run against the "nativist" part of Chomsky's thesis, but only against its purportedly necessary association with a "preformationist" stance, as customarily assumed within Chomskyan quarters (Jenkins 1979, Chomsky 1980, Lightfoot 1982, Thorton and Wexler 1999, Wexler 1999, Anderson an Lightfoot 2002, Guasti 2002, Hyams 2002, and Yang 2004b, just to mention some representative examples). In this section we argue that this "orthodox" form of Chomskyan nativism is problematic for reasons that go beyond those reviewed in the previous section and suggest the general lines of an "unorthodox" epigeneticist alternative. Some important consequences of this perspective, concerning the theory of UG and the integration of Theoretical Linguistics within the normal practice of Biology, are examined at the end of this section.

2.1. Poor stimuli, rich genes—ain't necessarily so

Chomsky's nativism derives from the "Argument from the Poverty of the Stimulus"—henceforth APS. The roots of the argument can be found in Chomsky (1959a), a critical review directed against the then dominant form of linguistic inductivism held by behaviorist psychologists (Skinner 1957). There, in addition to defeating behaviorism for both empirical and conceptual reasons, Chomsky made observations for the first time concerning the scarce, incomplete and episodic character of linguistic data available to children in many non-exceptional learning scenarios, and their promptness to understand and use previously unheard expressions. All these facts pointed to the idea that language acquisition looked like an "internally-driven" process more than a "stimulus-driven" one (Chomsky 1959a: 42)[8]. Chomsky did not commit himself to a particular theory as being capable of satisfactorily dealing with these facts then, but he had already pointed out—inspired by ethological literature (Tinbergen 1951) and some yet unpublished material from Lenneberg—that some form of genetic programming was an idea worth serious consideration. So, in 1959 the idea of a "language acquisition device" (LAD) was almost in place, yet not connected to any theoretical model as to how it worked. Actually,

[8] This fragment seems to have escaped Thomas' attention in her otherwise thorough historical reconstruction of APS (Thomas 2002). She points out some other fragments from Chomsky's earliest works in which he refers in passing to the "finite" (Chomsky 1955/1975a: 61, 1957: 15), "limited" (Chomsky 1955/1975a: 61) and "accidental" (Chomsky 1957: 15) character of children's exposure to linguistic stimuli.

Chomsky did not completely discard the possibility that it might consist of a system of bias, the maturation of which would enable the child to acquire a particular grammar by entirely inductive means (Chomsky 1959a: 44), but he seemed to favor the idea that such an extremely complex accomplishment would be out of reach of children were it not "based on a prior understanding of the structure of grammars" (Chomsky 1959a: 57).

The conjecture that all grammars share a universal component (UG), the knowledge of which does not obtain from experience—as witnessed by the fact that it mostly deals with superficially opaque properties of grammars—had to wait until Chomsky (1965), where the connection is definitely and firmly established between the concepts of LAD and UG—the latter seen as an inborn structure that eases children's path towards mature fully-fledged grammars. This is close to saying that UG is genetically encoded, an idea that became a common topic from Chomsky (1980) on, when UG was categorically presented as a particular aspect of the human genotype:

> The argument from the poverty of the stimulus leaves us no reasonable alternative but to suppose that these properties are somehow determined in universal grammar, as part of the genotype. (Chomsky 1980: 66)[9]

In the meantime, Chomsky limited himself to say that the basic principles underlying the knowledge of every particular grammar "are determined by the nature of the mind" (Chomsky 1967: 9)—clearly, not the same thing as seeing them as directly expressing particular fragments of the human genotype.[10]

This historical overview has an important moral: APS is in the background of Chomsky's genetic preformationism, but the connection was not always historically present in Chomsky's thought. From our present perspective, this should not come as a surprise, as there does not exist a necessarily conceptual bridge between poor stimuli and rich genes. This means that in order to establish such a connection some further

---

[9] Or, in a similar vein: "From the point of view I have adopted, universal grammar and the steady state are real. We expect to find them physically represented in the genetic code and the adult brain, respectively" (Chomsky 1980: 82-83).

[10] Some brief comments regarding the genetic determination of language as a mental organ can also be found in Chomsky (1975b). However, Chomsky's modest claim in such passages is that Biology can seriously attempt the task of finding "the genetic mechanisms that guarantee that the mental organ, language, will have the required character" (Chomsky 1975b: 91-92)—which, again, is not the same thing as seeing UG "as part of the genotype" (Chomsky 1980: 66).

assumptions have to be made, not always explicitly declared or sufficiently justified by generative grammarians. A critical examination of this state of affairs reveals that the conceptual foundations of Chomsky's grammatical geneticism are not as strong as customarily assumed.

In the following paragraphs we unfold this criticism. We conclude that the shortcomings of the "orthodox" implementation of linguistic nativism in a preformationist key can be overcome by the "unorthodox" variant of the innateness hypothesis, according to which innate features are not literally inborn but developmentally (or epigenetically) attained.

2.2. The truth of poverty

Linguistic stimuli are said to be "poor" in the sense that children would never attain complete knowledge of their mother languages, were they solely to depend on the data offered by their caretakers. A particular aspect of this "poverty" property—customarily seen as its more radical manifestation—is that children show very early competence regarding complex grammatical processes for which the environment does not seem to provide models required for a "data-driven" learning process (Hornstein and Lightfoot 1981). A hotly debated example of this sort is the formation of polar interrogatives by auxiliary fronting in languages like English, a process that children promptly master as governed by a "structure-dependent" rule ("front the *main* auxiliary"). In this particular case, stimuli are said to be poor because most samples offered to children are sentences in which the auxiliary in the corresponding declarative is the *first* auxiliary as well—as in (1), so it seems that the character of the corresponding rule (is it linear or structure dependent?) is left underdetermined by the data:

(1) a. *The dog is hungry*
b. *Is the dog hungry?*

Despite this fact, children never entertain the "linear-dependent" rule, as convincingly shown by Crain and Nakayama's (1987) experiments in which children from 3;2,[11] elicited the construction of interrogatives of the desired type in response to prompts such as *Ask Jabba if the boy who is watching Micky Mouse is happy*, never produce errors like (2)—which would correspond to the linear interpretation of the fronting rule:

[11] Children's awareness of the structure dependent character of rules at an earlier age is experimentally proven in Lidz *et al.* (2003).

(2) *Is the boy who watching Micky Mouse is happy?*[12]

The curious fact about this finding is that, as previously believed on just intuitive grounds[13], models of the corresponding correct construction—interrogatives with auxiliary fronting across a complex subject containing another auxiliary element, like (3)—are almost lacking from any known data-base of child directed speech (Legate and Yang 2002) and other sources of spontaneous adult speech (Sampson 2002):

(3) *Is the boy who is watching Micky Mouse happy?*

In any event, were we to admit that sentences of this type are used at higher frequency rates than previously thought—as argued by Pullum and Scholz (2002) on the basis of a particular corpus of written journalistic language the situation would not change in any significant way. As observed by Lasnik and Uriagereka (2002), the large majority of samples offered to children continue to be of the underdetermined type in (1), so data alone do not explain children's rejection of sentences like (2) from such an early age: they are clearly biased from the start in entertaining no other possibility than the "structure-dependent" rule.[14]

Children attain their grammatical competence very early in domains in which relevant models are scarce or even non-existent—(3), or frequent but hardly informative—(1b). So the conclusion is that they are

---

[12] They produce errors, of course, the most typical of which consists in repeating the auxiliary, as in: *Is the boy who is watching Micky Mouse is happy?* Such an example can be ambiguously interpreted as including a second instance of either the "first" or the "main" auxiliary. However, the fact that children produce sentences like *Can the boy who is happy can see Micky Mouse?*, but not like *Can the boy who can see Micky Mouse is happy?* (Crain and Nakayama 1987: 234-236), clarifies the question in favor of the idea they are applying a "structure-dependent" rule also in these cases.

[13] As witnessed by statements like the following one, frequently raised by Chomsky in the 70s and 80s: "A person may go through a considerable part of his life without ever facing relevant evidence, but he will have no hesitation in using the structure-dependent rule, even if all his experience is consistent with hypothesis 1 [the linear-dependent rule]" (Chomsky 1975b: 32).

[14] Besides, as observed in Crain and Nakayama (1987: 257) and Freidin (1991), sentences like (3) give support to the grammaticality of the 'structure-dependent' rule, but they are uninformative regarding the ungrammaticality of the 'linear-dependent' rule. So, at best, they would serve to introduce something like a 'disjunctive' rule into a grammar obtained by inductive means.

compulsorily biased to "see" stimulus as structured according to a certain pattern of hierarchical organization—(4):

(4) a. [[*The dog*] *is hungry*]
b. [[*The boy who is watching Micky Mouse*] *is happy*]

This conclusion points to another aspect of the "poverty" property, as the input processed by children is not transparent with respect to such a pattern of organization. This is not the same thing as saying that primary linguistic data (PLD) consists of sets of barely organized sequences, with no properties at all children can take advantage of. On the contrary, the input is rich in acoustic information—for example, prosodic information, to which children are highly sensitive from a very early age (Mehler *et al.* 1988). So, the intonation change that typically comes after a subject NP could be thought to serve as a cue announcing the main component from which auxiliaries are to be extracted in the relevant conditions—as represented in (5):

(5) *The dog # is hungry*

However, this is a very "poor" signal in order to guide the correct structural analysis, as it is also fully compatible with—and actually closer to—an analysis in which two constituents just happen to follow each one after the other—as in (6), which would only serve as the basis for a rule of the linear type:

(6) [*the dog*] [*is hungry*]

So, stimuli are "poor", irrespective of their frequency, because cues indicative of structural organization are meager and ambiguous—which seem to us the core aspect of the "poverty" property, as it pervades every single sample of linguistic stimulation.[15]

To be fair, the fact that children never produce (pseudo) sentences like (2) could also be interpreted differently as customarily done: Instead of concluding that children never entertain the linear hypothesis (*front the first auxiliary*), one can also conclude that what really happens is that they never entertain certain structural hypotheses (*front a non-main*

[15] In Chomsky's words: "Sentences do not appear with brackets, intonation boundaries regularly marking phrases, subscripts identifying the type of phrase, or anything of the sort" (Chomsky 1975b: 32).

auxiliary)— somewhat unexpectedly, as such a hypothesis also seems fully compatible with their structural talent. This latter interpretation seems to us the most reasonable one, in the face of children manifesting what we would like to call "linear blindness" concerning linguistic inputs from a very early age. So, both (7a) and (7b) below being "structure-dependent" rules, why do children never seem to entertain the second possibility?

(7) If forming polar interrogatives by auxiliary fronting:
a. Front the main auxiliary
b. Front a non-main auxiliary

This is a point in need of clarification: children seem to be equipped with an inborn instruction like "construct a structure-dependent rule, ignoring all structure-independent rules" (Chomsky 1975b: 33), but this leaves why they favor certain "structure-dependent" options over others in constructing grammar particular rules unexplained. To our knowledge, this is an aspect of the question that has not received special attention in the literature.

A possible answer to this puzzle is that (7a) happens to be part of the innate component of grammar (UG)—with children using or forgetting it according to the properties of the target language (to which the heading of the rule refers). This is not a possibility to be completely discarded, but it is clear that it corresponds to a very costly conception of UG,[16] in that it foresees a lot of detailed contingent information concerning particular grammars—most of which is disposed of by many languages.

Another very promising answer is that primary linguistic data (PLD) of the type children indisputably have access to—(1b), are enough to secure that (7b) is a rule not worth consideration. This means that data alone could fulfill the role that the previous hypothesis delivers to UG instead.[17] Actually, it is not difficult to construct a "data-driven" story to explain it:

(i) Children are "linear blind" concerning linguistic inputs, so they cannot refrain from "seeing" sentences like (1a) as obeying, roughly, the

[16] Examples of which are for instance, Baker (2001) or Yang (2002).

[17] We think that this is also the alternative most congenial to Chomsky's original case, which was not directed as a defense that something along the lines of (7a) was part of UG, but simply as a particular illustration of a universal principle concerning the general "structure-dependent" character of particular rules of grammars.

structure in (4a)—as well as not being able to refrain from "seeing" (1b) as a sentence in which the *main* auxiliary has been fronted;

(ii) (1b) is a type of sentence not uncommon in normal linguistic environments, so children easily obtain a model to be applied as a first hypothesis (the "main auxiliary" strategy) from available data if an auxiliary is to be fronted in a partially new syntactic context—as in Crain and Nakayama's experiments;

(iii) As no sample defies the first hypothesis, it becomes fixed as a rule of general application.

Points two and three above correspond to the kind of facts routinely captured by means of connectionist models of learning—the current paradigm of inductivist or "data-driven" learning. However, the most relevant part of the explanation above is that the joint action of PLD and general learning strategies—the connectionist part of the story—would be pointless were it not for the fact that they operate on the solid ground of the structural scanning of the input based on an "innate-primed" procedure.[18]

This analysis departs from other more extreme versions of the "innateness hypothesis" in an important aspect as it is commonly held that children cannot derive from "negative evidence" information about the extent to which rules apply (see Marcus 1993, for an overview). This is seen by most generative oriented acquisitionists as a corollary of the "finite" character of PLD, which prevents the application of the "as *no* sample defies the first hypothesis" strategy (iii. above) as a legitimate learning tool. However, it is also customarily assumed that if a particularly significant type of data is frequent enough within PLD,[19] it can be credited with the role of fixing certain rules—as in the case of *there*-type expletives in relation to the "non pro-drop rule" in English (Hyams 1986), somehow defying the expectancy that new data disconfirming the rule could be in the way of entering the sample. Something along these lines can be thought of as also being operative in the case of the "main auxiliary" rule, as *is the dog hungry?*-type questions are common enough to fix the rule despite the expectancy that further data could disconfirm it. A difference is that *there*-type expletives and similar devices (like the use of the OVS order as a cue to fix the V2 parameter; see Yang 2002) are thought to be operative in deciding between cross-linguistically attested competing

[18] This is in agreement with the conclusions previously expounded, for example, in Fodor and Pylyshyn (1988) and Marcus (1998) concerning the limited learning capabilities of associative machines devoid of innate bias.

[19] See Legate and Yang (2002) on the issue of "frequency sufficiency".

possibilities, which obviously is not the case in the situation at hand. Our contention is, however, that the mechanism can also be thought to be useful to block cross-linguistically unattested possibilities—like (7b)—in the absence of UG specific instructions—like (7a). The conceptual grounds of the idea are spelled out below.

Our personal journey into the classic "main auxiliary fronting" motive is a little bit different from other accounts of the same issue, as we conclude that it also serves to illustrate how "data-driven" learning procedures—the combined action of PLD and general data-processing strategies, seem to be enough to acquire large portions of particular grammar systems. It is in this way that children acquiring English grammar probably learn, for example, (1) that polar interrogatives obtain by fronting an auxiliary, and (2) that the auxiliary to be fronted is the main one. That much is part of the idiosyncrasy of the English language that children have to learn from scratch. It is our conclusion, however, that such general learning procedures operate on the base of certain "innate-primed" biases—like the "structure dependency" expectancy,[20] in the absence of which the former would remain completely inert.[21] The resulting model is something akin to a "third way" solution to the problems of language acquisition, somewhat in between strong versions of nativism and inductivism, which we believe fully compatible with the strategies of "minimalist reduction" put forward in Chomsky (1995) and subsequent works.

Chomsky's project implies connecting different aspects of the design of language to "three factors that enter into [its] growth in the individual" (Chomsky 2005: 6), namely:

> [factor 1.] "Genetic endowment, apparently uniform for the species, which interprets part of the environment as linguistic experience";

---

[20] To which should be added children's knowledge concerning the "substantive" categories present in the rule ("polar", "interrogative" and "auxiliary"), the origin of which is an important open question.

[21] A note of terminological clarification is in order. We are using the terms "data-driven" and "innate-primed" as in, for example, Pullum and Schloz (2002), with the former referring to general learning or data processing strategies—i.e., not restricted to particular domains of application. However, such strategies are in themselves "innate" even for the most recalcitrant of the empiricists, as they are not learned from experience but part of one's own endowment to learn (see, for example, Cowie 1999). The terminology is misleading, but we think that confusion disappears as soon as we integrate the terms within the model we are ready to erect.

> [factor 2.] "Experience, which leads to variation"; and
> [factor 3.] "Principles not specific to the faculty of language" [such as] "principles of data analysis that might be used in language acquisition and other domains" [and] "principles of structural architecture and developmental constraints". (Chomsky 2005: 6)

Waiting for some criticism to be put forward in the next subsection, a correspondence between Chomsky's tri-factorial model and our own analysis of the "main auxiliary fronting" case can be provisionally established, as our point (i) above—the "structure-dependency" expectancy, and "innate-primed" bias—can be connected to factor 1, while our points (ii) and (iii)—the combined action of PLD and general data-analysis procedures—can be connected to factors 2 and 3, respectively. Besides, our analysis also fulfill the core minimalist strategy of charging the weight of explanations on factors 2 and 3, at the expenses of factor 1, which in the case under scrutiny means prioritizing the "data-driven" component of acquisition over the "innate-priming" one. Our conclusion is that the workings of an innate component within the LAD—albeit a residual one—still seems to be a sharp conclusion deriving from the "poverty" property so far discussed, impeding a more radical minimalist recipe based on the "data-driven" model alone. This is however a provisional conclusion, as we claim in the next subsection that the picture presented thus far can still be reworked within a more minimalist orientation.

2.3. On UGenes—or lack thereof

The traditional generativist strategy has been that of overcoming the under-determination of stimuli by over-determining UG, conceived of as a genetically encoded collection of grammar specific instructions. As Smith puts it in the context of a discussion on the phenomenon of "structure dependency":

> Either the primary linguistic data which the child learning its first language is exposed to are sufficient to determine the form of the grammar the child is acquiring; or the universal properties of language have become encoded in the genes of the children. If arguments for the poverty of the stimulus are correct then the first of these alternatives cannot be adequate. (Smith 1999: 173)

Things have substantially changed with the advent of the Minimalist Program (MP) (Chomsky 1995 and subsequent works), which basically means the introduction of new factors capable of fulfilling the same explanatory role but from outside the genotype. UG refers, as in previous generative models, to "mechanisms specific to FL" (Chomsky 2007: 5), directly given in the "genetic endowment for language" (Chomsky 2009: 19), that Chomsky identifies with factor 1 above. The difference is that the new minimalist framework favors factors 2 and 3 over this genetic component, seen as an explanatory "last resort" and, ideally, as a "residue" (Chomsky 2007: 26; 2009: 25). Remember that Chomsky explicitly says that they are "factors that enter into the growth of language in the individual" (Chomsky 2005: 6), so an evaluation of the impact of this program on an explanatory theory of acquisition cannot be irresponsibly put aside (Longa and Lorenzo 2008, Lorenzo and Longa 2009).

We provisionally concluded in the previous subsection that something along the lines of the instruction spelled out in (8) seems to be a good candidate to be part of UG in the sense above:

(8) "Construct a structure-dependent rule, ignoring all structure-independent rules."

We derived this conclusion, following standard practice, from a close scrutiny of the properties of PLD on the basis of the APS. The minimalist methodology now leads us to ask whether such a statement can be derived from factor (3)—the preferred solution, or must it rather be definitely conceptualized as a genetically encoded instruction of a domain specific character. Chomsky himself includes what he dubs "principles of structural architecture" within his own list of putative third factor effects. So the question is reduced to this: Do we really have compelling reason to include (8) within UG? In the following paragraphs we show that we don't.

From the computational perspective on mind, the one traditionally embraced by Chomskyan linguistics (Putnam 1961; Fodor 1975), the way sequential inputs are processed is a function of the "working memory" space connected to a "sequencer", both devices jointly composing a "computational system" (Gallistel and King 2009). Thus, whether sequences are processed in a strictly linear way or in a structural fashion is a fact to be derived from the memory power of the corresponding computational regimes. In the former case, the system will have the capacity of a "finite state automaton" while the latter will have that of a "push-down automaton". Technical details are unimportant here (see Balari and Lorenzo 2009b, and references therein). The substantive question is: How natural systems (brains) can attain computational powers equivalent to that of abstract (or artificially implemented) machines?

A reasonable conjecture is that neural networks extending forwards and backwards in the cortex (Aboitiz and García 1997, Aboitiz *et al.* 2006), crucially converging in the frontal area (Embick *et al.* 2000, Grodzinky 2000, Lieberman 2006, Moro *et al.* 2001, Musso *et al.* 2003), probably perform the task of a working memory component similar to that of artificial systems, a possibility explored at length in Balari and Lorenzo (2008, 2009b, and submitted). Again, technical details are not important. The decisive point is that the capacity of such networks to retain and, eventually, to structure items while performing a task is also reasonably to be thought to correlate to the size and strength the said networks attain during the development of different types of organisms.

This brief sketch of the ultimate anatomical basis of principle (8) does not preclude the possibility that a particular gene—or collection thereof—could be specifically devoted to guide the neural implementation of one of these networks in relation to the linguistic function. In that case—and only in that case, (8) could be sanctioned as belonging to UG in Chomsky's sense. This is a very unrealistic picture however, taking into account that cortical growth seems to obey a more holistic pattern of development than that implied by the idea. Many genes are known to correlate with different phases of brain growth (Benítez Burraco 2009, for an overview), but as far as we know they seem to work in a way that justifies Kaskan and Finlay's (2001: 27) assertion that the brain is one of the few organs whose development appears to follow a set of rules affecting the organ as a whole.

Two clarifications are in order. The first one relates to a logical consequence of the previous comments. If the scenario put forward above is on the right track, then the most natural expectation is for such a computational apparatus to be used in a wide range of domains, as actually acknowledge by Chomsky himself:[22]

> It is interesting to ask whether this [structure building] operation is language-specific. We know that it is not. (Chomsky 2007: 7)

[22] In this and the following quotations from Chomsky (2007), he specifically comments on the operation "Merge", not on the "structure dependency" principle. However, as the former operation is, according to MP, the one ultimately responsible for the type of structure which linguistic expressions obey, it also shapes the character of the rules affecting them. So, Chomsky's comments can be read as statements on the latter principle under a minimalist perspective.

However, Chomsky seems to favor the idea that the application of these computational resources in domains other than FL is "parasitic" or "derivative from" its linguistic uses (Chomsky 2007: 7). Such an idea is against well-established facts in the field of developmental Psychology, where it has long been known that the capacity for hierarchical organization grows from a state in which it exists as an anatomically and functionally undifferentiated ability, the differentiation and diversification of which in specific domains of application is a relatively late outcome of development (Greenfield 1991).

The second clarification, actually an extension of the first one, has to do with Chomsky's further remark that even if the structure building operation ("Merge") underlying linguistic computations in the way depicted in (8) actually happens to be a general purpose one, applied in such disparate domains as, say, music, mathematics, social intelligence, navigation, and so on (Corballis 2007), its connection with FL is still to count as a fact of UG:

> The conclusion that Merge falls within UG holds whether such recursive generation is unique to FL or is appropriated from other systems. If the latter, there still must be a genetic instruction to use Merge to form structured linguistic expressions satisfying the interface conditions. (Chomsky 2007: 7)

We think that this is a very dubious contention, both on empirical and conceptual grounds. Connectivity among different brain components is known to be a property which obtains as a by-product of cortical growth (see Striedter 2005, and references therein), so the idea that there must be specific genetic instructions to connect the computational system and the peripheral components generating usable linguistic expressions—a way of conceptualizing Greenfield's differentiation processes in the domain of language, runs counter to both neurogenetic expectations and the minimalist method.

So the conclusion that we reach concerning the particular case of (8) above is that it is not a principle to be conceptualized as particular aspect of UG, notwithstanding APS considerations. Rather, we think that it just captures a natural consequence of how human brains are structured and how they develop. In the case under examination, the structure-dependent interpretation of the incoming input when extracting the rules of the underlying grammars is not to be seen as a property of either PLD or UG, but simply as a reflex of the computational resources put to work in this as in other tasks—its specialization for the linguistic function deriving from developmentally contingent episodes in which the said resources connect to other brain components, composing an integrating system.[23]

2.4. Development—a neglected fact vindicated

The structure-dependent character of the rules of grammars—captured in (8) above—has a developmental history, as is also the case of particular grammatical rules like the "main auxiliary fronting" of English—captured in (7a). The former is the natural consequence of (i) the growing of the cortical networks serving as a working memory space for the processing of sequential items, and (ii) the connection of such a memory device with the sensory-motor and thought systems that read these items as linguistic expressions—a process that we see as a by-product of cortical growth. The latter is the consequence of the resulting system being exposed to certain linguistic stimulation (PLD) under the influence of data analysis principles of general domain sort ("most frequent → first model", "first model → default model", and so on), also following their own developmental trajectory. The conclusion clearly is that second and third factor effect interactions are enough to explain the "growth in the individual" (Chomsky 2005: 6) of these aspects of English grammar—and corresponding and similar phenomena in other languages. As for the first factor—a specifically evolved genetic component codifying grammatical information, we did not need to make such an assumption, as Laplace would have put it. The conclusion we reach is that individual development seems to be rich enough to fill the gap between the poverty of stimulus and the complexity ultimately attained by grammar systems. Historically, this is a shocking conclusion, as development has been a fact traditionally neglected by generativists on two different grounds:

---

[23] A caveat. One thing is to say that such networks are mature enough to allow the structural computation of sequential items at an early age, and a different thing to say that children are "blind" to compute linguistic expressions as obeying linear patterns of organization. Remember that a computational device capable of operating within a certain regime of complexity is also capable of performing operations of a lower level of complexity. So, a deeper analysis of the case perhaps requires the specification that the system uses its capacity maximum as a "elsewhere condition". We do not see this point as particularly problematic, as it has been observed before that hierarchical patterns of organization tend to emerge whenever possible. See Simon (1962) for a general approach, and Longa and Reynares (2010) for a particular application to the case of language.

(i) Development is causally innocuous. This is the position put forward in Chomsky (1975b: 119-122), for example, where he notes that linguistic maturation is obviously a fact. He adds, however, that as no piece of evidence corroborates the idea that each particular intermediate state (S_i) has a formative influence in the immediately subsequent state (S_j), we better act on the assumption that learning is instantaneous.[24]

(ii) If any, the influence of developmental concerns on the process of acquiring a language is a negative one. This is a position defended, for example, in Hyams (1986: 168-171), where the contention is made that language acquisition would actually be an instantaneous process—under the assumption that UG is already implanted at birth in children's minds were it not for the difficulties and delays that operate on them until the systems by means of which they access PLD are mature enough.

Our arguments point to a very different conclusion. Development is not an innocuous or negative factor on the growth of language and the acquisition of particular linguistic systems. Quite the opposite: As in any other biological realm, the unfolding of a multifarious collection of factors and the interactions brought about by such processes of natural growth are causally sufficient to be able to explain the emergence of species typical features like FL. This view, which requires no "blueprint", no "plan", or no "program" of the outcome encoded in the genotype, is the one preferred on minimalist grounds and, above all, the most congenial one with the biological argument put forward in the first part of this chapter. Genes are obviously also crucial within this framework—as developmental processes could not take place otherwise, but they lack the purported function of serving as the symbolic code in which details of outcomes are represented from the start, as well as the alleged role of being the central controllers of the outcome's unfolding—a mission better thought of as distributed among the whole system of factors. In the case under discussion, this basically means that no appeal to UG is required, if understood as a set of genetically encoded instructions containing information about the design and developmental schedule of grammars (Borer and Wexler 1987, Wexler 1990). The following passage from Lenneberg (1967) epitomizes the resulting position:

> Our present capacity for language may well go back to species-specific alterations in genetic material (intracellular changes) which, however, affected certain rates and directions of growth during ontogeny (temporo-

[24] Chomsky elaborates on the same motive, for example, in (1980: 201) and (2000a: 140-141; fn. 11).

> spatial gradients), producing a peculiar ontogenetic phase of an optimal confluence of various abilities; thus a critical period for language acquisition might have come about. (Lenneberg 1967: 265)

This view runs against the "consensus view" shared by most generativists (Hornstein *et al.* 2005), but its conceptual and empirical underpinnings are solid enough to motivate a debate in which the voice of experts from many fields (biologists, developmental psychologists, acquisitionists, etc.) should be welcome. Such a debate will not be devastating in respect to the point around which, in our opinion, true consensus among generativists exists: Humans are endowed with an innate faculty to learn and use complex linguistic systems. In this we agree with Chomsky (1975b: 122), who, commenting on the fact that something along the lines of the ideas put forward in this chapter could turn to be true, defends that it would only "complicate" the innate hypothesis, without falsifying it.

The required "complications" are not so complicated and are worth taking into serious consideration. They basically amount to adopting a "dispositional" form of nativism—in the sense introduced in the first part of this chapter, closer to the rationalist antecedents of the nativist position, as witnessed, for example, by the following quotation from Descartes:

> This is the same sense as that in which we say that generosity is 'innate' in certain families, or that certain diseases such as gout or stones are innate in others: It is not so much that the babies of such families suffer from these diseases in the mother's womb, but simply that that they are born with a certain 'faculty' or tendency to contract them. (Descartes 1648 [1985]: 303–304)[25]

From this point of view, a feature—say, FL—is deemed "innate" not because it (or a schematic version of it) is somewhat given from the onset, but rather because its being present at some particular point of an organism's life is guaranteed (all things equal) by a system of development comprising both internal and external factors, robust and canalized enough to tolerate perturbations in any of them, and open to a certain range of plasticity. A peculiarity of this perspective is that—in agreement with minimalist expectations—factors are not required to be of a domain-specific sort, because what makes a feature species typical—and, eventually,

[25] See Cowie (1999) and Balari and Lorenzo (submitted) for some comments on this passage.

species specific—is the system as a whole and not any of its component factors.

The fact that "specificity" is something only to be expected at this systemic level is a welcome consequence of the perspective. As recently brought to the fore by Boeckx (2009b) and Hornstein (2009), the idea that UG consists of very detailed information concerning the properties of grammatical systems is biologically anomalous, as one cannot find anything similar in other organisms. Under such a view, the human genotype seems completely exceptional in this respect, which posits a serious difficulty in explaining FL as a normal outcome of natural evolution. However, if we adopt a "factorial" model of development (Minelli 2003) and assume that FL, instead of being the programmed outcome of an exceptional bunch of genes, results from an evolutionarily exceptional assembly of common developmental factors, the task of conceptualizing our linguistic capacities within the normal parameters of evolutionary explanation eases radically. This is an idea that has not escaped Chomsky's attention, as witnessed by the following passage:

> The less attributed to genetic information for determining the development of an organism, the more feasible the study of its evolution. (Chomsky 2007: 4)[26]

In closing this section, it is worth remembering that nativism has been the target of some criticism based on the idea that it merely signals the pitfalls of empiricism, without articulating an alternative substantive theory (Cowie 1999). We are convinced that incorporating nativism in a developmentalist or epigeneticist framework, along the lines presented in this chapter, gives us the opportunity of constructing such a theory for the particular case of language.

---

[26] Or as previously stated by Cowie (1999: 125-126): "I you're going to be a nativist about the mental, then, if you keep the inbuilt stuff relatively simple, you can get away with issuing a promissory note (or notes) on the biology. You can wave your hands around, claiming either that the mechanism in question was shaped in some relatively straightforward way by natural selection or else, perhaps, that it arouse fortuitously though some non-Darwinian process. The more you enrich the hypothesized inbuilt structure, though, the less trustworthy such notes appear".

2.5. A defense of UG—or half a defense

If the view sketched in the previous sections holds generally, then UG—factor 1 in Chomsky's (2005) model, is an empty factor. Nevertheless, a system of development now comprising factors of the other two types—factors (2) and (3)—can exert an identically strong influence in constraining the attainable systems of linguistic knowledge. So, languages can still be said to obey UG, in the traditional sense of being shaped within a restricted range of variation, without implying that a priori knowledge of certain particulars of grammars—UG in the sense of Chomsky (1965)—is required for children to accomplish the task of acquiring a particular linguistic system within this space of variation. So, a divorce is implied by our model between the role of UG in the theory of variation (a "descriptive" goal) and its role in the theory of acquisition (an "explanatory" goal, according to Chomsky 1965), thus breaking with a long-standing premise of generative linguistics, made explicit in statements like the following one:

> Saddle with the dual goals of descriptive and explanatory adequacy, the theory of grammar is primed to offer solutions to the problem of language variation and acquisition in a single package (Yang 2010: 1161)

The solution in question has been translating the principles putatively constraining the shape of possible grammars into a "theory of the mental mechanisms children employ to learn grammars" (Yang 2006: 147). Such a view, which has become consensual under the format of Principles and Parameters Theory (Hornstein *et al.* 2005), so defends that knowledge of these universal principles is genetically inherited, with options open to fixation through experience triggering.[27] This is, however, an anomalous solution, both on conceptual and biological grounds, which relies on the dubious contention that by knowing the essential properties of a phenomenon or having an accurate description of its pattern of variation one readily obtains an explanation concerning how it obtains these properties and attains its steady form (Cowie 1997: 27, Lewontin 2000: ix).

---

[27] As a matter of fact, the "principles and parameters" format was first proposed in an influential paper on comparative syntax by Rizzi (1978) and interpreted as a theory of acquisition only after that (Chomsky 1980: 66; 1981: 4). See Rizzi (2005) for an historical overview.

Our alternative does not imply the defeat of UG. It simply suggests, in agreement with current minimalist thought (Longa and Lorenzo, in press), that UG must be dispensed of the causal role on acquisition that generative linguistics has been crediting to it in the last few decades. Rather than this, a model can be envisioned in which UG refers to certain crucial properties of linguistic systems resulting from the epigenetic interactions of a set of heterogeneous developmental factors, the joint action of which describes a space of possible linguistic forms.[28] Within this view, the theory of acquisition seeks to uncover the factors comprising the LAD that causally intervenes in the development of linguistic systems, while the theory of UG seeks to establish the formal properties that draw up the boundaries of these systems, seen as emergent properties of the underlying developmental resources and interactions: Two related yet independent tasks for theoretical linguistics.

[28] See Alberch's (1989, 1991) "developmental morphospace" concept as a suitable biological model for this idea—Balari and Lorenzo (2008) for an application to cognition and language. See also Fitch (2011) for a congenial approach.

CHAPTER THREE

THE MYTH OF LANGUAGE DIVERSITY

JOSÉ-LUIS MENDÍVIL-GIRÓ

1. The diversity of languages as a myth

The title of this contribution is not intended to imply that diversity in languages does not exist. That languages are different is an objective fact, not a myth. What I call *the myth of language diversity* is the belief that the diversity of languages is profound and substantive enough to be considered the crucial phenomenon in the scientific study of language.

Undoubtedly, the fact that languages are different is a remarkable and interesting phenomenon in itself. Also, the study of such diversity affords a valuable perspective on the human faculty of language, and is not, as sometimes argued in certain areas of contemporary linguistic research, an argument to deny its existence.

Many readers will have noticed the implied reference in the title of this chapter to the influential paper by Evans and Levinson (2009a): *The Myth of Language Universals*. Evans and Levinson (E&L) do indeed believe that the existence of Universal Grammar (UG) is a myth that Chomskyan linguists have injected into cognitive science, something that constitutes an obstacle to the scientific investigation of language: "a great deal of theoretical work within the cognitive sciences thus risks being vitiated" (E&L 2009a: 429).

E&L react against Chomskyan universalism arguing that it distorts the reality of language, compromising realistic and meaningful research within cognitive science:

> The claims of Universal Grammar, we argue here, are either empirically false, unfalsifiable, or misleading in that they refer to tendencies rather than strict universals. Structural differences should instead be accepted for what they are, and integrated into a new approach to language and cognition that places diversity at centre stage (E&L 2009a: 429).

In the following pages I will try to show that, on the contrary, it is the exaggerated assessment of the depth of the diversity of languages what constitutes an obstacle to the ultimate acceptance of the study of language into the natural sciences.

Disdain of the diversity of languages can certainly be an obstacle to progress in our understanding of how the brain creates and uses language. But what E&L propose, in their rejection of universality, ignores important discoveries about common properties of language made over the past half century and risks throwing the baby with the bath water.

If one adopts a biolinguistic point of view, the relationship between language and languages is proportional to the relationship between life and life forms, organisms.

Of course, what we find in the physical world are organisms and not life itself. Life exists only as life forms, and it is quite true that these are notoriously diverse. But this is not to deny that life exists (e.g. in contrast to the inorganic world), or that we should consider it a secondary matter to ask which common properties define and explain it. Indeed, this is the central goal of biology as a science.

There is no reason to believe that this shouldn't also be the case in biolinguistics, unless we reject the comparison, that is, unless we reject that all existing languages are manifestations of the same faculty of language, sustaining the view that languages are essentially cultural objects, abstract tools that have evolved independently and whose essential structure comes from outside the mind and brain.

This is the most common conception of languages among researchers who belong to what we might call the *Functional-Cognitivist Paradigm* (FCP). This conception of languages, an anti-biolinguistic one, underlies E&L's approach and feeds what I have called the myth of the diversity of languages.

E&L explicitly formulate this myth as one of their conclusions:

> The diversity of language is, from a biological point of view, its most remarkable property–there is no other animal whose communication system varies both in form and content (2009a: 446)

While it is undeniable that the diversity of human languages is notable and relevant, the claim that diversity is the most remarkable property from the *biological standpoint* can indeed be challenged. This statement appears to be based on the belief that no other animal communication system varies in form and content as human language does.[1] However, this is doubly questionable. On the one hand, it is not indisputable that human languages vary in form and content as much as E&L assume. On the other hand, it is not clear whether there is indeed no variation in other animal communication systems.[2] In addition to the known cases of whales (Payne 2000), many songbirds have dialectal variation:

> Most birds do not simply mimic the song of adults exactly: In many species, individuals create new, novel songs that are built upon but not identical to the songs they heard as nestlings. This creative aspect of birdsong ensures that each generation hears slightly different songs from those of the previous generation. This process of song transmission across generations, with slight novelties introduced by creativity and or erroneous copying, leads to 'dialects' of birdsong: Birds in different regions sing quite different learned songs (Fitch 2009b: 289).

It is difficult to consider this description without reflecting on what happens in the case of human languages, unless we apply different criteria to humans than to other species. "Language" change and diversification, then, is not so rare in the natural world. However, naturalists have not been tempted to conclude that there are organisms of the same species that have radical and profoundly different systems of communication and knowledge, except in the case of humans (as seen in the quotation from E&L, above).

It is tempting to consider this as an example of the kind of suspension of scientific rationality when it comes to studying the human being that Chomsky has frequently noted when defending the naturalistic approach to human language:

> The basic properties of humans seem to be identical, close to identical, except for pathology, which you can find anywhere. So you could investigate every individual, just like you could investigate every apple to make sure it follows the laws of motion when it falls, but nobody does that because there is so much evidence that they have to be identical. In fact it's interesting that these questions only arise in the human sciences. They never arise in physical sciences. It's all the same. Why don't they arise

[1] In fact, according to Pullum and Scholtz, "Evans & Levinson's *major point* is that human languages are intriguingly diverse rather than (like animal communication systems) uniform within the species" (2009: 466, my italics).

[2] Of course, human languages are used for communication, but this does not mean that languages *are* communication systems (see Berwick and Chomsky 2011: 39).

> elsewhere? Because there is a kind of rationality that prevails in the study of the natural world, which is somehow cancelled when we study ourselves. At that point we become very irrational. So we ask the kinds of question that wouldn't arise in studying other aspects of the physical world —even studying other animals. (Chomsky 2010a: 30).

What Chomskyan naturalism means in this regard is that the burden of proof is on those who deny that human beings have a unique capacity for language and state that this ability varies greatly and deeply within the species. Certainly this would be a strange biological fact, unless of course we think that language has nothing to do with biology and nature.

The radical and deep diversity of languages is in this sense a myth, a myth coined on the old anthropocentric prejudice that human beings are essentially a matter of culture rather than nature, that human languages are cultural artefacts, not natural objects.

Indeed, the insistence on the plasticity of human cognition and on the power of our ability to learn cannot be interpreted but as an attempt to present human language as a phenomenon alien to the natural world:

> Structural diversity at every level is not consonant with a theory of fixed innate language structure, but instead suggests remarkable cognitive plasticity and powerful learning mechanisms. We pointed out that human communication is the *only* animal communication system that varies in myriad ways in both form and meaning across the species, and this must be a central fact that should never be lost sight of (E&L 2009b: 473)

Faced with a similar criticism of dualism by Smolensky and Dupoux (2009), E&L claim:

> Smolensky and Dupoux ignore the recent synthesis of biological and cultural evolution. Thus they assert "language is more a biological than a cultural construct." We would go further: "language is one hundred percent a biological phenomenon" (E&L 2009b: 479).

As I hope to show, this rhetorical statement hides a direct rejection of Chomskyan naturalism. The "number one enemy" of E&L and, in general, of the FCP, is the assumption that human beings have a natural capacity for language (i.e. what traditionally has been called UG). Yet, as I will discuss below in greater detail, this is a false problem, and mainly a terminological one (see also Chapter 2). The attack on UG simply disguises an assault on the consideration of human language as a natural object and, ultimately, on the idea that the human mind is also a natural object.

One possible explanation as to why a non-naturalistic conception of human beings is linked to the rejection of a naturally conditioned faculty of language (FL) would be that if we recognize the existence of such a capacity as a component of the human mind, what follows is the idea that the other components of the mind would be similar or largely identical to those of other organisms, something that is certainly made explicit in Hauser *et al.* (2002) and that collides with the anthropocentric idea that there are substantial differences between humans and other animals. As Penn *et al.* point out:

> At present, most comparative psychologists believe that the difference between human and other animal minds is 'one of degree and not of kind' (Darwin [...]). Among researchers willing to admit that the human mind might be qualitatively different, most argue that our species' cognitive uniqueness is limited to certain domain-specific faculties, such as language and/or social-communicative intelligence. We believe this view of the human mind is profoundly mistaken. In our opinion, the discontinuity between extant human and nonhuman minds is much broader and deeper than most researchers admit. We are happy to report that E&L's target article strongly corroborates our unpopular hypothesis (Penn *et al.* 2009: 463)

As we know, the hypothesis of a natural capacity for language is the Chomskyan answer to the "spectacular fact (...) that any normal human child can learn any human language, and no human language is learnable by any other extant species" (Penn *et al.* 2009: 463). As these authors ask, "why are human languages so easy for us to learn and so unthinkable for everyone else?" (*ibid.*). Chomsky's response that only humans are endowed with a human FL as part of their nature "fits nicely with the presumption that the rest of the human mind is more or less like that of any other ape" (Penn *et al.* 2009: 463). One way to reject such a conclusion, undesirable from an anthropocentric point of view, is to support the myth of the diversity of languages:

> But as E&L point out, the diversity of human languages suggests that our faculty for language relies largely on domain-general cognitive systems that originally evolved for other purposes and still perform these non-linguistic functions to this day. If E&L are right, there should be significant differences between human and nonhuman minds outside of language (Penn *et al.* 2009: 463).

E&L themselves point out, as a remarkable fact about our species, that the same set of organs (the same in all people) serve to learn and use a set of heterogeneous and radically different languages (2009b: 479). In fact, this is a serious problem for the contention that languages are radically diverse.

To use the well known Chomskyan terminology, this would be more a mystery than a problem. But we can try to make it a problem using the logic of rational inquiry, which always suggests that behind similar but apparently diverse and heterogeneous systems (be they physical phenomena as the behaviour of fluids, natural organisms or human languages) there underlies a set of principles and parameters of variation that can reveal the root causes of similarities and, of course, of differences.

2. Inductive and deductive approaches to language unity and diversity

At least since Borsley (1991) many authors open their discussion of the uniformity and diversity of languages by citing Martin Joos' claim that "languages can vary without limit" (Joos, 1957: 96, quoted from Borsley, 1991: 1) as a counterpoint to Chomsky's position that for human languages "there is only a computational system and one lexicon, apart from its limited kind of variety" (Chomsky 1995: 170).

Haspelmath (2008) has noted that unlike generativists, "functionalists do not assume that they will find the same syntactic categories and relations in all languages […], but they expect languages to differ widely and show the most unexpected idiosyncrasies" and, consequently, functionalists "tend to agree with Joos's notorious pronouncement that 'languages can differ from each other without limit and in unpredictable ways'" (2008: 93).

The difference between these two views on language and languages can be expressed as follows: The FCP involves an *inductive approach to language*, while the Chomskyan paradigm implies a *deductive approach to languages*.

The deductive approach deals with so-called *I-language* (persons' language organs, see Chomsky 1986, Boeckx this volume) as the object of study and, as in any scientific research, the objective is to build a theory on the properties and principles governing this language organ and to discern the factors involved in its development (e.g. Chomsky 2005). As more I-languages are examined and other cognitive systems are studied, the initial theory is adjusted, always trying to increase the explanatory power without neglecting descriptive adequacy. As in the study of the motion of bodies or cell development, it is assumed that the laws and principles formulated are maximally general, and that the observed peculiarities must follow deductively from the consideration of additional factors (whether friction, the presence of certain proteins, or parametric variation associated with certain functional categories).

The inductive approach proceeds otherwise. Here, the objects must be studied in themselves. A detailed description is made of each object on its own terms, and once the description is exhausted or completed, generalizations about possible underlying uniformity can be made.[3]

Although not in itself a conclusive argument as to how to judge the extent of diversity in language, there is no doubt that from the standpoint of modern science, the deductive model is the adequate one.[4]

The adoption of an inductive or deductive model to express the relationship between language and languages, it seems, correlates to a specific concept in science. According to the empiricist (inductive) approach, science aims to discover the causes and the nature of things; it involves obtaining mental representations of external reality. In a rationalist (deductive) approach, by contrast, science aims to translate nature into mathematical language, and involves building theories to determine the existence of things.

At least since the time of Galileo we have understood that the aim of science is not to find concepts or representations of the entities of reality, but to build mental realities (concepts and theories) and attempt to discover experimentally which of these have some basis in what we perceive.[5]

However, the differences that these two traditions show in the way they are applied to the issue of the diversity of languages is not ultimately based on different conceptions of science; rather, the different conceptions of science are inspired by different conceptions of the object of study.[6]

From a Chomskyan point of view, language is conceived as a natural phenomenon, and languages are understood as particular environmentally

---

[3] As E&L remark, "D-linguists [as opposed to C(homskyan)-linguists, JLMG] prefer Boasian 'methodological relativism' –first analyze a language in its own terms, then compare" (E&L 2010: 2734).

[4] As Haspelmath notes, for the functionalist model "description is separated strictly from explanation" (2008: 93), but such a separation is simply inconceivable from the point of view of natural science.

[5] "The standard that inspired the modern scientific revolution was abandoned: the goal is intelligibility of theories, not of the world" (Chomsky 2002: 68).

[6] However, E&L are explicit in this regard: "We have no quarrel with abstract analyses per se, but we would like to see these arise inductively, and not be derived deductively from a model based on English and familiar languages." (E&L 2010: 2754).

conditioned manifestations of that phenomenon. That is, we proceed *deductively from language to languages*.

The clearest manifestation of this procedure is parametric theory. Regardless of the specific formulations that it might take, and regardless of more or less literal interpretations of certain metaphors, the basic logic of parametric theory remains strong: from common design principles, the various emerging systems respond to variations in development processes that have systematic implications, just as happens in the development of natural organisms. The only way to deny parametric theory thus formulated is to demonstrate that there is no systematic grouping of properties in the structural typology of languages.[7]

On the contrary, from a functional-cognitive point of view, we proceed *inductively from languages to language*. This model implies that languages exist in themselves and that language is a secondary concept induced from the descriptive generalizations obtained from the study of languages. This, then, is not a strictly naturalistic approach, since languages are conceived primarily as cultural objects. Not surprisingly, from this point of view linguistic universals can be seen as a myth, since for the inductive model the diversity of languages is, by definition, irreducible, languages being primary objects with their own nature.

From a deductive point of view, the inductive program is not only empirically unattainable, it is also incompatible with a naturalistic study, since it is based on the hypothesis that human languages exist in themselves as purely cultural objects that colonise the human brain.[8]

Consequently, the inductive model described by E&L (2009a, 2010) is a perfect example of how the naturalistic research of language can be hindered. In positing the exercise of a description of each language in its own terms and with their own categories ("Every language, we have said, is a thing unto itself", E&L 2010: 2737), we would be including in the universe of data only the surface forms of languages (purely descriptive entities such as "dative case in Hindi", "double object construction in English" or "clitic climbing construction in Spanish"). If the theory of language must be made starting from such a set by induction, the expectations of success are severely reduced. As Pinker and Jackendoff

[7] Baker (2001) and Holmberg and Roberts (2010) are robust attempts to explain parametric clusters as effects of a hierarchical relationship among grammatical properties that determine other grammatical properties.

[8] To use Deacon's influential metaphor: "In some ways it is helpful to imagine language as an independent life form that colonizes and parasitizes human brains, using them to reproduce" (Deacon 1997: 111).

rightly pointed, "any survey that fails to consider the larger design space for conceivable languages is simply unequipped to specify how 'remarkable' the actual diversity of languages is" (2009: 465).

I have suggested that the inductive approach of the FCP is based on a conception of languages as cultural objects. This vision goes back to the dawn of the science of language, and is clearly reflected in the following statement by Edward Sapir:

> Speech is a human activity that varies without assignable limit as we pass from social group to social group, because it is a purely historical heritage of the group, the product of long-continued social usage. It varies as all creative effort varies—not as consciously, perhaps, but none the less as truly as do religions, the beliefs, the customs, and the arts of different peoples. Walking is an organic, an instinctive, function (not, of course, itself an instinct); speech is a non-instinctive, acquired, "cultural" function (Sapir 1921: 2).

It may be instructive to consider briefly in what ways Sapir was right and in what ways he was not. Although in other parts of his influential book Sapir speaks of language and not of speech, it is clear that if we limit ourselves to the surface, his claims are credible: if we define a language as a historically determined set of lexical items (sound/sense pairings), then he was right. In the following pages, Sapir develops a comparison between talking and walking, and he says that without a society around you, it would still be possible to learn to walk, but you could not learn to speak according to the conventions of that society, and again he is right. The point is that an I-language is obviously not simply a manner of speaking, it is not simply a collection of sound/sense pairings and a set of rules for combining them. But that's how the myth of the diversity of languages is born, from an incomplete picture of what languages are.

From a deductive point of view, I-languages are the language organs of persons. A remarkable property of the language organ of human beings (but shared with any other organ) is that it is sensitive to environmental information in the process of development and, therefore, it is susceptible to individual and group variation.

If an I-language is actually a specific and historically determined pattern of materialisation of the faculty of language, it makes no sense to rely solely on specific historical realisations to draw conclusions about the unitary or diverse nature of languages, and about the faculty of language in itself. If we move to biology, it would be tantamount to accepting the views of the traditional naturalist whilst dismissing what the molecular biologist has to say.

3. The False Problem of UG: Innatism and Emergentism

The functional-cognitive model recriminates Chomsky's postulation of a specifically linguistic and specifically human FL, arguing that it is more economical, more biologically plausible, and more consistent with standard scientific reasoning to assume that no such thing exists, and that languages can be explained as the result of general limitations of learnability, the recruitment of other cognitive systems, and functional pressures arising from their use for communication and thought. This is certainly a possibility, one which the minimalist program developed by Chomsky and followers has in fact been exploring over the last fifteen years (see chapters 1 and 2).

It could be said, then, that what has confronted the traditions in linguistic research during the last fifty years is the problem of the innate or emergent character of the FL. However, this is a false problem, a false controversy, because FL is, like any other human faculty, simultaneously innate and emergent.

The wisest way to deal with a false problem is to stop thinking about it, and I will do so here, with the exception of some brief notes to justify the apparently contradictory statement that FL is at the same time innate and emergent.

On the one hand, there is no doubt that human beings have an ability that enables them to learn and use a human language. Given that the rest of known organisms (whether natural or artificial) lack such an ability, it is fair to say that this ability is specifically human and, therefore, innate in any normal human being. As Chomsky points out,

> to say that "language is not innate" is to say that there is no difference between my granddaughter, a rock and a rabbit. In other words, if you take a rock, a rabbit and my granddaughter and put them in a community where people are talking English, they're all learn English" (2000b: 50).[9]

---

[9] Fitch follows this same elementary line of reasoning: "Clearly, immersion in a linguistic *environment* is not enough for spoken language to develop in most organisms. There must therefore be *something* about human children which differentiates them from other species, and this something provides one of our core *explananda* in biolinguistics. We might gloss this neutrally as 'the human capacity to acquire language'. In generative linguistics this capacity is traditionally called the 'Language Acquisition Device', and a characterization of its properties termed 'Universal Grammar'" (Fitch 2009b: 288).

On the other hand it cannot be false that human language (as a human ability) is emerging, that is, the result of the combination or the arrangement of elements that in themselves are not language. To return to our analogy between life and language, life is certainly an emergent phenomenon, but by no means less real because of this. As Stuart Kauffman puts it,

> life is not located in the property of any single molecule -in the details- but is a collective property of systems of interacting molecules. Life, in this view, emerged whole, […], not to be located in its parts, but in the collective emergent properties of the whole they create […] The collective system is alive. Its parts are just chemicals (Kauffman 1995: 18, 24).

In fact, Chomsky himself, when describing FL as a subcomponent of the brain that is specifically dedicated to language, notes: "As a system, that is; its elements might be recruited from, or used to, other functions" (Chomsky 2004: 124, fn 1). In a similar vein, Anderson and Lightfoot remark that

> even if it were to become clear that there is no real segregation between language-related and non-language-related brain tissue, it would still be useful and important to treat the language capacity as a discrete and specifiable human biological system in functional if not anatomical terms (2002: 236).[10]

And this is both useful and important here, because such "methodological modularity" is common in any scientific inquiry. Rothstein and Treves (2010) note that "evidence accumulated over the past century, but little noticed by linguists, indicates that there is no dedicated neuronal machinery to subserve complex cognitive capacities, least of all machinery specialized for language" (Rothstein and Treves 2010: 2717). In this sense, it seems that "our abilities to learn, produce and interpret language are supported by the same neural mechanisms, at the cortical microcircuitry

[10] As noted by Rooryck *et al.*, the Strong Minimalist Thesis explores precisely this direction, but this does not negate the notion of UG: "In terms of Hauser *et al.*'s contrast between FLB and FLN, the latter would then be empty and the uniqueness of UG would reside in its particular combination of traits each of which is shared with some other domain or organism. As human language differs from all other systems something (but not necessarily some thing) must be unique" (Rooryck *et al.* 2010: 2655). For an elaborated proposal of how to conceive this "something" see Chapter 2.

level, as any other cognitive function, including those (most) that we share with other mammals" (*ibid.*). Nevertheless, they add:

> While something is known about the neural mechanisms underlying, for example, visual and auditory processes, we know next to nothing about how the nervous system processes information about the relation between sounds and symbols and how complex meanings are represented compositionally in the brain (*ibid.*).

Although we can infer that cortical transactions behave the same way in language as in other, better understood functions, "how they sum up to produce the faculty of language, and why they should do so only in the human species, remains unclear" (*ibid.*).

The solution proposed to address this mystery is simply to accept that language, as an emergent system, may have its own properties, so that linguistic structures themselves will be the result of a particular organization and accumulation of neural processes that are necessarily (and inevitably) more basic:

> The mystery may stem precisely from the nature of language as an emergent property, its being more than the sum of its parts, arising, that is, by the system-level combination of elementary cortical operations. From such a perspective, the central question for cognitive neuroscientists is what neural mechanisms can facilitate compositional interactions, and how the range of grammatical structures emerges from a much narrower range of neural mechanisms. Further, what is the nature of the plasticity which allows children to acquire a native-speaker linguistic competence in a finite time; that is, how do children learn to manipulate a finite set of symbols in such way that they can produce and interpret an infinite number of novel strings and thus convey and comprehend new information (Rothstein and Treves 2010: 2717-2718).

A similar view is that of Marcus (2006), who suggests that the possible mental organization in cognitive (or functional) modules can be relatively independent of the modular organization in neurobiological terms, while the possible neurobiological modules may themselves be relatively independent of genetic modules (see Benítez-Burraco, this volume).

Of course, there is a lot of research to be done to determine the shape, nature, extent and evolution of the principles governing the origin, development and use of the language faculty (as the other contributions to this volume clearly reveal), but it does not make sense to focus the controversy in terms of *a priori* ideas about the false issue of innateness and emergence of FL.

It is reasonable to think that certain principles will be specifically linguistic (specifically human) at certain levels of discussion and knowledge, and that such principles will have to be reduced to other principles that, at a given time (and necessarily) will cease to be specifically human and specifically linguistic. It is also expected that there will be some dispute over what evidence we need to consider in addressing the question of whether we can go from one level to the subsequent levels of reduction (see Boeckx, this volume). But to infer that there is no human FL is an illicit step that would certainly be detrimental to any naturalistic study of language (unless of course one thinks that language is not is a natural object, which is the idea underlying the FCP).

If we define UG as the initial state of FL (the human capacity to learn language in Fitch's terms), that is, as the set of natural principles (whether genetic or not, whether biological or not) that determine the architecture of human languages and limit their ranges of variation, UG by definition exists. Of course, it could be argued that the very definition of UG, which claims that these principles are specifically linguistic (in fact, specifically grammatical, if we stick to the literal interpretation of the term) is the problem. This is Tomasello's objection:

> For sure, all of the world's languages have things in common, and E&L document a number of them. But these commonalities come not from any universal grammar, but rather from universal aspects of human cognition, social interaction, and information processing–most of which were in existence in humans before anything like modern languages arose (Tomasello 2009b: 471).

Here we face two separate issues: one mainly terminological and the other conceptual.

The terminological problem has to do, clearly, with the meaning of the expression *Universal Grammar*. Note that Tomasello does not reject the notion that there are language universals, but that the properties that explain them are specifically linguistic:

> Why don't we just call this universal grammar? The reason is because historically, universal grammar referred to specific linguistic content, not general cognitive principles, and so it would be a misuse of the term. It is not the idea of universals of language that is dead, but rather, it is the idea that there is a biological adaptation with specific linguistic content that is dead (Tomasello 2009b: 471).

But we can ask what is meant by "a biological adaptation with specific linguistic content": Does it mean that there would be genes that specify grammatical categories? Or that there would be parts of the brain dedicated to language and whose removal would leave intact the rest of cognitive and motor faculties? Or that there would be linguistic neurons and nonlinguistic neurons, just as there are pyramidal and spherical neurons? No one seems to have defended such a position. The idea of a "biological adaptation with specific linguistic content" is a straw man. Innatism refers to the bias that the body and the laws of nature underlying its anatomy and physiology impose on the systems of knowledge that are developed. If everyone agrees that this bias exists, then the dispute is meaningless.

Most certainly, UG is a heterogeneous set, since it may include restrictions derived from the principles that govern brain development and even purely formal principles that apply to every system of memory and computation (Chomsky's third factors). The notion of UG, although for historical reasons it alludes to grammar (in the sense used by philosophers of the eighteenth century), does not presuppose that each and every one of the principles that form it have to be specifically linguistic.

The question of which of the principles governing the formation of LF in each person are specifically human and specifically linguistic is an empirical matter that cannot be resolved prior to the determination of what such principles are. The distinction between FLN and FLB of Hauser *et al.* (2002), and the factorialisation formulated by Chomsky (2005) (which is common in developmental biology) are simply attempts to make this more explicit in the investigation of FL.[11]

E&L concede that it is more of a terminological problem than a real one, and they themselves suggest talking about the human capacity for language, avoiding the term UG:

> To make this crystal clear: UG/FLB must include all the infrastructure for language including the neuroanatomy for speech and cognition, theory of mind, statistical learning capacities, and all the communicative and cooperative motivations and interactional abilities special to the species [...] If all this is agreed, it is time to change the terminology and avoid all the theory-laden terms used so far. We would urge our colleagues simply to talk about *the human capacity for language* (including language-special

[11] In the programmatic terminology put forward in Hauser *et al.* (2002) the computational system and the interfaces would be part of the faculty of language in the narrow sense (FLN) ("only the core computational mechanisms of recursion as they appear in narrow syntax and the mappings to the interfaces" (Hauser *et al.* 2002: 1573) and SM and CI systems, along with other potential components and FLN, would form the faculty of language in the broad sense (FLB).

> aspects of cognition, if any) and the ingredients or elements that may contribute to that – the rest carries too much baggage, as the field advances rapidly beyond the creaky old baggage train (Evans & Levinson 2010: 2742, my italics).

However, as I noted above, the question is not whether linguistic principles and structures may be reduced to more basic principles and structures (and ultimately to neuronal structures), but when, how and at what level the reduction can be done without ceasing to account for the structure of languages. The model E&L support skips a stage, so to speak, the stage of linguistic theory itself. Among the detailed (albeit superficial) description of the structure of languages and the general cognitive principles that E&L mention, there is a vacuum of explanation, and this diminishes the scientific value of the model they defend and contributes to the unfounded myth of the diversity of languages.

Although it may seem paradoxical, in the study of language we must be both nativists and emergentists; in this there is no choice. The difference between those who claim one or the other is a matter of their optimism regarding our capacity to understand emerging systems and of the degree of compliance we have with the formal and abstract structure of human languages as systems of knowledge, which is not uniform among different traditions.

In fact, a real problem behind this false controversy is that a large number of linguists and cognitive scientists ignore or simplify the formal structure of languages, that is, the system of knowledge which is materialised in each particular language. One might ask why some linguists assume that the structure of languages is learnable from environmental data and others say that it is not. It cannot be because we do not know what is and what is not learnable. We know, at least since Post (1944), that there are things that cannot be learned from a sample by induction and others that can. The difference of opinion emerges from the fact that we see the structure of languages and the very nature of language in different ways.[12]

---

[12] As Rizzi notes, "much of the typological tradition discussed by E&L has chosen to stick to an extremely impoverished, non-abstract descriptive apparatus, a legitimate methodological decision, but one which severely restricts the possibility of capturing forms of invariance." (Rizzi 2009: 467). In a similar way, Safir maintains that "it is hard to see how E&L's commitment to superficial linguistic analysis would recommend them to be a better partner for the broader enterprise of cognitive science than those who are committed to discovering what people know

The shallowness of functionalist/cognitive models of the structure of languages (as systems of knowledge) is perhaps the cause of the confidence in the possibility that induction and analogy may be an adequate means of extracting the information needed to develop such systems of knowledge (for a critique of this position see Yang 2009).

But as I have argued, the problem is not only terminological, nor does it only involve degrees of optimism and professional rivalries. It may be instructive to return now to the final statement of Tomasello's cited objection, that is, that many of the factors that determine the unity of language (including universal aspects of human cognition, social interaction, and information processing) "were in existence in humans *before anything like modern languages arose*" (my italics). Note, however, that in this Tomasello is implicitly assuming that human languages emerged externally to such "previous factors", that is, that human languages are cultural collective inventions or discoveries, objects external to the mind and the brain.

In this externalist context it is conceivable then that there are profound and substantial differences among languages, within the loose external frame determined by such factors, which are considered extralinguistic by definition. If we deny the existence of UG (i.e. the initial state of FL), then we are denying that humans have a bias towards language, that is, a specific capacity to learn the language from the environment, and that this specific capacity inevitably determines its structure. This position would then involve the theoretical possibility that different languages can activate different general cognitive systems or use cognitive systems differently. In such a case, it would not even be justified, as E&L suggest, "simply to talk about the human capacity for language", as there may be various human capacities for language. Such a possibility is suggested, in fact, by McMurray and Wasserman (2009) in their laudatory review of E&L (2009a):

> Yet, just as there is no universal structure, there may be no universal learning mechanism responsible for language. Language is a behavior assembled by many processes, an assembly guided by the language being learned (McMurray and Wasserman: 2009: 459).

---

about the language that they speak, including what is abstract, and necessarily beneath the surface of linguistic phenomena" (Safir 2010: 2725).

As we discuss in more detail in section 6, this is the gateway to linguistic relativity as an unavoidable consequence of the externalist approach to human language.

4. A note on the minimalist UG and the universality of syntax

To some extent it could be said that the externalist approach of the FCP is converging with some developments of the Minimalist Program (MP), in the sense that the MP (see Chomsky 2007, Boeckx, this volume) in effect implies a severe restriction of the notion of UG, and an increase in the weight attributed to non-specifically linguistic factors in the explanation of language structure. This resemblance, I suggest, is only superficial.

Although in some areas convergence does occur, for example in the tendency to commit to language processing and use factors the explanation of structural typology and other aspects (Newmeyer 2005, Boeckx 2011a), the fact is that the MP remains fundamentally internalist and naturalist in its development and, contrary to what happens in the FCP, the MP also postulates a restrictive conception of what is a possible human language.[13]

According to current minimalist research, a language is not an inventory of morphemes and words and a set of grammatical rules, but a common computational system (what Hauser *et al.* 2002 call *Narrow Syntax*) associated through (at least) two systems of interface to respective components of the mind (at least the sensory-motor system, SM, and the conceptual-intentional system, CI).

We might call this approach the *Universal Syntax Hypothesis* (USH). According to USH, all languages would share the same (narrow, internal) syntax and the structural differences between them would be the consequence of differences in the externalisation of the computational system in connection with the SM component.

Chomsky has formulated (Chomsky 2010a, 2010b, Berwick and Chomsky 2011) the hypothesis that there is a crucial asymmetry in the relationship between FLN and the other essential components of FLB (CI

---

[13] Arguably, in terms of Chomsky's (2005) three types of factors that determine the development of language, the P&P model relied more on type 1 factors and the MP model on type 3 factors. In other words, the P&P model had endorsed a more geneticist "environment" of the biology of the time (years 1970-1990*)*, while the MP echoes the developmental biology revived in recent decades (see Benítez and Longa 2010, and Longa and Lorenzo, this volumen*)*.

& SM), in the sense that the evolution of FL in the species would have favoured the relative transparency in the connection of the internal syntax with the CI system at the expense of the connection between the internal syntax and the SM system. Such asymmetry would then imply that the main source of variation among human languages would be in the connection between the internal universal syntax and the SM system, i.e. in the externalisation of language[14].

The following quote from Chomsky, neatly—and informally—summarizes this point of view and sets it in relation to learning and language change:

> But what about the relation between this internal system and the sensory motor system? That's the externalization problem. Well, the sensory motor system had been around for hundreds of thousands of years. It's a completely separate system. It has nothing to do with this internal entity. So there is a hard problem to solve. How do I relate that internal system to the sensory motor system for externalization? Well, it's a hard problem and in fact if you look at language, that's where practically all the complexity of language is. When you study a second language, about all you study is externalization. You study the sounds, the particular lexical choices, which are arbitrary, the inflectional system, you know, how to conjugate verbs, some facts about word order, and so on. That's just about all you have to learn. You don't have to learn the syntax and the semantics because that's there already. That's part of your nature and probably it's part of your nature because that's the way physical laws work. It's meeting conditions of computational efficiency—or so we would like to show. The externalization systems are overwhelmingly—maybe, some day, we will discover *entirely*—where languages differ from one another. The wide variety of languages is almost entirely, maybe entirely if we know enough, in the externalization process, the secondary process of getting it out into the sensory motor system. That's also where languages are very susceptible to change, so say teenage jargon or invasion or something else. That's where languages change a lot. That's where they vary. (Chomsky 2010a: 20-21).

The crucial idea is that the interfaces between this universal (human) syntax and the rest of the components of language (mainly the S-M one)

---

[14] Sigurðsson puts it this way: "much as there are no direct mappings of morphological features onto phonological features or sound waves, there do not seem to be any one-to-one mappings from syntax onto morphology or onto PF in general" (Sigurðsson 2009: 179). See Boeckx (this volume) for a specific articulation in this line of reasoning.

are different for each human being, because they are sensitive to external experience during their development (the type 2 factor), and therefore they are grouped culturally. But note that if we claim that languages are essentially different, then we are identifying "language" with the historically modified interfaces, that is, we would be excluding from the definition of language that which (hypothetically) is common to all of them, hence making the claim that they differ sharply a trivial one.

If we substitute for the question *how deep is the diversity of languages?* the question *how deep is the diversity among species?* it is easier to perceive that there is no clear and objective answer.

If one focuses on the external phenotype (e.g. body size or nervous system) the difference between an orang-utan and an earthworm is enormous. If we continue down to the embryonic level, similarities increase and differences decrease; at the genetic level the degree of similarity is even greater, and if we jump into the biochemical mechanisms that drive cells, we reach indistinguishability. If we really are to take seriously the naturalistic approach, the same logic should be applied to languages.

It is quite possible that the human genome specifies very little about what is a possible human language (see Benítez-Burraco, this volume), but it is also true that the genome specifies little about what is a possible human pancreas. In spite of this, each human ends up having a pancreas essentially homogeneous in the species, because the development of this organ (as that of all the others), in addition to being genetically bounded, is strictly constrained by epigenetic and environmental factors and, of course, by principles of development, biochemistry and physical laws. The language organ does not escape this logic, although among the "environmental" factors there is plausibly much greater variety (see Longobardi 2003). However, when biologists talk about the influence of the environment on the developmental process, they are referring principally to the environment of the cell, not to the environment of the organism, which has an undeniable influence, but a far more mediated one. The same applies to language. The environmental (non-genetic) factors that regulate language development are also (but not exclusively) internal to the brain and internal to the language faculty, in the broad sense. It is precisely in these factors where we should look for language universals (and probably for typological tendencies), and not in the ultimately derived languages, that is, in the different patterns of materialisation.

5. The Naturalist Approach: Languages as natural objects

Nineteenth century historical linguistics was not unfamiliar with the identification of languages and natural organisms, or with the consideration of linguistics as a natural science. The best known and most relevant formulation is that of August Schleicher:

> Languages are organisms of nature; they have never been directed by the will of man; they rose, and developed themselves according to definite laws; they grew old, and died out. They, too, are subject to that series of phenomena which we embrace under the name of 'life'. The science of language is consequently a natural science; its method is generally altogether the same as that of any other natural science (Schleicher, 1863: 20-21).

Although this paragraph has subsequently been rejected and even ridiculed (see, for example, Keller 1990), in my opinion it should be considered a historical precedent for what we now call biolinguistic research.

In fact, when we address the problem of uniqueness, diversity and change in languages from the current biolinguistic standpoint, the comparison between languages and natural species is no longer the illustrative analogy it was for Darwin, but an important source of inspiration.

An I-language is a person's language organ, his/her faculty of language. Therefore, there are not around 6,000 languages in the world, but billions, as many as there are people (in fact many more, given that in the brains of bilingual people are more than one I-language).

The only real, existing thing, from the biolinguistic point of view, are those billions of I-languages. Everything else (varieties, dialects, languages, families, etc..) are abstractions that we make by grouping I-languages according to their resemblance or their historical origins. The same is true in the biological realm: what exist are the emerging states of matter that we call life forms, the organisms (the billions of animals, plants, fungi, etc. that live on the planet), whereas varieties, species, families, kingdoms, etc. are abstractions we make on the basis of genetic and morphological similarity and historical origins.

Thus, an I-language is a historically modified natural object. And this is exactly what natural organisms that are grouped to form natural species are: historically modified natural objects.

Darwin used the analogy between languages and species in order to illustrate the theory of natural evolution, using findings from the historical linguistics of the time, especially to strengthen the idea that the similarity between similar systems could be explained as an inheritance from a shared common ancestor (see Alter 1999, Mendívil-Giró 2006, 2009). In his use of this analogy, which is more explicit in *The Descent of Man* than in *The Origin of Species*, Darwin identified species with languages, and organisms with the components of languages (words, sounds). Most modern versions of the analogy between languages and species are based on this type of correlation between the two orders (language = species, component of a language = organisms). This is not so in the case of Schleicher. In his aforementioned review to the German version of *The Origin of Species,* he deals with the correlation as follows:

> The species of a genus are what we call the languages of a family, the races of a species are with us the dialects of a language; the sub-dialects or patois correspond with the varieties of the species, and *that which is characteristic of a person's mode of speaking* corresponds with the individual (Schleicher, 1863: 32, my italics).

What Schleicher mentions as "that which is characteristic of a person's mode of speaking" is the closest definition that could be formulated at the time of the concept of I-language. And, indeed, the most appropriate terms of comparison are those that identify, on the one hand, I-languages with the organisms that form a species, and, on the other, E-languages (understood as a populations of similar I-languages) with species.[15]

But this language organ, in addition to being a natural object, is also a historical object. The language organ of a person who speaks English is different from that of a person who speaks Spanish: both share (allegedly) a fundamental design, which conventionally we call UG, but differ because of contingent events that can only be explained historically. Migrations, divergent changes, borrowing and isolation have produced two different natural objects (actually, millions of them, as many as there are speakers of both languages). But it is important to note that the very fact that the I-language of an English speaker and the I-language of a Spanish speaker are historically different should not make us think that they are purely historical objects, just as a horse and a buffalo are different historical objects, but are not purely historical, but also natural.

---

[15] Of course, there are other definitions of E-language (see Chomsky 1986), but the only one relevant for our purposes is that of E-language defined as a population of similar I-languages.

A linguistic species (like a natural species) is a population, a group of individuals with enough similarities. The degree of this necessary similarity is determined by viable reproduction, in the case of natural species, and by mutual intelligibility, in the case of languages. And in both cases it is a diffuse and somewhat arbitrary border. In the case of species, fertile reproductive potential depends on a very pronounced morphological and genetic similarity between two individuals; in the case of languages, mutual and fluent intelligibility depends on a very pronounced lexical, grammatical and phonological resemblance between two language organs.

The comparison works because (as argued by Pinker 1994, Chapter 8) the evolutionary mechanism is based on the same principles: inheritance, mutation and isolation. Heredity explains that organisms resemble their descendants (lions breed lion cubs, not chicks). Mutation explains that descendants are not exactly identical to their progenitors. Replication is imperfect. Genetic mutation and other factors generate variation among organisms, and it is through the variation of organisms that natural selection operates. Isolation prevents natural selection becoming homogenous in all populations, and results in the unequal distribution of variants among different populations.

The case of languages is just the same; children speak the same language as their parents (assuming that they raise them), but not exactly the same. Sometimes reanalysis (or errors) generate variation, and it is variation that feeds the social selection that leads us to adopt new forms and transmit them, along with traditional ones, to successive generations. Physical and social barriers prevent the homogenization and levelling of the relative frequency of variants.

The meaning of the term *evolution* in evolutionary biology is not (or should not be) improvement, optimization or refinement, but simply the differential transmission of certain variants over others. And the same should be applied to languages. This is crucial, but has not always been well understood. Many of the comparisons between the evolution of languages and species that have been made after Darwin have been in the wrong direction because they were often inspired by an incomplete view of the theory of evolution, according to which adaptation to the environment is all we need to explain the actual structure of living beings.

Indeed, the comparison between I-languages and natural organisms could be (and has been) used from the anti-universalist standpoint to justify the myth of the diversity of languages, in the sense that there is a noticeable difference in the structure and way of life between (at one end) a bacterium and (on the other) a human being with a brain composed of about one hundred billion (10^{11}) neurons. And because of that I think the comparison is legitimate, in that the diversity among languages is as remarkable and significant for the study of language as the diversity of species is for the study of life. The way in which languages differ is, then, an excellent source of information on how the brain constructs, processes and uses language.

It is noteworthy that from the phenotypic point of view life forms are extremely varied, but it is also true that all forms of life use exactly the same chemical processes and biological mechanisms to develop, metabolise, reproduce, and die. Compared to inorganic objects, forms of life are fundamentally similar. To dismiss or underestimate the biological and natural foundation of language would be comparable to renouncing the discovery of the basic principles of life in favour of a descriptive analysis of living forms and of the induction of certain patterns or processes common to all of them. Nobody, of course, proposes this in the sphere of life sciences, but in language science this is indeed the case.

As we have seen, E&L (2009) propose as an alternative research program to the biolinguistic one, a model in which "the natural" is first separated from "the cultural", and then it is denied (rightly) that there are exceptionless universal properties in the phenotypic realization of languages. E&L accuse some of their critics of using a definitional notion of UG, rendering it unfalsifiable (2010: 2741), but they in fact do the same in the opposite direction: they consider a priori that human languages are cultural systems, that is, they exclude from the very concept of language the natural conditions that it can have, given that in their model such conditioning is by definition language external. Therefore, the notion of language with which they operate is partial: the result of segregating biological invariants from what is subject to historical development. It is hardly surprising, then, they fail in their attempt to find linguistic universals.[16]

Returning to our analogy between language and life, we might say that E&L do not find the common basis of life because they do not look for it in biochemistry, in the principles of self-organization, in the laws of development or in epigenesis, but in genes and phenotype, where they

---

[16] Baker (2009) notes that each language he studies is a source of amazing diversity and of striking uniformity, and points out that E&L fail to find true universals because they "are looking at only one side of this paradox" (Baker 2009: 448).

cannot exist, since this is the domain of the contingent historical events that have singled each species (and each language).[17]

In a way, we can say that the model suggested by E&L to address the problem of uniqueness and diversity of languages is an incomplete one that reproduces the same limitations that characterize the neo-Darwinist model of natural evolution. Neo-Darwinist biologists tend to downplay the sources of order and structure of organisms other than the adaptation by means of natural selection. And in the same vein, functional-cognitive linguists tend to downplay the principles that determine the uniformity of languages, and identify as sources of the observed uniformity those principles derived from the functions (adaptations) for which language is used.

6. The long shadow of neo-Darwinism: Languages as cultural objects

Stephen Jay Gould (1996, 2002) has described the fundamental difference between the neo-Darwinian model and its alternatives through the metaphor of the billiard ball against Galton's polyhedron. According to the neo-Darwinian point of view, an organism could be represented as a billiard ball in motion. Each time the cue hits the ball there is a variable and distinct movement; there is variability and it is a free variation that might go in any direction. The cue hitting the ball represents natural selection, and the ball goes where selection drives it. This constitutes, in Gould's terms, an *externalist*, *functionalist* and *adaptationist* theory. By contrast, the anti neo-Darwinian point of view presents the metaphor differently. The organism would not be spherical, but a polyhedron resting on one of its faces. Of course, it remains necessary for the cue of natural selection to hit the polyhedron (if not, it will not move), but once the cue hits it the prospects for change are very constrained: it is a polyhedron, which has a certain internal structure that limits variation, so that certain options are more likely than others and some are impossible, however interesting they might be from an adaptive point of view.

---

[17] As noted by Duffield in response to E&L (2009a), "no matter how deep one digs into mature grammatical systems, there is no logical reason to expect that one will excavate UG in any recognizable form, any more than one should discover universal principles of embryology through an in-depth study of mature organisms" (Duffield 2010: 2675)

Gould and numerous other authors do not deny natural selection and its importance as a mechanism, but suggest that we must also make a thorough study of the influence of negative restrictions (and/or positive channels) on Darwinian adaptations. As pointed out by Kauffmann, "biology since Darwin is unthinkable without selection, but may yet have universal laws" (Kauffman 1993: 25).

What is at issue is not whether Darwin was right, but whether he was absolutely right, that is, if the distribution of the morphology and structure of life forms is simply the result of adaptation through natural selection, or whether other factors that place limits or somehow channel the evident work of natural selection are also in operation. As suggested by Kauffman, the matter is that "we do not understand the sources of order on which natural selection was privileged to work" (Kauffman 1993: 643).

It seems clear that E&L's eulogy of diversity fits more in the conception of languages as billiard balls, and that the Chomskyan universalist approach equates more readily to a vision of languages as Galton's polyhedrons. Although not specifically an argument in this debate, it is important to note, as made in the previous chapters, that modern evolutionary biology tends to encourage the pluralistic view of evolution than that brought about by the neo-Darwinist new synthesis[18].

Obviously, from a functionalist point of view what matters is the function. The function to be satisfied is more important than the means employed to satisfy it. This predicts a scenario where linguistic changes are restricted only by the functions that must be met for communication and thought, which in turn would imply that languages can diverge freely provided that the functions entrusted to them are met. Thus, from the externalist, functionalist and adaptationist view (to use Gould's characterisation), a language is by definition a historical object that has been created by its development through successive adaptive changes.

On the contrary, from an internist, formalist and non-adaptationist viewpoint, a language is, of course, a historical object (affected by the Basque substrate, the Norman conquest, etc.), but it is also a natural object with an 'ahistorical' structure which marks boundaries, the roads opened

[18] "Pluralism, instead of the belief in the omnipotence of natural selection, is the norm in evolutionary biology today. (...) Obviously, natural selection only serves as a filtering condition on preexisting variations, and the primary question is how these variations first came into existence. In other words, arrival of the fittest, instead of survival of the fittest, is the core issue in any evolutionary study" (Narita & Fujita 2010: 364)

or closed to changes. This model predicts that linguistic changes are superficial, and therefore also predicts a restricted range of language diversity.

Given that for E&L languages are essentially cultural historical objects, the patterns of similarity among languages must be explained either by shared historical heredity, or by evolutionary selective pressures *external* to the evolving systems:

> To the extent that there are striking similarities across languages, they have their origin in two sources: historical common origin or mutual influence, on the one hand, and on the other, from convergent selective pressures on what systems can evolve. The relevant selectors are the brain and speech apparatus, functional and cognitive constraints on communication systems, including conceptual constraints on the semantics, and internal organizational properties of viable semiotic systems (Evans & Levison 2009a: 446).

Note that in the absence of further specification, this statement cannot be false. This amounts to saying that the explanation of the structure of organic beings must be the result of natural selection, the laws of self-organization of organic matter, development principles and physical laws. The central issue (both in biology and in linguistics) is to define what degree of restriction changes present, be they evolutionary (in the case of natural organisms) or historical (in the case of languages).

As regards evolutionary biology, Gould makes the central question clear:

> In what ways does the skewed and partial occupancy of the attainable morphospace of adaptive design record the operation of internal constraints (both negative limitations and positive channels), and not only the simple failure of unlimited number of unconstrained lineages to reach all possible position in the allotted time? (Gould 2002: 1053).

Gould argues strongly in favour of the first option, on the basis, among other arguments, of the discovery in modern genetics of a deep homology between taxonomical types separated by more than 600 million years. These types share many ontogenetic channels based on levels of genetic retention (e.g., so-called Hox genes) that a neo-Darwinian model would consider implausible, given the alleged capability of natural selection to modify any line in a single direction according to its long and contingent history.[19]

Not surprisingly, E&L (2009b) argue, contrary to the claim of Pinker and Jackendoff (2009) that languages actually occupy very partially the space of conceivable design, that such partial occupation could be a consequence of the fact that languages have not yet had time to develop certain ways, given the short time they have been evolving (about 100,000 years, according to the authors).[20] It is impossible to judge this argument, since there is no way of knowing objectively whether 100,000 years is a long or short time span. It is, though, instructive to see how E&L, perhaps inadvertently, align themselves conceptually with the neo-Darwinian point of view, according to which the evolutionary process is limited externally and not internally to the evolving system itself.

In fact, the neo-Darwinian idea that the methodology of evolutionary theory can be applied wherever any dynamic system exhibits random variation, selection between variants and differential inheritance, is probably correct and, as I have shown, is clearly applicable to languages. Yet this does not mean that there is only one way of conceiving of the methodology of evolutionary theory, or that there is only one way to think of languages as dynamic systems with random variation and differential inheritance.

In the tradition represented by Deacon (1997) and Hurford (2002)[21], to which E&L's programmatic article is definitely indebted, when they speak of a language as a dynamic system they are actually talking about E-language, while I-language is a mere reflection of the former in the brains/minds of speakers, who merely transmit it with changes.

---

[19] As Benítez-Burraco and Longa (2010) point out, Mayr noted more than forty years ago that "much that has been learned about gene physiology makes it evident that the search for homologous genes is quite futile except in very close relatives" (*apud* Benítez-Burraco and Longa 2010: 310). These authors identify this view as "different genes for different animals". They also note that "for more than a century, biologists had assumed that different types of animals were genetically constructed in completely different ways", while "Evo-devo has shown (...) that such a contention was unjustified" (Benítez-Burraco and Longa 2010: 310).

[20] "The earliest modern human remains date back to about 200,000 BP, and outside Africa date from only 100,000 years or so ago. If that is the date of the great diaspora, there has been relatively little time for diversification" (Evans and Levinson 2009b: 477).

[21] See the works collected in Briscoe, ed. (2002) for an overview of the model. Hurford (2002) calls this tradition as the "expression / induction" model.

Although in some ways the analogy is appropriate, in the sense that language change occurs through iterated learning processes, the problem is that the object of study is located *outside* speakers. E-language is identified with the species, but there is no clear equivalent of the natural organism, the individual forming the species.

A conclusion that the authors of this tradition share (Briscoe 2002: 10) is that languages evolve to adapt to acquisition requirements, a surprising way of seeing the issue that again illustrates that we are operating with a biologically inconsistent view of E-language, as an external object or as a set of sentences, and not as a population of I-languages. This approach may sound appealing, but in the end it is unacceptable, since it would be analogous to claiming that natural organisms have evolved to be expressed by DNA. Obviously, DNA is not part of the environment to which organisms adapt, and in the same way acquisition requirements imposed by human brains are not part of the environment to which languages adapt, unless we claim that languages are objects outside the brain and not properties/states of the brain itself. Indeed, such is the case with Deacon's influential theory of brain and language co-evolution:

> The extra support for language learning is vested neither in the brain of the child nor in the brains of parents or teachers, but outside brains, in language itself (Deacon 1997: 105).[22]

Let us imagine a stone channel through which water circulates. Of course we can say that water is adapted to the shape of the channel, but it would be surprising if we were to ignore that the channel structure also strictly determines the form water adopts. Notice that the problem is not whether learnability requirements are general or are specifically linguistic (a difficult and central issue, but ultimately an empirical one), but one of considering them part of the adaptive environment for languages. And this is precisely the usual point of view of the FCP:

> Human children appear preadapted to guess the rules of syntax correctly, precisely because languages evolve so as to embody in their syntax the most frequently guessed patterns. The brain has co-evolved with respect to language, but languages have done the most of the adapting (Deacon 1997: 122).

---

[22] E&L themselves suggest that the model offered is intended to determine "how much design must be in the organism and how much design can evolve on purely cultural grounds" (E&L 2010: 2734) and they definitely opt for the second option.

For this externalist and adaptationist conception, in which languages evolve independently, the explanation for convergent evolution is not a matter of the restrictions of evolving systems themselves, but, as in the neo-Darwinian model of natural evolution, an explanation which relies on the evolutionary analogy favoured by the similarity of adaptive environments:

> Grammatical universals exist, but I want to suggest that their existence does not imply that they are prefigured in the brain like frozen evolutionary accidents (...) they have emerged spontaneously and independently in each evolving language, in response to universal biases in the selection processes affecting language transmission. *They are convergent features of language evolution in the same way that the dorsal fins of sharks, ichthyosaurs, and dolphins are independent convergent adaptations of aquatic species* (Deacon 1997: 115-116, my italics).[23]

It is illuminating to observe how Deacon (also Briscoe, and E&L) seem to take for granted that the resemblance between legs, wings, eyes and fins in the animal kingdom can be explained via pure analogy, that is, as a product of independent (convergent) evolution. However, this is not the only conceivable path. It would certainly be unwise to rule out effects of this type in language change, but the recent development of evolutionary theory clearly makes it inadvisable to ignore the role of formal and invariant principles in the explanation of linguistic convergences.

Although it is a deep homology and not (at least apparently) a consequence of "design laws of organic matter", the fact is that the evolution of the examples mentioned by Deacon and Briscoe (dorsal fins, wings and eyes) is one of the central arguments that can be used to oppose the adaptationist model of the new synthesis. Thus, as shown by Sampedro (2002: 119 et seq.), Gehring's group showed that the gene Pax-6 (eyeless in Drosophila) is the *same* regulatory gene that controls the tens or hundreds of genes that form both Arthropod and human eyes. This makes clear that while evolution and selection have modified many of these

---

[23] In a similar vein, Briscoe states: "such historical pathways can be stereotypical responses to similar pressures arising in unrelated languages, *in much the same way that eyes and wings have evolved independently in different lineages many times*, without the need to posit a substantive theory of such changes or to see them as deterministic" (Briscoe 2002: 13, my italics). Of course, E&L share this view: "the distribution of attested structural types across the design space reflects the likelihoods of the evolutionary pathways that would engender them, rather than specific constraints on the structures themselves" (Evans & Levinson 2010: 2734)

genes to produce eyes so incredibly different as the compound eye of crustaceans and our own human eye, there is in fact a deep homology here. And the same can be said in general terms of the other body parts mentioned: wings, legs and fins have all traditionally been used as examples of evolutionary analogy, of convergent evolution, as clear examples of how the medium shapes the adaptation of organisms, yet all have been revealed, so to speak, invented at a time in nature.

The conception of languages as external systems whose evolution is (more or less loosely) constrained by the human brain and human societies is valid as a metaphor, but it certainly can be detrimental to the naturalistic study of our language faculty. As noted by Kauffman "with the onset of full-blown evolutionism and Darwin's outlook on branching phylogenies, the very notion that biology might harbour ahistorical universal laws other than 'chance and necessity' has become simple nonsense" (Kauffman 1993: 5). The same error could occur in linguistic theory if the neo-Darwinian model of the FCP is accepted. As Kauffman also argues, "evolution, while destroying the idea of fixed species, simultaneously swept away the impetus to seek ahistorical laws of organic form" (Kauffman 1993: 3). But this very impetus, encouraged by the Chomskyan model, has gained ground in recent years, and its direction in this regard is akin to that of the bulk of biological science (see Narita and Fujita 2010, Benítez-Burraco and Longa 2010, Boeckx, this volume, Longa and Lorenzo, this volume).

If there is, as suggested by the minimalist model, a universal syntax (derived from type 1 or type 3 factors, or a mixture of both), such a syntax may function as a source of "order" in the "organisms" that would be independent of (in fact immune to) natural selection (i.e. language change). As suggested by the formal biological model proposed by Kauffman, part of the order that exists in the natural world exists "not because of selection, but *despite* it" (Kauffman 1993: 16, original italics).

7. Co-evolution: When Linguistic change and language evolution are conflated

When speaking of co-evolution there is a tendency to identify (in fact to mistake) the evolution of language as a human faculty with the evolution of languages as historical objects.[24] This is not surprising

[24] Hereinafter (and to avoid ambiguity) I employ the term *language evolution* to refer to the evolution of the language faculty in the human species over geologic

because, as we have seen, in the FCP the conception of the faculty of language is purely inductive, and thus the faculty of language in the species would evolve as a consequence of the individual evolution of languages. However, the confusion between linguistic change and language evolution has serious consequences for the scientific and naturalistic study of language and sends an inappropriate message to the cognitive sciences.

As we have seen, the mere formulation of a theory of co-evolution between language and the brain implies a dualistic conception according to which language is external to the brain (and not a property or a state of it). It also implies (which is even more striking) that language and the brain evolve independently, although with interactions. This, indeed, is the point of view defended by E&L (although they do not mention Deacon):

> Coevolving sociocultural systems (languages) and their biological platforms (human brains) spur each other on [...] Once a coevolutionary approach is adopted [...] the presence of any universal feature calls for explanation in terms of two interacting components: the evolving semiotic system, and the evolving capacities of its users (E&L 2010: 2746).

The dualist logic of the co-evolutionary model implies that languages function as adaptive environments for the brain, and that the brain functions as an adaptive environment for languages, since the model states that the brain evolves to learn (and use) languages, and that languages evolve in order to be learned (and used) by the brain.

It is difficult not to perceive this process as circular. Note that since brains are, clearly, prior to languages, then languages must have been produced by brains in first instance (or less plausibly, they were copied form another species). After this, languages would have been evolving (changing over time) and acquiring new emergent properties that, in turn, served as a new adaptive environment for brains themselves. Whence comes the structure of a language, other than that resulting from constraints imposed by the brain itself, remains a mystery.

Since the FCP tends to reject the notion that human beings have a natural inclination towards language, and prefers to think of general

---

time (the issue discussed by Lorenzo, this volume), and the expression *language change* to refer to the change of languages in historical time, i.e. to processes such as those that led from Old English to current English or from Vulgar Latin to Spanish. This does not mean giving a qualitatively different value to the terms *evolution* and *change*.

learning and processing systems, it also favours a view according to which languages have done most of the adaptation in the co-evolutionary process while the brain has been limited to achieving a general state. As we have already seen, Deacon has noted that, "the brain has co-evolved with respect to language, but languages have done the most of the adapting" (1997: 22).[25]

However, apart from a problem of circularity, the theory of language-brain co-evolution has other potentially critical implications for cognitive science. If we assume some degree of co-evolution and we also assume the great depth of the diversity of languages (the central premise of the program presented by E&L), then we have no choice but to assume that different languages might have produced different types of brains, that is, different types of faculty of (or capacity for) language in the same species. Far from rejecting this conclusion, E&L seem to consider it seriously, even suggesting that the evolution of the "human capacity for language" is still changing:

> There are also fundamental differences in the way language is implemented in brains (...) reflected in considerable individual differences in language performance, providing the raw material for ongoing evolution. If we are at all interested in language diversity and language variation, or the instantiation of language in the brain, we will need to bear a coevolutionary model of language in mind (E&L 2010: 2742).

It seems to follow, then, that any differences in the localization of language in individuals or groups of individuals could be explained because those individuals evolved in different linguistic contexts. It is hard to imagine a more extreme relativism.

With their emphasis on the conception of language as a cultural tool ("processes of cultural evolution hone languages into the marvelous instruments they are", E&L 2010: 2742), E&L suggest that since "the anatomy of the hand has coevolved with tool using" (*ibid.*), human anatomy would have coevolved with languages. One might ask then why the use of different types of tools did not cause differences in the anatomy and physiology of the hand, which appears essentially uniform in the

---

[25] Note that otherwise it is not easy to distinguish this position from the traditional generative conception of UG as a *Language Acquisition Device*. Smith and Kirby (2008: 283) pose the dilemma clearly: "There are two possible sources of structure in language: biological evolution of the language faculty, or cultural evolution of language itself", and they make the second.

species. There are two options: either the structure of the hand is not the result of the co-evolution with manual tools (which seems the most reasonable), or the hand evolved when it used a single type of tool. In such a case we would have to assume that if there are not different capacities for language, it is because the FL evolved when there was only one language or a restricted type of languages. But in such a case, the evolution of the FL could not have been slow and gradual, as E&L (2010: 2742) stipulate.

Consequently, we can say that co-evolutionary theory predicts that humans should be grouped according to various capacities for language, just as they are superficially grouped by the colour of their skin, the shape of the eyes, or the size of the nose.

In fact, E&L consider this possibility:

> The interaction of genes and language continue to evolve, as shown by the recent finding that language change has been channeled by population genetics in the last few thousands of years (E&L 2010: 2742).

They cite as evidence for this "recent finding" the study of Dediu and Ladd (2007), according to which there is "preliminary evidence that gene pools with certain biases in allele distribution are more likely to harbour languages of specific sorts" (E&L 2009b: 480).[26] However, this study does not demonstrate that variation in certain genes has an influence on the development of tonal systems. Dediu and Ladd certainly suggest that these genetic differences might have an influence on language change, favouring tones:

> If differences in language and speech-related capacities are variable and heritable and if the genes involved have inter-population structure, it is likely that populations may differ subtly in some of these aspects, and that differences between populations could influence the way languages change through cultural evolution over time (Dediu & Ladd 2007: 3).

Their study, though, is based on a statistical correlation and nothing else (see Benítez-Burraco, this volume, for a thorough discussion on the criteria for the setting of correlations between genetic mutations and cognitive aspects). Mark Lieberman's commentary on this study suggests

---

[26] To which they add: "We are not dealing, then, with an invariant machine at all, but with a biological system whose evolution has relied on keeping variance in the gene pool" (E&L 2009b: 480).

that it presents merely that, a statistical correlation, and does not prove a causal relationship between genetic variants and the occurrence of tones in languages, which might have many possible explanations:

> I suspect (though I haven't shown) that by jiggering the parameters of the simulation, you could get a frequency-distribution of geographical correlations rather like the one the Dediu and Ladd found, without assuming any meaningful connection at all between genes and linguistic traits. And of course there really are some connections, if only because of linguistic endogamy.[27]

Even if we can bridge the theoretical (and empirical) problem of a diverse natural capacity for language in humans, a model in which linguistic structure and complexity is the result of the individual evolution of languages cannot avoid the prediction that some languages may be more developed and evolved than others. In other words, what underlies this kind of approach is (implicitly at least) a denial of the so-called *Languages Uniformity Hypothesis* (LUH). The LUH is a consequence of the belief that all human languages have the same degree of development or evolution, and that hence there are no primitive languages. This stance routinely appears in linguistics textbooks and is often presented as an achievement of modern linguistics. The typical ingredients of the formulation of the LUH are statements such as the following ones: There are no primitive languages; all languages have the same degree of complexity; all languages can perform the same function; all languages have the same dignity; all languages have at an abstract level the same

[27] In http://itre.cis.upenn.edu/~myl/languagelog/archives/004554.html (accessed 09.20.2010). Dediu and Ladd do not contradict these conclusions and, in contrast, recognise the lack of correlation (for what is known today), and declare that "it's certainly true, as Mark says, that our geographical correlations would mean more if they had proceeded from some experimental demonstration of some sort of genetically linked, language-related, cognitive/behavioral/perceptual difference". In the context of our discussion, the motivation to publish their study is not without interest: "But given the widespread assumption (rooted in the Boasian tradition, but with a significant contemporary boost from Chomsky) that the human language faculty is absolutely uniform across the species, it's very unlikely that we would have been able to get funding to look for such a difference first. So we started by doing something we could do on our own without such support, namely testing the apparent correlation" (in http://itre.cis.upenn.edu/~myl/languagelog/archives/004564.html, accessed 09.20.2010).

structure; all languages have the same basic components; all languages offer the same degree of difficulty of acquisition as native languages.

Of course, some of these claims might be false and others true, and authors, especially those aligned to the FCP, might share some of them and not others. Hereafter I will refer to the LUH in a broad sense as the assumption that all current languages have the same degree of evolution, i.e. that there are no languages that represent an earlier or less developed state of human language.[28]

Although they are connected, it is important not to confuse the LUH, which is an empirical hypothesis, that is, a statement about the existence or non-existence of certain objects, with the so-called *uniformitarian principle*, which is a fundamental methodological principle of historical linguistics. In fact, uniformitarianism is a general principle of any historical science, whether linguistic, biological or physical.[29]

The LUH raises two critical issues: (i) if it is correct or not, and (ii) if correct, why.

In general, most linguists accept the broad version of the LUH, but it is the case that they do not agree what the ingredients of the formulation are; it is easy to imagine, for example, that E&L would not subscribe to all the points listed above. This makes addressing problem (i) especially fruitless, since previously we must specify clearly what is meant by "primitive language" or what is meant by "a language that represents a previous evolutionary state of the language faculty", which would lead us to define "language faculty", and so on.

We will therefore focus on problem (ii) in order to show that although almost all linguists formally endorse the broad version of the LUH, the hypothesis does not follow from all approaches to the question of language diversity.

If we ask any of the countless advocates of LUH why there are no primitive languages, two types of responses are likely, which I will call

---

[28] A typical (and significant) example is that of Joseph Greenberg, who stated the following: "Certainly, then, the evolution of language as such has never been demonstrated, and the inherent equality of all languages must be maintained" (Greenberg 1957: 65).

[29] A particularly clear definition of that general principle is that of Roger Lass: "the principles governing the World (= the domain of inquiry) were the same in the past as they are now" (Lass 1997: 25). This general principle invites us to assume the maxim that nothing impossible today was, in principle, possible in the past, so we should rule out reconstruction proposals involving states or processes that are impossible today.

"homological" and "analogical" theories of language uniformity and which, not surprisingly, match the biolinguistic and the functional-cognitive paradigms respectively.

The homological theory of language uniformity would contest that all current languages are the result of a single language faculty, which is the consequence of natural evolution in the species; therefore, any natural language spoken by a human being is limited or conditioned by this faculty and cannot be primitive. Primitive languages disappeared with the disappearance of species that had primitive language faculties.[30]

The analogical theory, in contrast, would argue that there are no primitive languages because all current languages have been evolving for tens of thousands of years, constrained by human processing and acquisition systems and by the cognitive and communicative functions they must fulfil. During this time, either modern human languages caused the extinction of primitive languages, or the latter turned into the former. We can characterise this view as analogical in the sense that the explanation of the uniformity would be the long-term historical evolution of each language.

There is a crucial difference (not always marked) between these two approaches regarding the LUH: the relevance that is given to linguistic change in the process of language evolution. For homological theories, language change does not have functionally positive or negative effects, and is wholly independent of the phenomenon of language evolution, which concerns the biological evolution of species. However, for analogical theories, language change is typically sensitive to such external factors and, although not often explicitly recognized, it is not conceived of as independent of the phenomenon of language evolution.

In English, but not in other languages, the word *language* serves two functions: it represents particular languages (cf. French *langue* or Spanish *lengua*) and also language in general (cf. French *langage* or Spanish *lenguaje*). Thus, an expression such as *language evolution* can refer to both the historical changes in languages and the evolutionary process that created our special ability for language. However, this "unfortunate ambiguity" (Hurford 1992: 273) is not the only cause of frequent uncertainty regarding the use of expressions such as *language evolution* or

[30] The term *homological* is used metaphorically, in the sense that I do not mean that from this point of view structural uniformity of language is the result of common inheritance from a protolanguage, but rather that the source of such uniformity is the faculty of language as a biological attribute of every human being.

language development. For some authors, there is an authentic vagueness to the process of language evolution and the process of language change, which is of great importance in predicting the LUH. From this viewpoint, language evolves across the evolution of languages, which in turn are influenced by cultural evolution, whereas for 'homologists', languages do not evolve but simply change.

When we define a language as a historically modified mental organ, we are assuming that linguistic change is the main cause of linguistic diversity, a conclusion identical to Darwin's when he suggested that species' mutability explained the existence of different species. Therefore, the question regarding the extent of linguistic diversity is tantamount to the question of the transforming capacity of the change process. Can processes of linguistic change produce anything that is not a language? The answer seems clear: "no". On the other hand, can linguistic change produce a language from anything that is not a language? Again, the correct answer seems to be "no". This is the answer from a biolinguistic point of view, but it is not universally accepted. In fact, many authors think that linguistic changes were a central part of the evolution of language (e.g. Heine and Kuteva 2007 and references there).

Consider, for example, Comrie (2003):

> As a result of the recent development of grammaticalisation as a tool in historical linguistics it has been possible to develop a more general variant of internal reconstruction (…) that does enable us to come up with plausible hypotheses concerning earlier states of *language development* (Comrie 2003: 249, my italics).

What is Comrie referring to, linguistic change or language evolution? This short extract, which includes one reference to the method of internal reconstruction and another to the concept of grammaticalisation, may suggest that Comrie refers to linguistic change in historical time, but this is not the case. He really refers to both, simply because a purely historical conception of languages admits no precise way of differentiating between the two phenomena. This, in my opinion, weakens the LUH notably.

Later, in the same article, our suspicion is confirmed:

> We can take grammaticalisation and base on it a kind of generalised internal reconstruction that gives us access to hypotheses concerning earlier stages of the language in question and by generalising our conclusions to earlier stages of language in general (Comrie 2003: 249).

The reference to "earlier stages of language in general" only has meaning as an allusion to the development of the faculty of language in the species. However, note that we are then admitting that linguistic change is directional and capable of changing the faculty of language (if recognised as such at all). Insofar as languages change at different rates and in different directions, we must also admit that the faculty of language is not uniform across the species and that, in fact, this faculty continues evolving among different groups.

In the homological scenario there is a point in human evolution that produces a modern faculty of language. After this point all languages will be modern languages.

In the analogical scenario human biological evolution is, of course, a condition for language development; however, it does not specify the structural properties of languages, but only a new environment for the evolution of languages. Thus, once humans crossed this frontier of evolution, human languages began to develop, through linguistic evolution, the structural complexity that now characterises them.

The critical difference is that in this scenario the imaginary line that separates cognitively modern humans and their ancestors does not coincide in time with the imaginary line that separates primitive languages and modern languages. That is, the historical moment at which every historical lineage reaches, through the effect of linguistic change, the status of non-primitive language may be (must be?) different for each linguistic lineage. This scenario envisages the possibility that certain human languages, perhaps now extinct and unrecorded, did not successfully cross that frontier.

Notably, the LUH also follows from this scenario, but in a rather different way. Under this approach, it is conceivable that some languages, although spoken by anatomically modern humans, continue to be primitive languages. Actually, this analogical model predicts that it is highly probable that during long time spans, some anatomically modern human groups would speak primitive languages while others spoke modern languages.

According to this model, if there are no primitive languages today, it is because they have either become extinct or evolved historically towards modern status. In other words, in the analogical scenario, the LUH is contingent even though it remains a highly probable outcome.

Now it is easier to understand why Comrie (2003) asserts that current studies on grammaticalisation (which conceive of it as unidirectional) can offer insight into early states of human language, and why he proposes to relativise the uniformitarian principle, excluding from it ancestral language states.

On this approach, the transition between modern and primitive languages is gradual and merely historical. Language evolution (beyond a general biological endowment) limits itself to the evolution of languages; therefore, the reconstruction of the past of languages is equivalent to the reconstruction of primitive language.

As we have seen, in Deacon's tradition, the ambiguity of expressions such as *language evolution* entails not only vagueness but also a deliberate mixture of the two meanings (this is precisely the meaning of 'co-evolution'). Recall that if we assume this model to be correct, we need also to assume that all languages have reached a sufficient level of complexity to affect their users' brains (that is, to function as adaptive factors for the brain). However, the model then predicts that it is possible that there exist human groups with pre-human cognitive abilities: those whose languages did not evolve in the proper way.[31]

The main difference between the analogical model and the homological one is that the former lacks the boundary provided in the latter by the assumption of a qualitative jump between classes of languages. The postulation of a biological evolutionary endowment that gives rise to a modern FL allows one to establish a historical discontinuity between the classes of languages generated by evolutionarily differentiated FLs. The absence of this boundary in the analogical model may cause the attribution of the LUH to the hypothesis that languages have converged into uniformity under the effects of the passing of time and external pressures. But note that in this case the LUH is not predicted (let us say it is a possibility, not a necessity) and henceforth we must entertain the possibility that this confluence has not occurred. In such a case, some languages could be primitive, that is, languages that because of diverse historical ups and downs failed to reach the aforementioned boundary. Everett (2005) proposes just this about the Amazonian language Pirahã, which according to him lacks one of the (supposedly) central traits of human languages: recursion. Everett claims that in this language, phrases cannot be embedded into other phrases (so that there are no subordinated clauses). Everett interprets this as a cultural constraint on grammar and presents it as an empirical argument against a naturally determined FL and in favour of relativism. Nevins *et al.* (2009) forcefully refute this analysis. In his reply, Everett (2007) admits that he does not mean that the Pirahãs'

[31] Nevertheless, I know of no one who has claimed this, not even Daniel Everett, the champion of modern relativism.

minds are limited in the use of recursion or the use of recursive languages, which orientates us here to the analogical model; what Everett is implying, then, is that some languages analogous to those of the previous stage of human evolution did not become extinct, but rather still exist today. This is a denial of the LUH.

It is true that the homological model smacks of circularity insofar as modern languages are defined as those spoken by modern humans. However, in the final analysis, this is the crucial matter: is the boundary between pre-language and language just cultural, or is it biological?

To answer that it is both things (as in co-evolution theories) amounts to saying that it is cultural insofar as if the historical process does not occur, then neither does the transition. If we make this claim, however, we assume that there exist modern human beings that nevertheless speak primitive languages, that is, classes of languages produced by ancestral language abilities that for contingent reasons (social or cultural) have not had sufficient time to evolve in the expected way.[32] I think that this is the only foundation on which the emerging new and radical relativism can be constructed, but it is a rather weak one. The empirical robustness of the LUH, then, can be seen as an argument in favour of the biological theory of the uniformity of languages.

8. Neo-relativism and human nature

It is not surprising that the centrality that the FCP confers on the diversity of languages and the sharp perception of its depth have meant a revival of linguistic relativism (e.g. Lucy 1992, Levinson 2003). In fact, as we have seen, the perception of how deeply different languages may influence cognitive processes of human beings is even increasing.

Questions such as to what extent the languages we speak determine how we see the world, how we think, and how we live our lives have always fascinated people and have been addressed by a multitude of philosophers, anthropologists, linguists, and with much more zeal recently by psychologists and psycholinguists. Precisely because of the close attention it has received from various fields, linguistic relativity is not easy to define. Instead of trying to do so, I shall characterise *linguistic neo-*

[32] Even Comrie, when he suggests distinguishing between "the human language potential" and the "realisation of the human language potential" (Comrie 2003: 250), is obliquely introducing an anti-unformitarian stance.

relativism by citing the prestigious experimental psychologist Lera Boroditsky:

> Linguistic processes are pervasive in most fundamental domains of thought, unconsciously shaping us from the nuts and bolts of cognition and perception to our loftiest abstract notions and major life decisions. Language is central to our experience of being human, and the languages we speak profoundly shape the way we think, the way we see the world, the way we live our lives (Boroditsky 2009: 129).

This strong relativistic statement has at least three implications, all of them problematic: (i) if language influences thought, then language is different from thought; (ii) languages influence the way we see the world, and it is not our way of seeing the world what influences languages; and (iii) there is empirical evidence confirming the first two statements.

As regards the first implication, note that the mere question about the relationship of determination that exists between language and thought necessarily implies previous definitions of *language* and of *thought*, but these definitions are not to be found in the works that have addressed this issue, both past and present; this is little surprise, since we simply lack of such definitions. Of course it is entirely possible that language does not only influence thought, but that language is part of thought. Even in such a case the relativist hypothesis is not necessarily correct, and is so only if language is identified with languages (i.e. in an inductive sense). Thus, from a point of view according to which the diversity among languages is relatively superficial (that is, confined to the manner in which homogeneous computational processes are materialised), the discovery that human thought makes extensive use of language does not in any way lead to the conclusion that human thought is fragmented into groups coinciding with languages. Quite the contrary, it would support the hypothesis that human thought is essentially homogeneous within the species. In fact, any theoretical model that accepts a certain degree of natural bias towards language acquisition should reject Boroditsky's reativist hypothesis[33]. Only a radically empiricist conception of the human

[33] As Bolender notes, "if some sort of Whorfian hypothesis turns out to be true, as these works suggest, this should not only mean that linguistic differences account for some cognitive differences. It should also mean that linguistic similarities account for some cognitive similarities across cultures. If linguistic differences so crucially enter into cognition, as a Whorfian would claim, it is unlikely that similarities would be cognitively irrelevant" (Bolender 2010: 2662).

mind and brain could accept that external cultural objects (human languages in the FCP) could "profoundly shape the way we think".

One can infer that Boroditsky (2009) and Deutscher (2010), to cite only recent work of great social impact, accept that languages influence thought because they operate with a rather vague notion of thought (and, obviously, because they believe in the myth of the diversity of languages). In a review of Deutscher (2010), published in *The New York Times* (September 5, 2010), Derek Bickerton notes that the aspects of language Deutscher deals with "do not involve 'fundamental aspects of our thought', as he claims, but relatively minor ones". Thus, Bickerton points that issues such as location, colour or grammatical gender (which are also the main topics dealt with experimentally by Boroditsky to give support to her proposal) probably have little conditioning on our thought in the daily management of our lives, and far less still when developing political, scientific or philosophical thought.[34]

Let us address the second and third implications of the relativist hypothesis: that languages determine our way of seeing the world (and not vice versa), and that we have empirical evidence for this. In fact, Boroditsky argues that, compared to discussions about the relationship between language and thought in the past, empirical work is currently being developed that could solve these ancient disputes. Specifically, she states that research conducted in her laboratories at Stanford and MIT has produced a large collection of data from around the world (China, Greece, Chile, Indonesia, Russia and Aboriginal Australia), and she concludes that

> what we have learned is that people who speak different languages do indeed think very differently and that even flukes of grammar can profoundly affect how we see the world (Boroditsky 2009: 118).

This is not the place for a detailed review of the ingenious experiments and remarkable discoveries in this field of research, but it may be sufficient to consider, for example, what kind of experimental evidence supports the claim that "flukes of grammar can profoundly affect how we see the world".

Boroditsky (2009: 127) asks whether the fact that in Russian grammar *chair* is masculine and *bed* is feminine makes Russian speakers think that in some way chairs are more like men and beds are more like women. Her

---

[34] Thus, Bickerton concludes in the aforementioned review: "Moreover, with the possible exception of color terms, cultural factors seldom correlate with linguistic phenomena, and even when they seem to, the correlation is not causal".

conclusion is that it does. To provide empirical support for this assertion, Boroditsky and colleagues (Boroditsky *et al.* 2003) asked speakers of Spanish and German to describe in English opposite gender objects in these two languages (for example *key*, which is feminine in Spanish, *la llave*, and masculine in German, *der Schlüssel*). Thus, Spanish speakers were more likely to use words like *golden*, *intricate*, *small*, *beautiful*, *bright* or *very small*, while the Germans tended to use more words like *hard*, *heavy*, *irregular*, *jagged* or *useful.*[35]

Regardless of the details of the experiment and of the criteria for determining which terms are more characteristic of men or women in certain societies, it does not seem that these results show that the language we speak "profoundly" determine our world view, but only that speakers of languages with masculine and feminine gender marking may be prone to extend sexual stereotypes to sexless objects, an extension based on the analogy of grammatical gender that is not available for English or Japanese speakers. We simply cannot infer from these experiments that a Spanish speaker conceives of the bridges in a "profoundly" different manner than a speaker of German (or English). There does not appear, at least, to be systematic differences in how bridges (or keys) are built and designed in Germany or in Spain; in such tasks aspects of physics, engineering, strength of materials, economics and (of course) aesthetics come into play, and these do not seem to be grouped linguistically.

Of course, it is clear that languages interact with culture and that, to a large extent, are part of it. Another example discussed by Boroditsky and colleagues is the expression of space in languages, a topic addressed in depth by Levinson (2003). According to Boroditsky, Thaayore Kuuk speakers (Pormpuraaw, Australia) do not express space in reference to an observer (left, right, front, back), but use the cardinal points (north, south, east, west). It seems to be the case that having to speak a language in which instead of saying "my left leg" it must be said "my west leg" makes it essential that one always knows where west is. According to Boroditsky,

> the result is a profound difference in navigational ability and spatial knowledge between speakers of languages that rely primarily on absolute references (like Kuuk Thaayorre) and languages that rely on relative reference frames (like English) (Boroditsky 2009: 121).

---

[35] Another example used by the authors is *puente* ('bridge', masculine in standard Spanish) compared to *Brücke* (feminine in German), with similar results. Interestingly, *puente* is a variable gender word in Spanish, being feminine in some dialects.

She goes on to claim that

> speakers of languages like Kuuk Thaayorre are much better than English speakers at staying oriented and keeping track of where they are, even in unfamiliar landscapes or inside unfamiliar buildings. What enables them—in fact, forces them to do this—is their language (ibid.)

But if the *cause* is the language istelf, one might wonder what is the cause of a situation where in some languages people use "left", "right", "ahead" or "behind" and in others they use "east", "west", "north" or "south". It seems clear that the answer cannot be other than that the speakers of the second group pay much more attention to the cardinal points. The argument, then, is circular.

At some point, certain aspects of the culture or lifestyle of people led to those decisions, so it is unclear how one might conclude that it is the language which influences the world view, and not the world view which influences the language. All languages, especially in their lexicon, have a dimension which is sensitive to people's culture and lifestyle and, in effect, transmit culture and a view of life. A community that loves mushrooms is likely to have terms for many varieties of mushrooms, a richer lexicon in this respect than a language spoken by those living in deserts. But it would be unwise to say that having many words for different varieties of mushrooms causes a greater appreciation for such a food.

To put it in clearer terms: when Deutscher and Boroditsky (following numerous authors) argue that languages determine or constrain our vision of the world, what they are really saying is that those parts of languages sensitive to culture determine or constrain our culture, something that can not be disputed.[36]

---

[36] Boroditsky (2009: 124) asks "how we do know that it is language itself that creates these differences in thought and not some other aspect of their respective cultures?" (2009: 124). Her answer is based on experiments in which English speakers are trained to express time, for example, as Greek speakers do, and once they are trained, they show that "their cognitive performance began to resemble that of Greek [...] speakers" (2009: 124-5), which, according to Boroditsky, "suggest[s] that patterns in a language can indeed play a causal role in constructing how we think" (ibid.). But note that assuming these individuals have changed their way of thinking implies a very vague and imprecise definition of "thinking". Likewise it could be argued that when the subjects are trained they acquire a new "culture" that affects their performance. Boroditsky adds that "in practical terms, it means that when you're learning a new language, you're not simply learning a new

Undoubtedly, good orientation in space is part of the culture or the lifestyle of the Kuuk Thaayorre. It is also clear that their culture has made their language (like those of other aboriginal Australians) express in this way both spatial location and, by analogy, temporal location. Indeed, this is shown in experiments that Boroditsky (2009) reviews in which speakers of this language were asked to order some drawings temporarily, and instead of doing from left to right, they did it from east to west. But what the experiment really shows is not that language determines thought, but that Kuuk Thaayorre speakers are sensitive to their culture both when speaking and when performing certain experimental tasks.

Boroditsky formulates the key question as follows: "are languages merely tools for expressing our thoughts, or do they actually shape our thoughts?" (Boroditsky 2009: 118). Note, however, that the alternative offered is too narrow and blocks a lot of options that are discarded without consideration. In fact, as we have seen, we have every reason to believe that languages are not tools to express thoughts or shape our thoughts. There is the option that languages are different materialisations, variable and historically conditioned, of a single system of knowledge. Certainly, it would be surprising that a computational system that allows us to create new expressions without theoretical limit, without scope restriction and free of conditioning stimulus (and which is at the heart of every human language), were not also involved in the way humans think about the world and try to understand it in a specifically human way, beyond evident (and interesting) cultural differences.

I have argued above that psycholinguistic experiments designed to demonstrate the influence of languages on thought actually reveal that culture can affect languages. This is no doubt the case, and languages are important vehicles in the transmission of culture from one generation to another.

It could then be argued, however, that another path to linguistic relativism is opened up. Indeed, if we apply the analogy with natural evolution, one would expect that languages somehow adapt to the worldview of their speakers, in the same way that organisms adapt to their environment. That is, we might expect some type of correlation between the structure of languages and the cultural environment in which they are spoken, which in turn would help validate the relativist hypothesis.

It is true that, although in a blind and random manner, natural organisms adapt to the environment. Think of a hippo or a whale. It is

---

way of talking, you are also inadvertently learning a new way of thinking" (2009: 125), which in this context is the same as saying you are learning a new culture.

clear that they are adapted to live in the water, but only when compared with a cow or an antelope and not, for example, with a tench or a bluefin. The environment clearly influences the selection of variants, but it is difficult to delimit the extent of this. Speaking of culture, ideology or worldviews as adaptive frameworks of languages is too loose and general.[37]

It is therefore likely that the effect of culture on languages is very limited, although superficially very visible. In the biolinguistic model reflected in this chapter, the effect of culture (understood as the way of life, traditions and social, political and religious institutions of a human community) is limited to only those domains of language that are not driven by naturally conditioned principles (be them 1 or 3 type factors), mainly in the sphere of lexicon and, of course, phraseology.

As regards the deepest aspects of the structure of languages, if they adapt to the environment, then we can conclude that the environment is nearly invariable. This would not be surprising if such an environment were mainly internal to the brain and mind. As pointed out by Pinker, there are languages in which the direct object follows the verb and languages in which it precedes the verb, but there are no left-object cultures and right-object cultures.[38]

---

[37] Even in the natural world the environment may be rather complex. If we think of a whale, water comes to mind, but if we think of a peacock, we find that its physical environment is not very useful in explaining its hyperthrophic tail unless we include peahens' mating preferences as part of their "physical environment". Peacocks' wonderful tails do not help them to nimbly escape predators or go unnoticed, but it seems that peahens select this kind of tail, perhaps simply because they like it. Delimiting the environment for natural selection is not as easy as it seems, and the exercise is exponentially harder for "linguistic selection". The quantity of different factors that can affect the destiny of a linguistic variant is so complex and varied that the concept of adaptation to the environment ("culture") is simply not concrete.

[38] "One of the great findings of linguistics, vastly underappreciated by the rest of the intellectual world (and probably not highlighted enough by linguists themselves) is that the non-universal, learned, variable aspects of language don't fit into any meaningful, purposive narrative about the surrounding culture. Linguists have documented vast amounts of variation, and have a good handle on many of its causes, but the causes are internal to language (such as phonological assimilation and enhancement, semantic drift, and syntactic reanalysis) and aren't part of any symbolic or teleological plan of the culture" (Steven Pinker, in http://www.edge.org/discourse/recursion.html).

In the review of Deutscher (2009) mentioned above, Bickerton offers an exaggerated (almost cartoonish) hypothetical correlation that, unlike the expression of space or morphological gender, would be relevant to the relativist hypothesis:

> Suppose relative clauses appeared only when a society entered the market economy. Any such finding would revolutionize our understanding of the interface between language and culture. But not only has no such relationship ever been demonstrated, nothing remotely like it has ever been found (Bickerton, *loc cit.*)

I share this observation, but one may wonder if Bickerton has ever read Everett (2005).

9. Conclusion: The true value of linguistic diversity for Biolinguistics

I suggested at the beginning of this contribution that from a biolinguistic point of view the relationship between language and languages is comparable to the relationship between life and living organisms. For this reason I think E&L are right to compare the importance of species diversity as a way of understanding life with the importance of language diversity as a way of understanding language:

> In all these fields, variation is the lever used to discover underlying systematics: you can't find a gene without phenotypic variation, and you can't find a fundamental building block for language without comparing languages (E&L 2010: 2735).

However, a different conception of what languages are (and even of what natural science is) has led us to very different conclusions. I have here criticised E&L's proposal of seeing the diversity of languages as the essential feature of human language from the biological standpoint. But this does not mean that the diversity of languages should not be analysed, explained and used as a relevant factor at the heart of cognitive science.

We have seen that from a biolinguistic point of view languages are historically modified natural objects, just as living organisms are. We have also seen that from the standpoint of evolutionary theory, there are different ways of understanding what the organisms themselves are. As George C. Williams has put it, "mechanistic biologists assume an *organism-as-crystal* and adaptationists an *organism-as-artifact* concept" (1992: 6). The first view focuses on discovering, through variation, the general principles that influence and determine the structure of living beings, while the second focuses on explaining variation as a result of adaptation through natural selection.

If applied to the study of language variation, the two views produce conflicting and to a certain point irreconcilable approaches (the deductive and the inductive ones). However, Williams adds a third perspective: "An *organism-as-document* approach should also be recognized for biologists interested mainly in unique evolutionary histories" (Williams 1992: 6). In my opinion, this is the right perspective to understand the diversity of languages and convert it into a source of privileged information to understand the human faculty of language.

The FL is an attribute of the human species to the same extent as, for example, the capacity of vision or memory. But there is an important difference: a person's capability to see, with its inevitable peculiarities, is essentially the same for all people, whether socially proximate and distant. However, the FL of a person who speaks Spanish is different from that of a person who speaks Russian, although both languages are manifestations of the same natural faculty. From this point of view, the most important information that linguistic diversity gives us about the FL is precisely that the language organ of humans, while a mental organ like any other, is peculiar in a crucial sense: it is a mental organ that is influenced to a comparatively deep degree by culture and history. Thus, although vision, memory or the capacity of facial recognition are mental organs such as language (and they are all the result of the interaction between nature and environment), the structure and properties of the FL of each person not only depend on natural conditioning, but also on historical aspects that do not seem relevant to other mental or physical organs.

As noted by Longobardi (2003: 102), memory and vision appear to lack a relevant cultural history and, unlike language, do not allow a comparative study as in the case of languages.[39] The true value of the diversity of languages for cognitive science is that it provides what is

[39] "These cognitive systems are also subject to variation, but the variation is less rich, essentially individual rather than dependent on cultural transmission: it is not acquired from an environmental stimulus produced by similar cognitive structures ('other minds') and therefore modelled after the latter. Therefore, it does not obviously present the aggregation of properties changing with each other and correlating with external variables (e.g. space/time coordinates) which characterizes languages. In short, it hardly makes sense to talk about modern Spanish memory or medieval Russian vision" (Longobardi and Roberts 2010: 2702)

probably a unique opportunity in the field of cognitive science for addressing the same faculty from various historical solutions (documents) restricted by the same principles.

The deductive or parametric approach to the structural diversity of languages has resulted in work that is clearly beneficial to the future of cognitive science. On the one hand, in the wake of the linguistic typology of the second half of the twentieth century, this approach has led to the accumulation of a wealth of information about delimited regions of the grammar of languages, allowing a greater understanding of the many variables that can affect cross-linguistic comparison. On the other hand, the minimalist conception of UG has made possible a further refinement in approaches to the structural diversity of languages, leading to the hypothesis that structural variation is limited to the materialisation or externalisation of syntactic computations in morphological systems sensitive to environmental information and, hence, historical products themselves. As Longobardi and Roberts note, "this suggests that parametric analysis reflects actual processes of cognitive history, explaining the present distribution of syntactic properties on the grounds of their past" (2010: 2702).

The diversity of species is simply the result of change, and the same is true of the diversity of languages. Whenever a human being constructs from environmental data the interfaces between the computational system and the IC and SM systems, he/she is adding to his/her FL a historical record of how these interfaces were made in previous generations. In the discovery of regularities and systematic clusters of these processes, historical and comparative linguistics has a significant contribution to make to the investigation of the ultimate nature of the principles that characterise the human faculty of language.

In this regard, the following words from Rizzi may well serve as a conclusion to this chapter on the myth of the diversity of languages:

> The cognitive sciences should not overlook the results of half a century of linguistic research which has seriously attempted to identify the limits of variation: it is simply not true that languages can vary indefinitely (Rizzi 2009: 486).

PART II

LANGUAGE AND THE BRAIN

INTRODUCTION TO PART II

LOOKING FOR LANGUAGE INTO THE BRAIN

MARÍA DEL CARMEN HORNO-CHÉLIZ

The study of Language and Brain relationship, as a proper study, has no real tradition in our Faculties. In fact, language studies are usually included in human science, with arts and culture studies, while neurological processes are being faced by scientists who are not usually really interested in languages. Only some neuro-cognitive institutes and some brave researchers have decided to face this kind of studies systematically.

Nevertheless, there are lots of good reasons to study the biological basis of language. Firstly, as language is a human property we must study its biological basis for the same reason as we must study biological basis of digestive or respiratory systems. Besides, we have several practical benefits of this kind of studies. From a clinical point of view, understanding the biological basis of language is the best way to find the most suitable treatment for aphasic patients. It's also important for surgery, to inform the surgeon about the principal zones implicated in language faculty in order to avoid them.

More over the clinical benefits, studying language biological basis can give us some keys to afford theoretical studies. For instance, both the neuroimaging techniques and the study of word processing have shown that the level by level analysis of language is the correct one (see Cuetos' and Carreiras' and Baus' chapters) and studies about multilingualism in the brain has proved the relevant differences between lexicon and grammar or the equal nature of human languages (see Laka's contribution). In the other hand, it can also be useful to the deep knowledge of the brain and the psychological processes. In fact, linguistic ability is one of the most important functions of human brain, and it needs to be connected with other psychological processes (such as memory, attention, etc). In cognitive linguistics there is a tradition to use language input to understand better cognitive processes, such as memory (for example, Filipovich 2011). In sum, the study of language and brain relationship has clinical and theoretical benefits, but it also has practical applications. Its results could be useful to practical linguistics, such as second language learning or computational linguistics, among others.

But what do we mean with "language and brain relationship"? Mainly, two kinds of studies: the localization of different linguistic abilities in the brain (which is the main interest of Cuetos' chapter) and the different linguistic processes we must separate in linguistic ability and the way we have to understand the relationship among different kinds of linguistic information (see Baus & Carreiras' chapter). Both kinds of studies are applied in healthy and damaged brain, and both in monolingual or multilingual participants. As Laka explains in her chapter, this last possibility helps us more to know the real basis of the brain and language relationship.

1. Methodology

The biggest problem this kind of studies must resolve is methodological. As language is a human distinctive property, we cannot use control animal experiments to investigate. In the other hand, ethic limits restrict us to use human beings in this kind of experiments. In other words, language and brain relationship cannot be investigated with the experiments used in other fields of psychological or neurological studies.

This kind of researches is carried out in different ways. Firstly, we can work with people with or without damage in the brain. In the first situation, the development of the research is obviously aimed to clinical results. Nevertheless, the results of the investigations could help us to know more about the brain-language investigation (just in the case we don't generalize freely these results to healthy population). In these situations, more techniques are allowed, in order to help the patient to preserve the linguistic faculty from surgery. If the participants have no damage, in the other hand, more restrictions are limiting the techniques we can use, but the results are more easily generalized.

Due to the importance of the existence (or non existence) of brain damage and linguistic deficiency, the first step should be to test the linguistic faculties of the participants. In order to know the linguistic deficiency the patient has, it's common to use a series of neuropsychological tests. Usually, this kind of studies begins with the *Wechsler Adult Intelligence Scale* (WAIS), which has six verbal subscales (Wechsler, 1958): information, digits, vocabulary, arithmetic, understanding and similarities. Another possibility is the *Token Test* (De Renzi and Vignolo,

1962). This test consists of understanding and following some orders such as: "take the big red card". Sometimes, the orders are read instead of listened.

As we can see, this kind of tests is general, and it doesn't use all the information known about language, its levels (Safran, 1997) and its features. As Cuetos says in his chapter, the brain and language relationship must be studied using all the information linguistic science offers about language nowadays. So, it's necessary to improve these tests with more specific ones.

As we have said before, one of the interests of the study of the language and brain relationship is the localization of the different linguistic processes in the brain. There are several ways to afford it. The three most important ways are: (i) autopsy of aphasic patients; (ii) neuroscience technical of visualization and stimulation of the brain and (iii) neuropsychological tests.

The methods based on the autopsy of aphasic patients are fully explained in the first chapter of this part of the book, so we invite the reader who is really interested in these kinds of studies to find more information there. We just want to say in advance here that the traditional studies are based on the detailed description of the language deficits of these patients during their lives (see also Benítez-Burraco's chapter). After death, a proper autopsy of his/her brain should show the damaged zone. In this way, the linguistic damage must have a relationship with the damaged zone. As Cuetos explains in his chapter, the success of these studies remains in two aspects: (i) a proper register of the linguistic damage, based on a good knowledge of how human language works (that is, a level by level study of different linguistic features); and (ii) a detailed autopsy, with the inclusion of sub cortical levels.

The new techniques of viewing and stimulating different areas of the brain are a new way to work with human beings. The following are the most common techniques which give information about the brain activity:[1]

Positron emission tomography (PET): in one of the most common version of this technique, the patient must be injected with a radioactive 2-DG (2-Dexosi-glucose) solution in the left carotid artery. Once this substance is in the left brain hemisphere, it is absorbed by the active neurons just as fast as if it were real glucose. The advantage of 2-DG is double: firstly it is a radioactive substance, so it can be seen by PET; secondly, this substance cannot be metabolized, so it remains visible for

[1] The information given here about these techniques is technical. In Cuetos' chapter the reader can find a detailed critic about their limits and uses.

longer. With this technique we can ask the patient to make some activities (such as reading, repeating words, talking, listening, etc.) and we can observe which area of the brain is active in each case (Cabeza and Nyberg 1997).

Functional Magnetic Resonance (fMRI): this technique is better than the previous, because no substance is needed to be injected into the participants. It consists of measuring the level of oxygen in different areas of the brain. The more oxygen a neuron consumes, the more active it is. Another advantage of this technique is the images we can obtain: by better resolution, and 3-dimensional pictures, it gives information not only of the activity but also about the structure of the brain (Cabeza and Nyberg, 2000).

Electro-encephalography (EEG): this technique consists of several electrodes all around the patient's head. These electrodes show a signal about the electrical activity in the brain. Some waves (called Event Related Potentials) can be related with some neuronal activities. The most important objective in this case is being able to separate this Event Related Potential to the other waves (the nuisance). For that, the Averaged Event Related Potentials must be calculated. Nowadays, this technique has really improved its spatial resolution (Gevins, Leong, Smith, Le and Du 1995). Nevertheless, combining EEG with fMRI is still very useful to understand the relationship between language activity and human brain.

Magneto-encephalography (MEG): this technique measures the magnetic field changes in the brain caused by brain activity. The advantage of this technique is the temporal resolution (Hämäläinen, Hari, Ilmoniemi and Lounasmaa 1993).

Trans-cranial Magnetic Stimulation (TEM): this technique is different from all the rest described above. As we have seen, the previous techniques are based on controlling the areas of the brain which are active during linguistic activities. Using this technique, some areas of the brain are artificially altered by the development of a magnetic field on the surface of the cranium. This magnetic field stops the brain activity of this area of the brain, so it consists of observing the possible limits in linguistic activities without this area (Pascual-Leone, Walsh and Rothwell 2000).

As we can understand, this technique is the most useful one, in the sense that we can prove a causal relationship between the activity of one area of the brain and one specific linguistic task. Nevertheless, nowadays it has not been proven to be harmless, so it's not recommended unless there are good additional reasons to use it. For example, it's used before a surgical intervention. In this sense, all the people who are involved in this technique are patients with brain damage, and therefore, the results of this investigation is not freely generalized.

Apart from autopsy of aphasic patients and these kinds of techniques we have just mentioned, the brain-language relationship is investigated by different neuropsychological and psycho-physical tests:

Sodium Amytal (Wada 1949): this test follows the same logic as the stimulation of the brain and it is also reduced to clinical context. The Sodium Amytal is injected in the carotid artery in the neck. This anesthetic affects the homolateral hemisphere and then the researcher tests if the patient's linguistic behavior is normal. Later, the same process is done in the carotid artery of the other side of the neck. With this test, relevant information for surgery is found: which hemisphere is dominated by language. Although a great percentage of people has language localized in the left hemisphere, the surgeon must prove it in each case, in order to not affect the language ability of the patient.

Dichotic Listening (Kimura 1973): this test has the same objective as the previous one, but it could be used on healthy people, out of clinical context. Through stereophonic headphones, the subject hears three digits through each ear and s/he must repeat it. As each ear is connected with the contralateral hemisphere, digits will be better remembered by the ear contralateral to the hemisphere dominated by language.

Up to now, the methodology talked about has the aim of localizing the areas of the brain implicated in linguistic activities. From now on, our attention will focus on the study of the different linguistic processes and the relationship maintained by different kinds of linguistic information (e.g. phonological, semantic, lexical, etc.). These studies also profits from neuroimaging techniques. An example is the use of *Eye Tracking* technique. In the most common version of this technique, participants are asked to read some texts and the researcher controls the place the participants' eyes stop as well as the movement of their eyes. Duration, direction and speed are specifically evaluated. With this method, for example, the importance of the discourse markers has been proved (Loureda Lamas 2010). Nevertheless, this field of study has been mainly investigated by psycho-physical tests, such as the following ones:

Reaction-time: the measurement of the time participants need to begin to answer to a stimulus is the "reaction-time", used in mental chronometry. This dependent variable is the most useful test used in experimental psychology, and in psycholinguistics (Posner 2005). Donders' subtraction method consists on the premise that the more time the participants use in one test, the more mental processes are implicated in it. So, using reaction time as a dependent variable implies comparing at least two different tests.

As the result of this comparison, a hierarchical order (from difficult to easy) can be made. The more time a test needs to be done, the more processes it seems to imply. In Baus & Carreira's chapter the reader will find several examples of this technique.

The study of (natural or induced in the laboratory) mistakes: the study of mistakes (both natural and induced in laboratory) is a method that leads to make hypothesis about how the linguistic information is stored. If two words compete in their use, they are supposed to be stored together in mental lexicon. This method of study is also found in Baus & Carreira's contribution.

All this methodology we have just presented here is currently used in interdisciplinary groups of work. As Laka says in her chapter, only the interaction of diverse areas of knowledge (linguistics, psychology, neurology and neuroscience) can afford the development of this kind of research, overcoming present limits and offering new perspectives. Achieving this goal implies getting over the traditional mutual ignorance. In future, these studies will be collaborative or will not be at all.

2. Presentation and summary of Part II Chapters

This second part of the book is made up in three contributions. The first one presents the main studies of the localization of the language processes in the brain, using for this objective mainly participants with brain damage. The second chapter faces up to the other side of language and brain studies and it explains the main mental processes implicated in the production and the understanding of words. The last chapter includes the multilingualism variable in all these studies.

Firstly, Fernando Cuetos offers in Chapter 4 a presentation of "the role of aphasic disorders in the study of the brain-language relationship". His chapter begins with a brief introduction, where he justifies the benefits of the study of aphasic patients for the brain and language relationship knowledge. Although nowadays we have new techniques to study the biological basis of language, Cuetos claims the necessity to study aphasic patients. An interesting reflection about the limits of new techniques and the necessity of a complementary aphasic studies is offered here.

In section two Cuetos offers a historical view of these studies since the nineteenth century and the Wernicke-Geschwind model is presented. More over the historical view, this section offers a rigorous presentation of different aphasics, used in clinical context. Cuetos adds a table where they are compared by linguistic deficiencies and by implicated brain areas.

The third section explains some problems and limits of this traditional model. Although nowadays it is used in clinical context, this model is not accepted in the same way it used to be in the last century. Sections four and five offer a new way to face these studies, explaining both its benefits and its limits. In this part of the chapter a great amount of bibliography references are given to present the last discoveries about different linguistic disorders. These discoveries are important, due to their different way of understanding language, linguistic disorders and biological basis of language.

The chapter finishes with a general conclusion: Language and Brain relationship is a complex object of study, so it needs a good methodology which combines different kinds of study. In this context, the aphasic studies are very important.

Chapter 5, by Cristina Baus and Manuel Carreiras, deals with word processing, both in oral natural languages (sections 1 and 2) and in sign languages (section 3). The first section is about word recognition and the second is about word production. Both include a critical comparison of different theoretical models, using the results of various experimental works. The objective of both sections is showing relevant variables in real use of natural language. The third section studies in depth the sign language processing. As it is just another natural language possibility, it shares with verbal language processing some features. But, as it has modality differences, it has particular features, different from verbal languages. The chapter ends with a short conclusion.

Finally, Itziar Laka talks about multilingualism in Chapter 6 *More than on language in the brain*. She introduces multilingualism and its cognitive benefits (sections 2 and 3). In these two first sections the reader can find a definition of multilingualism and a speech about how normal multilingualism is in our species. Besides, there are some proves about advantages of knowing (and using) more than one language. In fact, these advantages go over linguistic abilities and reach cognitive abilities.

In sections 4, 5 and 6, Laka explores the relationship between multilingual brain and mental processes. Studies about the activation of linguistic areas show that multilingualism is an important variable to take into account. It is especially important to study the way language knowledge is inhibited in real linguistic use.

Sections 7 and 8 face different kinds of multilingualism, including the feature of age of acquisition. Some final remarks conclude the chapter: multilingualism is an absolutely relevant variable to study for understanding language and human brain nature.

CHAPTER FOUR

THE ROLE OF APHASIC DISORDERS IN THE STUDY OF THE BRAIN–LANGUAGE RELATIONSHIP

FERNANDO CUETOS

1. Introduction

The brain-language relationship is an enormously complex subject that must be approached from different angles and with different methodologies in order to lead to successful results. In the past two decades, great strides have been made thanks to technological developments, especially modern neuroimaging techniques. The availability of techniques such as PET or fMRI to observe brain activity while a healthy person carries out certain linguistic activities, or the ability to monitor processes on a millisecond-by-millisecond basis, as with electrophysiological recording techniques such as EEG and MEG, has certainly made possible a qualitative leap forward in the study of the brain–language relationship. Even so, neuroimaging techniques pose some important problems that limit what they can contribute. Most importantly, both neuroimaging and electrophysiological recording methodologies provide information about the correlation between the execution of certain tasks and the patterns of activation that appear in different areas of the brain, but one cannot conclude from this that the activated areas are wholly responsible for the linguistic activity being performed. A causal relationship cannot be established because the detected activity might simply be related to an epiphenomenon or resonance of the process under study. A second limitation of these techniques is that, in any linguistic activity, the areas of the brain responsible for language are not the only ones activated; so are areas that support other processes intervening in that linguistic activity, such as attention, memory, or task-planning processes (see also Benítez Burraco, this volume). Therefore, it is hard to determine which part of the cerebral activation corresponds to language and which to the remaining cognitive processes. A third limitation of the neuroimaging techniques, especially the electrophysiological ones, is that they do not provide information about what happens at the level of the subcortical nuclei (thalamus, basal ganglia, etc.) and especially the subcortical tracts that join the different zones of the cortex and play an important role in language processing. It is quite certain that recent developments in the diffusion tensor imaging (DTI) tractography technique are making possible virtual dissections of the tracts in living persons (Glasse & Rilling, 2008). However, DTI provides anatomical information about white matter tracts but not about the functional aspects of connectivity (Duffau, 2008). In summary, all these limitations make it necessary to complement the neuroimaging studies with other methodologies if one wishes to obtain precise and complete information about the neurological bases of language.

A methodology that has long been widely used in research into the brain-language relationship is the study of aphasic patients—that is, individuals who have lost part of their language following a cerebral injury, whether it was caused by trauma, stroke, viral infection, or any other cause. In fact, long before the modern neuroimaging techniques were developed, a considerable amount of information about the areas of the brain involved in language was already available from this type of study. The detailed analysis of the language disorders of these patients, followed by confirmation of the lesion's localization in the brain through autopsy once the patient has died, has provided valuable information, allowing researchers to deduce that certain zones of the brain are responsible for specific linguistic activities or at least take part in them. The assumption on which this is based is simple: If, for example, a person shows difficulties understanding spoken words and he or she has a confirmed lesion in the medial zone of the left temporal lobe, it can be concluded that that area of the brain is somehow involved in word comprehension.

The tradition of studying aphasic patients began in the second half of the nineteenth century. As we shall see in the following section, it underwent major development a century later, especially with the works of Soviet neuropsychologist Luria (1976), who examined a large number of World War II soldiers with shrapnel wounds to the head. With the investigations of Geschwind in Boston in the 1960s and 1970s, that tradition was resuscitated after several years in the doldrums (Geschwind 1965).

Although this methodology has numerous limitations, which we will describe later, there is broad-based consensus that the study of aphasic patients constitutes a very important source of knowledge about the neurological bases of language. It was extremely useful at the time when neuroimaging techniques did not exist, since aphasic patients were the only window through which the brain could be closely examined. Thus, aphasic patients were a type of experiment provided by nature, unfortunately all too frequently. But even today it is a fundamental methodology, and researchers do not doubt that results with brain-damaged patients constrain the interpretation of data obtained with the neuroimaging techniques (Binder & Price 2001). Hence the need to continue using this methodology and combine it with the neuroimaging techniques if one wishes to make progress in this complex subject.

2. The Wernicke-Geschwind model

The examination of aphasic patients in the scientific study of the neurological bases of language began a century and a half ago, in 1861, when French anatomist and anthropologist Paul Broca presented the famous case of Monsieur Leborgne, familiarly known as "Tan" because *tan* was practically the only sound he could utter. This patient's capacity for oral production was completely damaged, yet he understood everything that was said to him. This patient's post-mortem analysis showed a considerable lesion in the third gyrus of the left frontal lobe. Shortly thereafter, Broca published another case (Monsieur Lelong) with very similar characteristics and a lesion in the same zone. In the following two years, he collected data from six more cases, all with oral production problems and right hemiparesis. In contrast, one patient who had a similar lesion in the right hemisphere had no speech disorders. These findings led Broca to conclude that the frontal zone of the left hemisphere (later called *Broca's area*) was the zone responsible for language production.

According to Shallice (1988), Broca's investigation was of crucial interest because it served to emphasize two important points: first, that language is a function that can be damaged independently of the rest of the cognitive processes; and second, that this function is localized in a specific part of the brain.

A few years later, in 1874, the young German physician Carl Wernicke described two patients who also had language disorders, but with characteristics which were completely opposite to those described by Broca: Wernicke's patients spoke fluently but did not understand what others said to them. When he later autopsied the brain of one of these patients, he found damage in the posterior superior part of the left temporal lobe, behind the primary auditory cortex. Therefore he ascribed to this area (which would later be called *Wernicke's area*) the function of language comprehension.

Wernicke's work also had important consequences. First, it implied that language cannot be considered a monolithic faculty, since it would have to be subdivided at least in terms of receptive and expressive functions; and second, that focal lesions can cause specific deficits (Stirling & Elliott 2008).

Furthermore, Wernicke's contribution went beyond a description of these language disorders and their associations with certain zones of the brain; he also developed a theoretical model that explained the difference between his patients and Broca's. According to this model, there would be two word representation centers: the auditory center in the left superior temporal gyrus, and the articulatory center in the left frontal zone. These, in turn, would be connected to each other by nerve fibers. Thus, a lesion in the auditory center would produce comprehension disorders, whereas a lesion in the articulatory center would involve production disorders. Wernicke further predicted a third type of aphasia—conduction aphasia—which would result when the fibers joining the perceptive and articulatory centers are damaged, leaving them disconnected. The principal characteristic of this disorder would be the inability to repeat words, even when comprehension and production remain intact.

That disorder of disconnection between the two language centers predicted by Wernicke was set out by Lichtheim a few years later, in 1885, in a patient whose only disorder was effectively that of word repetition. Lichtheim also proposed a new conceptual center where meanings would be stored, which would be essential for word comprehension. Thus Lichtheim expanded Wernicke's model to three centers with three interconnections. Consequently, the number of possible aphasic disorders was also expanded as a function of damage to one of the centers or one of the connections, as shown in Fig. 4-1.

A patient with a lesion in the connection between the auditory and the conceptual centers would have all the traits of Wernicke's aphasia except that repetition would be preserved. This disorder was called *transcortical sensory aphasia*. A lesion occurring in the connection between the conceptual and the motor centers would have all the characteristics of Broca's aphasia except for the preservation of repetition. This disorder was called *transcortical motor aphasia*.

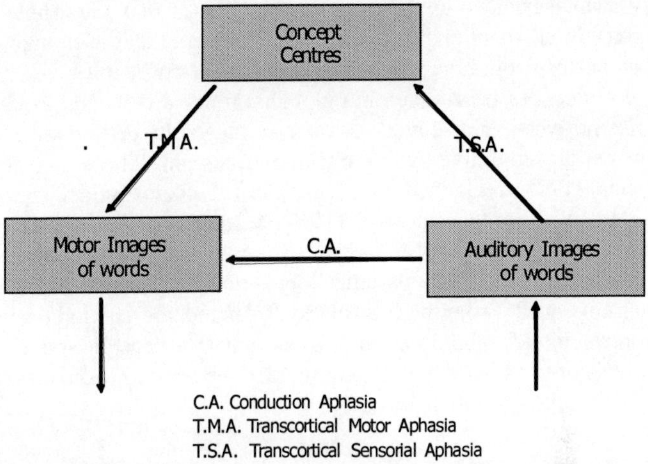

Fig. 4-1 Litchtheim's model

This way of studying the relationship between cognitive faculties and the brain prevailed throughout the latter half of the nineteenth century. The procedure consisted in studying each patient meticulously while he or she was alive and, when the patient died, performing an autopsy on the brain to establish the exact site of the lesion. With this methodology, the language model expanded to other domains, such as reading and writing. Thus, in 1881 Exner established a relationship between the posterior zone of the second left frontal gyrus (Brodmann's area 6) and handwriting; in 1892 Dejerine found one between the angular gyrus (area 37) and spelling; and other such relationships between specific areas of the brain and specific linguistic functions were discovered as well.

Nevertheless, at the end of that century, and especially in the early 1900s, strong criticisms against this trend were put forward, putting a stop to its development. The first came from Freud and Bergson, but the most critical authors were Marie in France, Goldstein in Germany, and Head in England (Shallice 1988). According to these authors' conception, called the globalist position, the different types of aphasias were nothing but different degrees of severity of cognitive dysfunction and might or might not have been associated with some type of motor disorder, rather than with a specific location in the brain. According to those authors, language is much more complex than the Wernicke-Lichtheim model indicate, and it depends on the entire brain rather than on specific zones. This globalist position was reinforced by studies from the American psychologist

Lashley, which seemed to support the hypothesis that the whole brain participates in all functions (mass-action theory) and that each area of the brain can perform any function (the equipotentiality principle).

Several decades later, when the globalist trend was losing ground due to the incontrovertible discovery of the role played by certain areas of the brain in certain cognitive functions,[1] the American neurologist Norman Geschwind (1965) reclaimed the Wernicke-Lichtheim model and added new evidence. Thus, Geschwind (1967) located the conceptual center, whose zone of the brain Lichtheim had not specified, in the inferior posterior part of the left parietal lobe—that is, in the angular and supramarginal gyri. From the information gathered, Geschwind designed a neurological model of the cerebral areas that intervene in spoken word comprehension and production and in reading written words aloud, as shown in Figures 4-2 and 4-3.

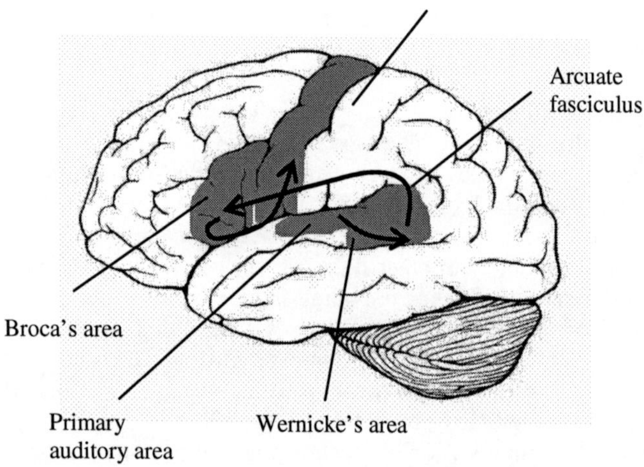

Fig 4-2 Speaking a word

[1] For example, the role of the hippocampus in memory, as showed by Scoville and Milner's (1957) study of patient HM.

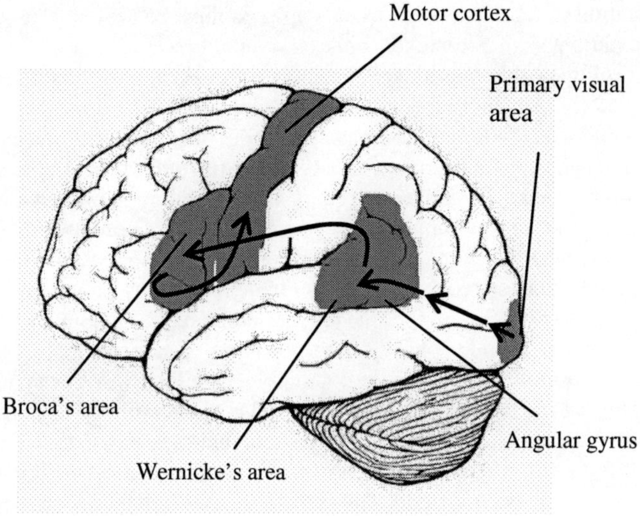

Fig 4-3 Reading a word

According to Geschwind, the brain areas that play a part in spoken word comprehension and repetition do so as follows: speech sounds coming from the ears through the auditory nerves are analyzed in the primary auditory area, located in the superior gyrus of the left temporal lobe. From there the information passes to Wernicke's area for lexical processing. From Wernicke's area, via the arcuate fasciculus, a beam of fibers that originates in the temporal lobe curves around the sylvian fissure and projects into Broca's area in the frontal lobe for phonological processing, followed by articulation through the motor area. In the case of reading, processing begins in the visual area in the occipital lobes, where information coming from the retina is analyzed. From there the information is projected to the angular gyrus to be transformed into phonological information, and from that area the same steps are followed as for oral language.

This model, which has been in use in recent decades and continues to serve as a reference, especially at the clinical level, has been tremendously useful both for stimulating research on the brain–language relationship and as a clinical model in the diagnosis of aphasic patients (Damasio 1992). For clinicians, it is of fundamental importance to have a taxonomy of disorders with which patients can be classified and diagnoses made, since this facilitates rehabilitation by making it possible to design different intervention programs as a function of syndrome. Obviously, the rehabilitation of a patient with Broca's aphasia must be very different from that of a patient with Wernicke's aphasia or conduction aphasia.

According to this Wernicke-Geschwind model, aphasic disorders can be grouped into seven major aphasic syndromes: the five described above (Broca's aphasia, Wernicke's aphasia, conduction aphasia, transcortical motor aphasia, and transcortical sensory aphasia) plus global aphasia (in which very severe language disorders affect both comprehension and production) and anomic aphasia (in which the only disorder is difficulty with word recall). Broca's aphasia, transcortical motor aphasia, and global aphasia are characterized by a lack of fluency, whereas patients with the other types present fluent speech. On the other hand, all but those with transcortical and anomic aphasias have problems with repetition.

This methodology also made it possible to infer from the symptoms which zone of the patient's brain had been damaged, which was extremely useful at a time when no neuroimaging techniques existed and it was impossible to tell whether or not an individual had a cerebral lesion and where it might be located. The only possible method was clinical assessment of the behavioral and cognitive disorders presented by the patient. The following procedure was used: First, all the patient's symptoms were detected through neuropsychological tests. On the basis of those symptoms, the patient was classified into one of the aphasic syndromes. Finally, the damaged cerebral zone was inferred. Thus, according to the Wernicke-Geschwind model, when a patient exhibits difficulty in spontaneous speech with non-fluent, telegraphic language, grammatically anomalous expressions, and repetition problems, but his or her comprehension is good, the diagnosis is Broca's aphasia and the third gyrus of the left frontal lobe is believed to be damaged. If a patient presents the same symptoms but preserves repetition, transcortical motor aphasia would be inferred, with the lesion probably located in the supplementary area of the left frontal lobe just in front of Broca's area, or in the connection between the supplementary motor area and Broca's area. If, on the other hand, the patient presents fluent language but uses numerous paraphasias and neologisms and shows difficulties with comprehension and repetition, this would be Wernicke's aphasia, with a lesion probably in the posterior zone of the left superior temporal gyrus. If symptoms typical of Wernicke's aphasia appear except that repetition is preserved, transcortical sensory aphasia is indicated, and the lesion is probably located in the angular or supramarginal gyrus. If the patient's main problem is with word and phrase repetition, conduction aphasia would be implicated and the lesion is likely to be affecting the fibers (arcuate fasciculus) connecting Wernicke's area and Broca's area. If the patient presents severe language disorders that affect both comprehension and production, the diagnosis is global aphasia and the patient has probably suffered a massive lesion affecting a large part of the left hemisphere. And if his or her only problem is nominal aphasia, perhaps he or she has a mild lesion in the left temporal lobe, although any lesion in the language zone can produce nominal aphasia. In fact, all types of aphasias are marked by some degree of problems with naming (Cuetos 1998).

The characteristics and cerebral localizations of the seven major aphasic syndromes are summarized in the following table:

Syndrome	Compre-hension	Speaking	Repetition	Naming	Fluen-cy	Brain Area
Broca's	+	-	-	-	-	Third gyrus of left frontal lobe
Wernicke's	-	+	-	-	+	Posterior left temporal gyrus
Conduction	+	+	-	-	+	Arcuate fasciculus
Trans-motor	+	-	+	-	-	Supplementary area
Trans-sensory	-	+	+	-	+	Angular or supramarginal gyrus
Nominal	+	+	+	-	+	Any language area
Global	-	-	-	-	-	Large part of left hemisphere

3. Some problems with the Wernicke-Geschwind model

Although the model was widely accepted by researchers and clinicians throughout the second half of the twentieth century, in recent years a series of problems have appeared that have cast doubt on its validity. Stirling and Elliott (2008) point out three:

1. Lack of specificity of the symptoms of aphasia. Syndrome taxonomy poses the serious problem that many symptoms do not fit easily into the syndromes. For example, Broca's aphasics not only have production disorders but comprehension disorders as well, principally when faced

with grammatically complex sentences such as those containing passives or relative subordinate clauses; Wernicke's aphasics have both comprehension problems and production disorders, as is demonstrated by their outcomes in naming tasks or their numerous errors, including neologisms (Cuetos *et al.* 2010).

In addition, it is difficult to cover the entire range of existing aphasic disorders with the proposed syndromes. Many patients do not fit neatly into any one syndrome because they present a variety of symptoms, sometimes corresponding to more than one syndrome. Some patients present disorders typical of Broca's aphasia, yet they also have symptoms that could pertain to Wernicke's, and vice versa. In fact, this is why a new syndrome called "mixed aphasia" had to be proposed, to include those patients with a variety of symptoms that do not square with just one syndrome. Unfortunately, syndromes are actually complex entities with very varied symptoms that may originate in different areas of the brain, as is shown by the numerous dissociations between symptoms within a single syndrome. For example, within Broca's aphasia dissociations between articulatory disorders and agrammatism have been found. Dissociations have even been found within agrammatism, between morphological and syntactic disorders (Berndt 1987), indicating that those symptoms have different neurological bases and can become damaged independently of each other.

On the other hand, very few patients display all the symptoms of a determined syndrome, which could be called a *pure syndrome*. In a study on 444 aphasic patients, Benson (1979) found that only 59% could be classified into one of the major syndromes. Marshal (1985) maintained that only 20% to 30% of patients fit neatly into the syndromes. A system that can correctly classify only one fourth of all patients hardly seems the best option.

2. The Wernicke-Geschwind model does not do justice to the vast richness and complexity of language processing (Poeppel & Hickok 2004). There is more to language than listening to and uttering words. Language has several different levels (phonological, lexical, morphological, semantic, syntactic, and pragmatic) and rules for combining the components of those levels: phonemes into words, words into sentences, sentences into discourse, etc. The Wernicke-Geschwind model did not take those levels into account, which made it difficult to interpret some symptoms, such as the comprehension problems in Broca's aphasia, which were overlooked. Broca's aphasics have trouble understanding grammatically complex sentences, such as passives and relative clauses, not because of semantic alterations but syntactic ones (Caramazza & Zurif, 1976). For this reason

they fail to understand reversible sentences of the type *The boy is chased by the girl*, where rules of syntax must be applied; but they have no trouble with sentences such as *The orange is eaten by the girl*, in which the application of semantic criteria facilitates comprehension. Therefore, a good model must take processing (and alterations) into account at the phonological, morphological, syntactic, semantic, and pragmatic levels. But each of those levels involves a series of processes that depend on different cortical areas and are therefore dissociable from one another.

3. Poor correspondence with neuroanatomical observations. When neuroimaging techniques, especially computerized tomography scanning and magnetic resonance imaging, were first introduced, it became possible to find the precise localizations of the lesions. Some studies using these techniques began to show the lack of correspondence between the syndromes and the areas of the brain that had traditionally been assigned to them. One of those early studies was that conducted by Basso *et al.* (1985) with a group of 207 patients with focal vascular left hemisphere lesions. They found that seven patients with lesions in anterior areas presented Wernicke's rather than Broca's aphasia; on the other hand, six patients with posterior lesions had Broca's aphasia. In a more recent study (Dronker, Redfern, & Knight, 2000), in which a sample of more than 100 patients was analyzed using behavioral and neuroimaging data collection as well, the authors confirmed that just over 16% of the patients classified by the aphasia assessment batteries as Broca's aphasics did not have lesions in Broca's area. What is more, only 50% to 60% of the patients with lesions in Broca's area showed symptoms of Broca's aphasia, one presented symptoms of conduction aphasia while the majority presented symptoms of anomic aphasia. In the case of Wernicke's aphasia, the correspondence was even weaker, only 65% of those classified as Wernicke's aphasics had lesions in that area and only 35% of patients with lesions in Wernicke's area showed symptoms consistent with that syndrome. As far as conduction aphasia is concerned, the majority had no lesions in the arcuate fasciculus, but rather in the superior temporal gyrus and the inferior left parietal region. They further confirmed that a lesion in the arcuate fasciculus resulted in a severe production disorder, with patients managing only to repeat syllables or words (Dronkers 2000).

These results should not be at all surprising in that current morphological studies are beginning to show clearly that the anatomical limits of the areas of the brain responsible for the symptoms are neither uniform nor clearly defined, since there are well-differentiated zones within them (Scott & Wise 2004).

Thus, there is no clear consensus about what is understood by Broca's area. Originally it referred only to area 44, but many researchers today include area 45 as well. From the cytoarchitectonic point of view, area 44 is a dysgranular premotor area, whereas area 45 is granular and belongs to the prefrontal lobe. This implies that area 45 is more cognitive whereas area 44 is more phonetic and more motor (Ardila 2010). Some authors even include area 47, the premotor cortex, and the frontal operculum within Broca's area as well (Kertesz, Harlock, & Coates 1979). In fact, articulatory disorders now tend not to be associated with Broca's area (Dronkers 1996) but seem to depend on the premotor cortex, whereas agrammatism depends on areas 44 and 45.

The boundaries of Wernicke's area are even less precise, since it comprises five or more different architectonic areas (Scott & Wise 2004), with a good number of functional regions involved not only in speech perception but also in transmodal integration (Wise *et al.* 2001). Area 22 is definitely a fundamental part of Wernicke's area, but there are others as well, such as the inferior parietal lobe and the middle and inferior temporal gyri (Glasser & Rilling 2008).

In the classical model, conduction aphasia was assumed to be produced when the lesion affected the arcuate fasciculus. Nevertheless, many current data cast doubt on this hypothesis, principally because lesions that do not produce conduction aphasia have been found in the arcuate fasciculus (Tanebe *et al.* 1987). On the other hand, cases of conduction aphasia have presented with cortical lesions and no damage to the arcuate fasciculus (Damasio & Damasio 1980). Cases of conduction aphasia caused by lesions in the left superior temporoparietal zone, involving the superior temporal gyrus, the supramarginal gyrus, or both, have been found (Hickok & Poeppel 2004). Cases of conduction aphasia caused by lesions in the subcortical zones, such as the white matter lateral to the insula or the putamen, have also been found (Parker *et al.* 2005).

On the other hand, recent tractography studies have shown that the arcuate fasciculus consists of two tracts, one that starts in the superior gyrus and the other in the middle gyrus. Conduction aphasia (difficulties only in repetition) is produced when the superior gyrus is damaged (Glasser & Rilling 2008), but it can also result from a lesion in the supramarginal gyrus (area 40), since the tract that starts in the superior gyrus passes just below the supramarginal gyrus.

Transcortical sensory aphasia results from a lesion in the middle and inferior temporal gyri, sometimes extending to the angular gyrus and occipital lobe. It can also be produced by a disconnection between Wernicke's area and area 37 (Stirling & Elliot 2008). The purest cases of transcortical sensory aphasia are those caused by carbon monoxide poisoning, because it damages areas 21 and 37. This entails the destruction of the lexical-semantic system, as occurs in semantic dementias, which accounts for their similarity. Nevertheless, the phonological circuit (Wernicke's area, Broca's area, and the supramarginal gyrus) remains intact, so repetition is unaffected.

Transcortical motor aphasia is produced by a lesion in the cortex anterior and superior to Broca's area, in the supplementary motor area, or at this area's connection to Broca's area (Stirling & Elliot 2008). However, transcortical motor aphasias caused by a lesion at the end of the tract that originates in the middle temporal gyrus are also found.

Finally, global aphasia is generally produced by a lesion in the medial artery that damages both anterior and posterior areas as it extends through a broad zone of the left hemisphere. Nevertheless, it has been confirmed that global aphasia can also be produced by a lesion that is limited to the temporal isthmus (Andrewes, 2001).

4. Toward a new neurological model of language

With all the problems with the Wernicke-Geschwind model and syndrome taxonomy that we have enumerated above, it seems clear that this is not the most suitable framework for studies of the brain-language relationship. On one hand, the model is overly restrictive of the brain areas involved in language. Current studies clearly show that Broca's area, Wernicke's area, and the arcuate fasciculus are only part of a wide-ranging neuronal network involved in language processing. This network is considerably more complex than the model proposed by Geschwind would suggest, given that it includes a large number of areas, even ones far from the language-typical areas. Broca's and Wernicke's areas continue to be the fundamental zones, but so are the dorsolateral prefrontal area, premotor cortex, supplementary motor area, superior temporal gyrus, middle and inferior temporal gyri, angular and supramarginal gyri, cerebellum, anterior cingulate, and thalamus (Price, Indefrey & van Turennout 1999). On the other hand, syndromes are overly complex categories that do not correspond to specific areas of the brain.

Consequently, the approach must be altered by acting on two different yet complementary fronts. First, new models must be developed that account for all the components of language and can explain all the alterations that aphasic patients experience. And of course, the new models must accommodate all the data being provided by the new neuroimaging techniques, including tractography. These models can be achieved only from a convergence of the various cognitive neuroscience disciplines whose object of study is the neurological bases of language—most notably linguistics, psycholinguistics, cognitive neuropsychology, and neurology—as well as from the results of neuroimaging studies. On the other hand, it is important to stop focusing on syndromes and talk instead about smaller, more uniform, more compact categories—that is, a type of subsyndromes (Willnes & Poeck, 1993) that combines symptoms that normally go together, are not dissociable, and have a clear neurological correlation and a single cognitive interpretation.

Most syndromes include dissociable symptoms, whereby they appear in some patients and not in others and therefore cannot be used to define the syndrome. This means that the syndromes can be subdivided into more specific categories. Thus, there is no longer any doubt that Wernicke's aphasia is actually a complex disorder involving several different symptoms that are also dissociable, since they can appear in some patients and not in others, precisely because they depend on different areas of the brain. In fact, Wernicke's aphasias tend to appear as a result of lesions in broad zones of the left hemisphere, including the left posterior superior temporal gyrus, left posterior middle temporal gyrus, supramarginal gyrus, or angular gyrus. When the lesion reaches only the superior gyrus, the patient cannot process phonemes, so the main symptom is phonemic paraphasias. In these cases recovery tends to be good because the corresponding zone of the right hemisphere can take over the function. On the other hand, when it affects the middle and inferior temporal gyrus, patients have trouble with word processing. Their most frequent errors are semantic paraphasias and nominal aphasias. In these cases the disorder is more severe and the prognosis poorer.

With these characteristics, and according to Hickok and Poeppel (2004), Wernicke's aphasia could be considered a syndrome consisting of two disorders: (i) alterations to comprehension (as in transcortical sensory aphasia) and (ii) alterations to repetition (as in conduction aphasia). There is a reason why the errors these patients commit also correspond to both disorders, since patients with the former disorder commit principally semantic paraphasias as in transcortical sensory aphasia, and those with the latter commit phonemic paraphasias as in conduction aphasia. In short, Wernicke's aphasia could therefore be considered a combination of those two disorders due to a more severe lesion that extends to more areas of the brain (and consequently to more linguistic processes).

Likewise, conduction aphasia can result from damage to the speech processing system in the superior temporal gyrus or from damage in the supramarginal gyrus at the point of interface between speech processing and articulatory processing (Hickok & Poeppel 2004). For this reason, two types of conduction aphasia can also be conceived: reproduction conduction aphasia and repetition conduction aphasia.

The number of subtypes of Broca's aphasia is even greater, since Broca's aphasia is a complex syndrome that includes many very different symptoms (e.g., agrammatism, verbal nonfluency, dysarthria). In fact, for years it has been known that many of those symptoms are dissociable (Berndt 1987, Parisi 1987), indicating that they depend on different areas of the brain. This should not be the least bit surprising, since Broca's aphasia is produced by lesions in very different zones. It is generally produced by a lesion in the left inferior frontal gyrus (areas 44 and 45), but it can also be caused by a lesion in the subcortical connections and even the basal ganglia. Some of the symptoms have very clear neurological foundations. Thus, dysarthria is produced when the lesion reaches the premotor cortex, and agrammatism appears when it reaches areas 44 and 45. As for comprehension and production problems, when the lesion destroys the ends of the tract coming from the superior gyrus, it produces phoneme processing disorders related to both discrimination and production. When it destroys the ends of the tract that comes from the middle gyrus, it produces disorders in both word comprehension and word production.

In short, rather than continue to resort to syndromes and the dissociations produced within them, it would surely be more efficacious to dispense with that taxonomy completely and base our arguments only on language processing models that have clear neurological correlates. In oral comprehension, for example, very detailed cognitive models have been developed on the basis of behavioral data obtained in laboratory experiments with healthy subjects and aphasic patients (Ellis & Young 1988). In these models, several processes or stages of word comprehension are differentiated, the first being devoted to phonological processing or phoneme identification. From the starting point of phoneme identification spring two alternative pathways: the sublexical pathway, in which the acoustic-phonological conversion mechanism transforms input phonemes to output phonemes (fundamental to learning new words); and the lexical pathway, through which word form and word meaning are accessed in the phonological input lexicon and the semantic system, respectively.

The recent neurological models clearly seem to support these cognitive models. According to Hickok and Poeppel's (2004) model, auditory information is first processed bilaterally in the primary auditory cortex. Functional neuroimaging studies have shown quite convincingly that phonological processing depends on the left superior temporal gyrus

(Binder *et al.* 2000). Phonological decoding later takes place in Wernicke's area. From there the auditory information can be transmitted directly to Broca's area, in what would be a simple repetition task (which would correspond to the sublexical pathway in the model of Ellis & Young 1988). Alternatively, it can be sent to the middle temporal gyrus for lexical-semantic processing, and from there to the frontal lobe for oral production (which would be the lexical-semantic pathway). The middle and inferior temporal gyri of the left hemisphere, specifically areas 21 and 37, are involved in lexical-semantic processing. On the other hand, recent tractography studies have revealed two pathways that go from the left temporal to the left frontal lobe: the superior temporal gyrus (STG) pathway, which connects the superior temporal gyrus (area 22) to the frontal lobe (areas 6 and 44), and the middle temporal gyrus (MTG) pathway, which connects the middle and inferior temporal gyri (21 and 37) to the frontal lobe (areas 44, 45, 6, and 9). The first pathway is for phonological processing and the second for lexical-semantic processing (Glasser & Rilling 2008). In other words, the first would be the equivalent of the sublexical pathway of the cognitive models, and the second corresponds to the lexical-semantic pathway.

This cognitive-neurological model perfectly predicts the type of aphasic disorders that can be expected as a function of the process that has been damaged. Thus, a stage 1 lesion of phoneme identification (superior temporal gyrus) produces *pure word deafness*—that is, the inability to recognize speech sounds while still being able to identify environmental sounds. If the lesion is produced in the acoustic-phonological conversion mechanism (STG pathway), the result is an incapacity to repeat unknown words and pseudowords even though the patient can repeat familiar words via the lexical route. This disorder is known as *auditory phonological agnosia.* If the lesion is produced in the phonological input lexicon (superior temporal posterior gyrus), *word meaning deafness* results. Patients can repeat the words they hear but do not recognize them in oral form. Even so, if they see them in written form they understand them perfectly, which means the problem lies with accessing meaning through the oral route. A patient of Kohn and Friedman (1986) was incapable of following the therapist's order to "point to your knee." All he could do was repeat, "Knee, knee, knee, what does knee mean?" Nevertheless, when the patient wrote the word and read it, he said, "Oh! Knee," and pointed to it with no difficulty. If the lesion affects the semantic system (middle and inferior temporal gyri), patients can still recognize words; that is, they know whether a word is real or invented but cannot access its meaning. This disorder is known as *semantic agnosia.* And if the lesion completely destroys the acoustic-phonological mechanism (STG pathway) and partially destroys access to word meaning (Wernicke's area and/or middle and inferior temporal gyri), *deep dysphasia* results. Patients with deep dysphasia are incapable of repeating pseudowords and unknown words and commit semantic errors in word repetition.

Something similar happens with production in that, according to cognitive models of oral production (for example, Levelt's 1989), several levels of processing can also be distinguished, including processing at the levels of planning, syntax, lexicon, phonology, etc. At least three processes—conceptual, grammatical, and phonological—exist only at the word level. The meanings one wishes to express are activated at the conceptual or semantic level; the lemmas corresponding to those meanings —that is, the lexical and syntactic forms of the words—are activated at the grammatical level; and the phonemes corresponding to those words are activated at the phonological level. The neuronal substrates responsible for each of these processes are different and therefore can be damaged independently of each other, giving rise to different types of disorders. The semantic level, given its complexity, depends on broad cerebral zones, although it depends fundamentally on the middle and inferior temporal gyri, the angular gyrus, and the superior frontal gyrus (Binder & Price, 2001). The grammatical level depends on the left posterior superior temporal gyri, although there is probably a cerebral overlap in the processing of the lexical and semantic levels. Finally, the phonological level depends on the left frontal and posterior temporal gyri (Binder & Price 2001).

Consequently, there are various types of nominal aphasias, in addition to the disorders at sentence level. The term *semantic anomia* is used when the patient fails to access meaning because of a lesion in the middle and inferior temporal gyri (which are observed in some Wernicke's aphasics but not in others). Patients with the neurodegenerative disease called *semantic dementia* also suffer from atrophy in those zones and are also marked by a loss of concepts that translates into content-poor language. Both patients with semantic nominal aphasia and those with semantic dementia mainly commit semantic errors by confusing concepts that are close in meaning (*lion* and *tiger*, *washing machine* and *refrigerator*, etc.). *Pure anomia* is produced when the lesion affects the superior posterior temporal zone, in which case patients can activate the meanings they want to express but cannot find the right words. The most common errors in this type of nominal aphasia are circumlocutions (the use of phrases to express a word that cannot be accessed—for example, *the thing you use to sweep* to refer to a broom), although they also commit semantic errors. Finally, *phonological anomia* is produced when the lesion affects Broca's area or the connection between the temporal zone and Broca's area, thereby affecting phonological processing. Patients utter the right word, but make pronunciation errors in one or more of the phonemes (Cuetos 2003).

In summary, the key issue is to implement language models in which each component process has a clear neurological correlate, so we can predict what will happen when a lesion that damages one of those components is produced; or, alternatively, so each aphasic disorder can be interpreted precisely on the basis of those cognitive and neurological models. This is the true path to understanding the brain-language relationship.

5. Contributions and limitations of studies with aphasic patients

Studies with aphasic patients will always be an important source of research in the brain-language relationship. However, in order for them to be really advantageous, certain assumptions must be changed. One is that language depends on centers of the brain (Broca's area, Wernicke's area, angular gyrus, etc.). Today, with the application of modern neuroimaging techniques, it is known that there are no brain centers that are responsible for cognitive functions; or, to put it another way, linguistic processes are not localized in specific anatomical areas but are the result of a pattern of interactions between different regions. Given that—as neuroimaging shows—as certain linguistic tasks are carried out, various areas of the brain are simultaneously activated those areas must be connected by neuronal networks; hence the superposition of symptoms that the majority of patients display. Obviously, these networks can have more connections in a particular zone (Pulvermuller 1999), and so a lesion in that zone is more likely to damage the network, causing certain symptoms to appear. But it can also be that lesions in zones far from the network may produce similar damages and, consequently, the same symptoms. For this reason, patients with similar characteristics are sometimes found to have lesions in different zones of the brain. Consequently, we most move away from the search for language centers or areas involved in language processing, and toward the search for neuronal networks—that is, networks responsible for the different linguistic abilities (Ardila 2010).

But most especially, when conducting studies with aphasic patients it is important to be very careful with methodology, since many errors of the past can be attributed to a lack of methodological rigor. Studies with aphasic patients must meet a series of strict criteria in order to yield reliable results. Mainly, these three principles must be followed: (1) conduct detailed and exhaustive patient assessments, (2) precisely locate the lesion, and (3) quickly establish the correlation between the language disorder and the lesion.

Classic studies sometimes led to misinterpretations precisely because these criteria were not taken into consideration. To begin with, patients were not studied in depth because the batteries for language exploration that we have today were not yet available, among other reasons. Nor were the language processing models that served as guides for exploration as complex and detailed as those we have today. Furthermore, information about the locus of the lesion was not very rigorous, even if it was collected through autopsy, partly because only the classic language areas were studied. Moreover, sometimes post-mortem studies were conducted long after the behavioral study, which compromised the correlation between the neurological data and the symptoms, either because the patient might have suffered new strokes or because a neurodegenerative dementia or other type of brain tissue alteration may have taken place. Of the two famous patients Broca studied and autopsied, one (Leborgne) had a history of epilepsy and multiple strokes, so he may have suffered important lesions in other zones of the brain that Broca did not examine. The other patient (Lelong) had presented neurodegenerative dementia for several years and so would have had many atrophied brain zones (Dronkers, Plaisant, Iba-Zizen & Cabanis 2007). As for Wernicke's patients, only one was examined in an autopsy; this patient had suffered from dementia with atrophy in the superior posterior temporal gyrus (Dronkers 2000). Therefore, neither of these cases can be called good examples for studying the brain-language relationship, despite the enormous influence these studies had throughout the history of neurolinguistics.

The second important factor is to have a reference of the lesion taken from the neuroimaging studies that is as complete as possible, although a reference taken from the DTI is also desirable. This is because most cerebral lesions, especially cerebrovascular lesions and viral infections, tend to destroy not only the gray matter but also the subcortical nuclei and tracts that connect remote areas of the cortex. If only the lesions in the gray matter are taken into consideration, erroneous conclusions can be drawn in attributing alterations produced by subcortical lesions to those areas of the cortex. Lesions often affect subcortical zones or tracts that join the areas of the brain, fundamentally from the basal ganglia and thalamus. Not all subcortical connections are known, but the thalamus is an important center of connection. Metter and Hanson (1994) confirmed that lesions in the thalamus can produce Broca- or Wernicke-type symptoms, depending on which area is damaged: if they are localized in the zone of anterior connections, they produce disorders very similar to those presented by Broca's aphasics, and if they are in the posterior connections the disorders are similar to those of Wernicke's aphasics.

The third factor to bear in mind is how much time has elapsed since the lesion occurred. Important changes take place in the damaged brain as time passes. If the patient is studied by a scanner or by magnetic resonance in the acute phase—that is, shortly after the lesion was produced—the results will be affected because there are clearly areas of the brain that are not functioning, which therefore produces behavioral changes (Price & Friston 2001). Nevertheless, these nonfunctioning areas are not necessarily damaged but simply reacting to the shock of the lesion or experiencing a momentary shutdown as a consequence of the loss of blood or cerebrospinal fluid generated by the lesion, also called *diaschisis*. When the diaschisis begins to abate, those zones begin functioning again and many of the patient's symptoms begin to disappear. This is why some patients who present Broca's aphasia in the acute phase end up with simple nominal aphasia or may even fully recover, and why many patients with Wernicke's aphasia move on to conduction aphasia. When a functional magnetic resonance or positron emission tomography is applied to a patient in the acute phase, a large hypoperfused zone of the brain will be seen. Yet that doesn't mean the entire zone is damaged, but simply that it has not yet fully recovered its functioning. Thus, if the goal is to establish a relationship between the damaged zone and behavioral disorders, one would have to expect that the images of hypoperfusion corresponding to the functional tests accurately reflect the damaged zones. But this can take time to appear because, although the diaschisis tends to diminish rapidly in the first few days, it can take months to disappear completely.

It would therefore be desirable to wait for the diaschisis to disappear before attempting to correlate behavioral alterations with the functional neuroimaging data. However, if one waits for the situation to stabilize and the patient to enter the chronic phase, one runs the risk that cerebral restructuring will be produced as a result of either the compensatory strategies the patient adopts to overcome his or her difficulties (cognitive reorganization) or compensatory changes in the anatomy of the brain (functional reorganization) as other zones take on the role of the damaged zone. This phenomenon tends to be more marked in young people whose brains are highly plastic, especially if they are engaged in a rehabilitation program.

6. Conclusions

The brain-language relationship is tremendously complex and can be studied successfully only with a combination of a variety of perspectives and methodologies. Neuroimaging techniques are certainly very important tools, but they must be complemented with other methodologies, including the study of aphasic patients. Language processing models that guide the explorations are especially necessary. In this sense, disciplines as varied as linguistics, psycholinguistics, cognitive psychology, neurology, computational sciences, and others must collaborate within the framework of the cognitive neurosciences in order to provide a global response to the problem.

Although syndrome taxonomy has been widely used in recent decades, it is too problematic for clinical use because it leaves many patients unclassified. It is also not particularly useful in the search for the neurological bases of language, since the syndromes are complex categories of symptoms that depend on broad zones of the brain. Nevertheless, the use of smaller, more homogeneous categories (such as sub-syndromes), the study of language alterations, and the study of aphasic disorders in people who have suffered brain damage can provide valuable information about how language is organized in the brain.

Some of the problems described with the methodology of studying patients do not arise when healthy subjects are studied with neuroimaging techniques. On the other hand, neuroimaging techniques help identify the neuronal networks that are distributed throughout the brain and are responsible for each linguistic process. But neuroimaging techniques also present problems, one of them being that until not so long ago they only detected cortical lesions, overlooking the subcortical ones, which are generally much more crucial. Thus the rupture of a tract that joins two areas involved in language will certainly have severe consequences for the patient's language, yet this will not be detected in neuroimaging. It's certainly true that, thanks to the techniques of DTI tractography, subcortical structures are being studied more and more. Some studies are now discovering the role of the subcortical structures in language, since some aphasias arise as a result of subcortical lesions. The thalamus and the basal ganglia are especially important centers of connection, and lesions in them produce severe linguistic alterations. Thus, some tractographic studies have discovered other connective pathways between the temporal and frontal lobes, in addition to the arcuate fasciculus (Catani, Jones & Ffytche 2005, Parker *et al.* 2005). Glasser and Rilling (2008) describe three tracts, each of which intervenes in the processing of a different linguistic level: One tract connects the left superior gyrus with the frontal lobe and is responsible for phonological processing; one connects the middle temporal gyrus with the left frontal lobe and is responsible for lexical-semantic processing; and the third connects the middle temporal gyrus and right frontal lobe and is responsible for prosody.

In summary, studies with aphasic patients serve as a guide to neuroimaging experiments to determine what should be looked at in the brain and where to look for it. Therefore, the combination of the two methodologies is no doubt providing data that illuminate the complex world of the brain-language relationship in a way neither of the two methodologies could achieve on its own. Proof of the importance of studying aphasic patients is that methodologies are being developed that make it possible to produce a type of temporary and reversible "virtual lesions" in healthy experimental subjects. With these virtual lesions, the specific area to be studied can be selected. One of these methodologies is the transcranial magnetic stimulation (TMS) technique, applied directly on the cranium, producing magnetic fields that influence the neuronal activity of the zone to which it is applied. When high frequencies are applied, the magnetic fields have excitatory effects; they are now being used in the rehabilitation of psychological (depression, schizophrenia, etc.) and neurological (Parkinson's disease, aphasias, etc.) disorders. When low frequencies are generated, neuronal activity is inhibited, producing a type of virtual lesion (obviously reversible once the application is terminated). The other methodology is direct electrical stimulation of the brain areas of persons who are about to undergo surgery, generally for removal of a tumor or an epileptic focus. In order to know which areas are fundamental to language processing so that the neurosurgeon can avoid removing them, he or she stimulates zones of the brain while the patient carries out certain linguistic activities. When a linguistic activity is shown to be affected, that means that zone is involved in the activity. Both TMS and electrical brain stimulation are therefore methodologies for pinpointing the roles of certain areas of the brain that are similar to, but more direct than, the methodology of studying patients. They have the advantage of making it possible to select the area to be studied, given that it is not always easy to find patients with focal lesions precisely in a specific zone.

CHAPTER FIVE

WORD PROCESSING

CRISTINA BAUS AND MANUEL CARREIRAS

Language is one of the most astonishing human abilities. It is a very powerful representational tool that allows a fast exchange of ideas between individuals, facilitating extremely efficient communication. Spoken and signed utterances, as well as written texts are composed of words or signs that are the basic bricks of meaning. The aim of the present chapter is to provide an overview on the cognitive processes involved in processing words/signs during comprehension and production in spoken, written and signed language modalities.

Word comprehension and production are parallel in some important aspects. Common mechanisms may underlie both sides of word processing, suggesting the importance of considering both comprehension and production in understanding language (Pickering & Garrod 2004). However, the most common approach has been to study comprehension and production processes separately. The main difference between word comprehension and production is the temporal course of information flow through the different levels of processing. Word recognition goes from phonology to semantics while word production goes from semantics to phonology. Following this processing time course, phonological, lexical and semantic effects will be ordered differently in the two processing directions. We will focus specifically on spoken and visual word recognition and production, as well as on sign recognition and production, reviewing the most relevant evidence on the influence of sublexical, lexical and semantic variables on lexical processing.

1. Word recognition

Language comprehension can be conveyed through different input modalities, some of them acquired naturally as speech or signs, while others such as reading, Braille or cued speech need to be acquired through instruction (Alegria, Leybaert, Charlier & Hage 1992). In this section we will focus on spoken and visual word recognition, specifically how comprehension is accomplished through these two modalities and how information at different levels of processing influences lexical access during comprehension.

1.1. Word recognition: From phonemes/letters to meaning

Reading is historically very recent in terms of the human evolution as compared to speech. Nonetheless, research on word recognition has been mainly conducted in the visual modality (mainly because of experimental limitations on the presentation of speech) under the assumption that basically the same processes were involved in both comprehension modalities. The speech/visual signal is decomposed in both modalities and mapped onto lexical representation to finally access meaning. However, there are some important differences between visual and speech comprehension: reading is self-paced and involves discrete eye fixations, while speech is speaker-paced and unfolds sequentially.

During spoken word recognition, the listener decodes the continuous speech signal to identify words. Despite the fact that spoken utterances contain very few cues indicating word boundaries and co-articulation exists (phonemes carry information about the previous and the following phoneme), listeners experience little difficulty in recognizing words. From the acoustic signal, discrete phonological representations are activated and sequentially map onto lexical entries stored in memory. Every time a listener hears a word, a number of lexical candidates is activated that may vary in their level of goodness with the input (Frauenfelder & Tyler 1987). Models of spoken word recognition agree that lexical selection takes place through competition between the activated lexical candidates (Luce & Pisoni 1998, McClelland & Elman 1986, Marslen-Wilson & Tyler 1980, Norris 1994). For instance, the TRACE (McClelland & Elman 1986) and the Shortlist models (Norris 1994) assume that there is inhibition between the lexical competitors. The two proposals, however, posit a different flow of information, that is, whether there is top-down and bottom-up feedback interaction (TRACE model) or only bottom-up interactions (Shortlist). Models differ also on the question of the lexical candidates that compete during lexical access. The Cohort model (Marslen-Wilson & Tyler 1980), which emphasizes the "left to right" nature of spoken recognition and the input signal as it unfolds over time, narrows the pool of lexical candidates by reducing the level of activation of those lexical candidates that do not match with the input. The process continues until the only candidate that matches with the input remains and then the word is recognized. In contrast, the Shortlist and the TRACE models allow a broader set of lexical candidates to enter into competition, although onset-matching competition influences word recognition processing to a great extent (Frauenfelder & Peeter 1990). Finally, the Neighbourhood Activation Model (NAM, Luce & Pisoni 1998) assumes that all the words sharing all the phonemes except one with the input would be activated (phonological neighbourhood), no matter the position of that phoneme in the word. As the input unfolds, the correct lexical candidate receives a larger amount of activation surpassing the rest of the candidates and reaching the threshold for selection. Thus, the target word *cat* would activate in the framework of the Cohort model the words *calm* and *candle* etc. In contrast, *cat* will activate in the NAM model words like *rat* and *cart* (but never *candle*).

During visual word recognition, the reader decodes the stream of letters into their orthographical/phonological representations that, in turn, will activate those lexical candidates stored in memory that match with the input. Selection of the best candidate for identification is then required to finally access its meaning. Most models of visual word recognition agree that lexical selection, as in spoken word recognition, takes place by competition between the activated lexical entries.[1] As with the models of spoken word recognition, models of visual word recognition differ on the exact mechanisms proposed for selecting the appropriate candidate to be recognized. They also differ on how phonology kicks into the whole visual word recognition process (see for instance the DRC vs. the IAM).

1.2. Variables influencing lexical access and selection during comprehension

Despite differences between the two modalities, research on spoken and visual word recognition has provided evidence that sublexical variables, such as the syllable, and lexical variables such as lexical frequency or neighbourhood density guide lexical access and selection during word recognition. In the following, we review the most relevant experimental evidence.

Regarding the influence of sublexical units on word recognition, syllables have been suggested to play an important role in lexical access,

---

[1] E.g., Dual Route Cascaded (DRC) model, Coltheart, Curtis, Atkins, & Haller (1993), Coltheart, Rastle, Perry, Ziegler, & Langdon (2001); interactive activation model, (IAM), McClelland & Rumelhart (1981); connectionist accounts, Plaut, McClelland, Seidenberg, & Patterson (1996); MROM, Grainger & Jacobs (1996).

both in spoken (Mehler, Dommergues, Frauenfelder, & Segui 1981) and visual word recognition (e.g., Conrad, Stenneken & Jacobs 2006, Perea & Carreiras, 1998). Most of the evidence comes from studies exploring *syllabic congruency* and *syllable frequency*.

One of the first studies showing that the syllabic congruency of words guides lexical access was conducted by Mehler *et al.* (1981). In a monitoring task in French, listeners had to detect a string of phonemes (*pal*) in words in which this string of phonemes constituted the first syllable of the word (*pal*.mier) or did not (*pa*.lace). Although the phonemic overlap was identical, the syllable congruent prime reliably led to shorter monitoring times, thus confirming the importance of the syllable during lexical access (but see Cutler, McQueen, Norris & Somejuan 2001). The same facilitatory *syllable-congruency* effect has been obtained in visual word recognition with a priming paradigm: participants were faster recognizing a target word (e.g., *ba*.sis) when it was preceded by a prime that was congruent with the first syllable (e.g., "ba") than when it was not (e.g., "bas") (e.g., Carreiras & Perea 2002, Carreiras, Ferrand, Grainger & Perea 2005).

The role of syllables during lexical access was also investigated by manipulating the frequency of syllables. In the lexical decision task (LDT) participants are presented with words and they have to indicate whether a visual or auditory string is an existing word or not. In visual word recognition an inhibitory effect of first syllable frequency has been repeatedly observed: it takes more time to identify words starting with a high-frequency syllable than words starting with a low-frequency syllable.[2] This has been called the *syllable frequency effect*. Assuming lexical selection by competition, the inhibitory effect of the first syllable frequency results from the fact that high-frequency syllables occurring in many words in first position will activate all those words that share this syllable in first position, causing an increased competition for selection. Low-frequency syllables, on the other hand, will be easier and faster to select because less competition is expected.[3]

Taken together, *syllable*-congruency effects and *inhibitory syllable-frequency effects* strongly suggest that the syllable plays a functional role

---

[2] E.g., Álvarez, Carreiras & Taft (2001), Perea & Carreiras (1998), Conrad, Carreiras, Tamm & Jacobs (2009), Conrad & Jacobs (2004).

[3] Different studies have shown that the inhibitory syllable-frequency effect cannot be due to other confounding variables such as the ortographic or morpheme frequency (Alvarez *et al.* 2001, Carreiras *et al.* 1993, Perea & Carreiras 1995, 1998).

during word recognition processes. However, despite the fact that the syllable effect has been repeatedly replicated, most models proposed for visual and spoken word recognition do not incorporate explicitly a level of processing for syllables,[4] but rather the phonetic level connects directly with the lexical level during word recognition.

Another source of evidence showing that words which are phonologically/orthographically similar compete for selection is the *neighbourhood density effect*: neighbourhood density has been defined as the number of words differing from the target word by one phoneme (e.g. Luce, Goldinger, Auer, & Vitevitch 2000, Luce & Pisoni 1998) or one letter (e.g., Colheart, Davelaar, Jonasson & Besner 1977) by substitution, deletion or addition in any word position. That is, words with a large number of similar words have a *dense neighbourhood,* whereas those with a low number have a *sparse neighbourhood*. In relation to neighbourhood in word recognition, two effects have been considered: *neighbourhood density* (ND) and *neighbourhood frequency*. A basic finding both in visual and spoken word recognition has been the inhibitory effect of ND: words with dense neighbourhoods are recognized slower than words with sparse neighbourhoods.[5] Results from spoken word recognition have been considered to reflect that words in high-density phonological neighbourhoods activate a set of phonologically similar words that will compete for selection (NAM; Luce & Pisoni 1998), challenging the notion that only the onset of the word may have a special status in word recognition (Marslen-Wilson & Zwitserlood 1989). However, results from the visual modality have been more mixed. Some studies found a facilitatory effect of orthographic neighbourhood density (e.g. Andrews 1992, Perea 1993, Ziegler, Muneaux & Grainger 2002), some others a null effect (Forster & Shen 1996) thus making it difficult for the models to account for the observed results. For instance, Grainger, O'Regan, Jacobs and Segui (1989) observed an inhibitory effect of neighbourhood density but driven by the presence of at least one high frequency word within the neighbourhood, suggesting the importance of the frequency of the neighbourhood in word recognition. *Neighbourhood frequency* refers to the frequency of occurrence of the neighbours. Participants tend to

---

[4] But see Conrad *et al.* (2010); Ferrand, Segui & Grainger (1996), Ferrand & Grainger (1994); see also Taft (1991), Taft & Radeau (1995).

[5] E.g., Vitevitch & Luce (1999), Luce & Pisoni (1998), Carreiras, Perea & Grainger (1997); but see Andrews (1989, 1992), Vitevitch & Rodriguez (2005) for a facilitatory effect of ND.

respond slower to words with higher frequency neighbours that words with no higher frequency neighbours.[6]

Lexical frequency (phonological and orthographical), defined as the number of occurrences of a word in a corpus (e.g., Kucera and Francis 1967), has been considered the most influential variable on word recognition (and in production as we will see). Both spoken and visual modalities have showed a robust lexical frequency effect: words used more frequently are recognized faster than words used less frequently.[7] Importantly, lexical frequency effects have been considered to arise at the lexical level. A variation on the threshold or resting activation levels of lexical representation drives lexical selection. Thus, high-frequency words are recognized faster than low-frequency words because they have lower activation thresholds (e.g., Morton 1969) or higher resting levels of activation (e.g., Forster 1979).

Finally, a main question that has been addressed by the different models of word recognition is whether meaning is only accessed once the word is selected, or whether semantic variables can influence word recognition (see Balota, 1990). Some models assume interaction and top-down feedback between the conceptual and lexical levels (e.g., McClelland & Rumelhart 1981) and can easily account for results as that obtained by Meyer and Schvaneveldt (1971). The authors reported that in a lexical decision task participants were faster and more accurate responding when prime and target were semantically/associatively related (e.g., *bread/butter*) than when they were not (e.g., *doctor/butter*). This *semantic priming effect* (see Neely 1991 for a review on the semantic priming effect) was interpreted as that the semantic information might also influence the ease with which a word is recognized. The semantic priming effect has been interpreted in the framework of a Spreading Activation Principle (Collin & Loftus 1975) according to which activation within and between levels of processing is expected. Thus, in priming experiments, semantically related words are processed faster than non-related words because the prime activates a network of semantically related words and thus, when the target word is presented, its conceptual representation is activated to some degree.

In sum, despite differences in the models and variables influencing lexical access in the two modalities, the core processes and mechanisms

---

[6] E.g., Grainger (1990), Grainger & Jacobs (1996), Carreiras *et al.* (1997); but see Sears, Hino, Stephen (1995).

[7] See e.g., Forster & Chambers (1973), Balota & Chumbley (1984); Murray & Forster (2004), Rayner (1998), Gollan, Slattery, Assche, Duyck & Rayner (2011).

underlying spoken and visual word recognition seem to be similar regardless of the modality.

2. Language production

Speech production is probably one of the most complex cognitive-motor skills, but human beings are very efficient speakers. In normal rate conversations, speakers produce between 2 and 4 words per second (Levelt 1989) and do not make more than 2 or 3 errors every 1,000 words (Bock 1991). One of the most explored aspects in language production has been lexical access and word selection: how words are retrieved from the speaker's memory system.

In the following section we will review the most relevant literature on speech production and how lexical and sublexical information influences lexical access processes.

2.1. Word production: From meaning to phonemes

Most of the speech production models agree that during speech production at least three distinct levels of representation might be accessed (Caramazza 1997, Dell 1986, Levelt, Roelofs & Meyer 1999) before speakers can articulate their ideas: firstly, speakers must decide which concept they want to communicate (concept selection). For instance, if the speaker wants to talk about a dog, s/he could refer to the concept DOG as "dog", but also as "puppy", "animal" or "mongrel". The decision of selecting one concept or another is mainly influenced by contextual factors, such the audience to which the speakers wants to communicate the concept ("*perspective taking*"; Clark 1997, Levelt 1989). In this sense, speakers would not use the same words to speak to an audience in a conference (where more formal language is required) as to friends in a bar (where slang and taboo words would be permitted). Once a concept is selected, the corresponding lexical entry must be selected from the mental lexicon (lexical selection), which has been calculated to have around 50-100,000 lexical entries in a literate adult (Miller 1991). Once lexical selection is accomplished, the corresponding phonological units (phonological encoding) of the intended word must be selected (/d/ /o/ /g/) to further accomplish the final goal of the speaker that is, to articulate the idea. In sum, in order to articulate the word "dog" speakers might access at least three levels of representation: concepts, words and phonemes.

To understand the architecture and functioning of the language production system, different methods have been used, ranging from natural observation of speech malfunctions, such as tip-of-the-tongue states and speech errors, to experimental techniques (reaction time analysis). Tip-of-the-tongue states are described as the feeling of knowing a word in spite of being temporarily unable to retrieve it.[8] Different studies have provided evidence that speakers experiencing a tip-of-the-tongue state have access to the conceptual representations and to the lexical-syntactic features of that word (e.g., grammatical gender), but are unable to access the appropriate phonological information that corresponds to that word, especially last phonemes of the word.[9] This state could be interpreted as reflecting a failure in the second stage of the lexical access process, thus supporting the two levels of representation, semantic and phonological.

In addition, slips of the tongue (speech errors) involving word and sound exchange speech errors are also congruent with the distinction between a lexical and a phonological level of representation and support the notion that different types of linguistic structure are processed at different stages of speech production. Word exchange errors involve words of the same grammatical class but different phonological structure; whereas the sounds that enter in an exchange error typically come from words of different grammatical classes but similar phonological environments (Garret 1980, Dell & Reich 1981).

The selection of words from the mental lexicon that match the communicative intention of the speaker is a crucial step between the semantic and phonological levels of processing. As in speech comprehension, it is widely assumed that the process of lexicalization during production follows a spreading-activation principle (Collins & Luftus 1975) that operates within and between levels of representation. For instance, in the course of naming a "dog", not only the concept DOG is activated, but also semantically related or associated concepts will be activated (e.g., CAT, BARKING, etc.), to a lesser degree than the intended concept. All activated concepts will spread part of their activation to the corresponding lexical nodes ("dog", "cat", "barking", etc.). At this level, a lexical selection mechanism is required to ensure the correct selection of the intended word ("dog"). Selection is considered to be a competitive process, based on the levels of activation of the lexical nodes. In this

[8] E.g., Brown & McNeill (1966), Brown (1991), Burke, MacKay, Worthley, & Wade (1991), Harley & Brown (1998).

[9] See among others Miozzo & Caramazza (1997), Caramazza & Miozzo (1997), Vigliocco, Antonini & Garret (1997).

sense, the higher the level of activation of the competitor lexical nodes, the more difficult will be the selection of the intended word.

2.2. Variables influencing lexical access in word production

One important issue for speech production models has been to understand what factors influence lexical selection. Two major effects have been observed with picture naming paradigms: *picture-word interference* and *lexical frequency effects*.

The picture-word interference paradigm (PWI) is a Stroop-like task (Stroop 1935), so that a line-drawing picture is presented in combination with a visual or auditory distractor word. The task of the participants is to name the picture while ignoring the word. From its first application in the seventies, the PWI paradigm has been considered a powerful tool to evaluate language production processes. One of the most replicated effects found with the PWI paradigm is the *semantic interference effect*: participants are slower to name a picture when is presented with a semantically-related word (DOG/cat) than when is presented with an unrelated word (DOG/table).[10] This effect evidences that not only the intended concept but also semantically-related concepts are activated and propagate a portion of their activation to the lexical level, supporting the notion that lexical selection is influenced by semantic activation.

Another effect observed in the picture naming task (without distractor presentation) that has received much attention has been the *lexical frequency* effect (see Garret 2001 for a review), from the observation that words with high lexical frequency are named faster and more accurately than those with low lexical frequency,[11] fewer TOT are observed for high frequency words (Stemberger & MacWhinney, 1986). The *lexical frequency effect* is supposed to arise at the point at which lexical representations are accessed. High frequency words are processed faster because they have a higher level of activation than low frequency words. As a consequence, less activation is needed for a high frequency word to reach the selection threshold and thus it will be retrieved faster.

In contrast to the widely assumed notion that semantic and lexical information influences lexical selection, much less agreement exists between models of speech production on whether lexical selection is also

---

[10] Glaser & Düngelhoff (1984), La Heij (1988), Lupker (1979, 1988), Rosinski, Golinkoff & Kukish (1975), Underwood (1976).

[11] E.g., Olfield & Winfield (1965), Jescheniak & Levelt (1994), Gollan *et al.* (2011).

influenced by phonological information. According to feed-forward discrete models, only the lexical node selected for production activates its corresponding sublexical (phonological) nodes (Levelt 1989, Levelt *et al.* 1999). Therefore, sublexical information cannot affect the selection process of the target lexical node, given that phonological activation always comes after lexical selection. In contrast, interactive models (e.g., Dell 1986, Harley 1993, Rapp & Goldrick 2000), and cascade models (Caramazza 1997) assume that sublexical information may affect the process of lexical selection. In this context, before lexical selection takes place, all the activated lexical representations send a proportion of their activation to their phonological units, which, in turn, feed back some activation to those lexical representations sharing sublexical information. For instance, in the course of naming the picture of a "dog", the lexical node "dog" sends activation to the phonemes /d/,/o/,/g/, which in turn send activation back to all lexical representations containing such phonemes; "doll", "door", etc. As a consequence, sublexical information may affect the process of lexical selection.

Three observations have been taken to suggest the presence of feedback: the *phonological facilitation effect* in the PWI, *the lexical bias effect* and the *neighbourhood density effect.*

The *phonological facilitation effect* refers to the observation that in the PWI participants are faster naming the picture of a "dog" when is presented with a phonological-related word (e.g. "doll"), or a picture, than when is presented with an unrelated word ("table").[12] This effect has been interpreted as the result of the extra activation that phonemes of the target receive from the picture and distractor lexical nodes (cascade hypothesis) which in turn send activation back to their corresponding lexical nodes, including the target lexical representation (feedback hypothesis) thus facilitating lexical selection.

The *lexical bias effect* refers to those errors (often called spoonerisms) in which phonological substitutions tend to result, above chance levels, in real words (e.g., Baars, Motley, & MacKay 1975, Dell & Reich 1981). The lexical bias effect arises because the probability of mis-selecting a phoneme depends on its activation level. Feedback from the phonemes

---

[12] E.g. Cutting & Ferreira (1999), Jescheniak & Schriefers (1998), Morsella & Miozzo (2002), Peterson & Savoy (1998).

with the highest activation level will activate the lexical representations for an existing word.[13]

Regarding the *neighbourhood effect* different studies have shown that words from dense neighbourhoods facilitate production, with respect to both speed (e.g. Vitevitch 2002, Baus, Costa & Carreiras 2008; but see Vitevitch & Stamer 2006) and accuracy (Stemberger 2004). The neighbourhood effect is thought to arise because the activation of the phonological properties of the target lexical node will be higher for those words that have a dense neighbourhood than for those that have a sparse one. This is because in the former case, there will be many words activating such common sublexical representations, while in the latter, only a few words will do so. As a consequence, the retrieval of the phonological properties of the target lexical node will be faster for words with dense than sparse neighbourhoods. Moreover, the frequency of such neighbourhoods also has a positive influence on speech production: words from high frequency neighbourhoods are named faster (Vitevitch & Sommers 2003; Baus *et al.* 2008), and suffer fewer slips of the tongue (Vitevitch 1997) and TOT states (Vitevitch & Sommers 2003).

For interactive models, these two effects are supposed to be a consequence of feedback between sublexical and lexical information. However, it may be the case that, as discrete models would predict, these effects are the result of a bias in a monitoring system.

Syllables have also been found to play a functional role during speech production. However, their specific role in production is still controversial as to whether syllabic units are represented in the mental lexicon (Dell 1986) or are computed online during post-lexical syllabification, with syllables the articulatory units (Levelt 1989).

Two main experimental effects have been used to investigate the role of syllable units in language production: The *syllable-congruency* and the *syllable-frequency* effects. In the syllable-priming study, participants name the target word after being presented with a prime that is congruent or incongruent with the word's first syllable. Experimental results from this paradigm have been mixed, making it difficult to determine whether the syllable is a functional unit in speech production. Only a few studies showed that *syllable-congruent* primes facilitated target production (Ferrand, Segui, & Grainger 1996, Ferrand, Segui & Humphreys 1997, Costa & Sebastián-Gallés 1998). In contrast, numerous studies showed a

---

[13] Hartsuiker, Roelstraete & Costa (2006); Costa, Anton-Mendez, Roelstraete & Hartsuiker (2006); but see, Viso, Igoa & García-Albea (1991), Perez, Santiago, Palma & O'Seaghdha (2007).

null effect of *syllable-congruency* and only an segmental length effect, that is, the longer the prime, the larger the priming effect.[14] The second source of evidence comes from exploring the *syllable frequency effect* that facilitates production, in contrast to what happens in comprehension (see above). Different studies have shown that words (or non-words) with high-frequency syllables (such as first syllables of disyllabic words) are named faster than words with low-frequency syllables (Carreiras & Perea 2004 Cholin, Levelt & Schiller 2006, Laganaro & Alario 2006). In contrast, less clear have been results with aphasic patients, with some of them showing fewer errors for high frequency words (Aichert & Ziegler 2004, Laganaro 2005) and others a null effect of syllable frequency (Wilshire & Nespoulous 2003). Thus, data on syllable representations seem to suggest, although not conclusively, that syllabic units play a role during speech production. Further research is needed, however, to provide evidence about the specific locus of syllabic representations during production.

In sum, we have outlined evidence of how information flows during speech production and how variables at different levels of processing influence lexical access and selection during speech production.

3. Signed language processing

The ability to convey our ideas and to understand ideas from others is universal, no matter the modality in which language is presented. Both spoken and signed languages are naturally evolved languages of hearing and deaf people respectively, acquired during infancy without formal instruction.

Signed language presents a unique research opportunity to differentiate linguistic universal processes and their underlying brain substrates from those processes specific to the modality in which the language is presented. In this section, we will review the similarities and differences between spoken and signed languages to later provide an overview of the most relevant evidence of the influence of lexical and sublexical properties on sign processing, both in comprehension and production (see also, Carreiras 2010 for an extensive review)

As is the case of spoken languages, there is no universal signed language, but rather hundreds of signed languages used by deaf communities around the world. Moreover, the use of those different signed

---

[14] Baumann (1995), Evinck (1997), Schiller (1997, 1998, 1999, 2000), Schiller & Costa (2006), Schiller, Costa & Colomé (2002).

languages does not depend on the existence of the spoken language used in the same geographical area (American Sign Language is different from British Sign Language).

The linguistic properties of signed languages, as in the case of oral languages, can be characterized according to different levels of processing: phonological, morphological, lexical, syntactic and semantic (Stokoe 1960). It may seem strange to use the term "phonological", with its definition as sound, when speaking about signed languages, but the term "phoneme" is used to refer to the minimal language unit without meaning that combines in a rule-governed way to form meaningful signs (Stokoe 1960, Battison 1978, Sandler 1986, Brentari 1998). In this sense, signs can be decomposed into different phonological parameters (Stokoe 1960, Stokoe, Casterline & Croneberg 1965): *location* of the sign in relation to the body, *movement* of the hands/fingers involved in the sign and *handshape* (that involves also the *orientation* of the hand, phonological parameter included by Battison 1978). These parameters are allowed to combine to form lexical signs with meaning. As in spoken languages, minimal pairs of signs can be created from changing one of the parameters. Moreover, signed language also present phonotactic constraints which determine the possible combinations of the three main parameters and this also varies from one sign language to other (for instance in the inventory of *handshapes* that can be used).

Above these phonological units, sign language research supports the existence of syllables in sign language (Brentari 1998, Corina & Sandler 1993, Wilbur 1993). Although there is no clear agreement on how syllables should be characterized, all of the models propose that syllables in a sign must include some type of movement. For instance, in most of these models (e.g., Brentari 1998, Perlmutter 1992, Sandler 1986, Wilbur 1993) *Movement* and *Location* are the skeletal structure from which the syllables are formed in signing. For instance, in the *Hold-Movement-Hold* structure proposed by Perlmutter (1992) the syllabic unit in signed language would be analogous to the CVC (*Location-Movement-Location*) syllable in oral languages, where the *Movement* would correspond to the peak of sonority and hence would be analogous to the vowel, while *Location* would be analogous to the consonant in oral languages.

3.1. Sign language comprehension

Both spoken word and sign recognition entail the segmentation of the acoustic/visual signal from a continuous input onto discrete phonological units that map onto the lexical representations stored in memory.

However, both modalities differ in the degree of sequentially/simultaneity with which words and signs are articulated and recognized. As previously mentioned, spoken languages are articulated sequentially (although co-articulation exists) by the vocal tract and recognition is also a sequential process in which acoustic-phonetic representations, extracted from a continuous input, are mapped onto lexical entries. Sign languages, in contrast, are articulated mostly by the simultaneous combination of various upper-body articulators (for instance, hands, arms and fingers) and the phonological parameters of handshape, location and movement are mainly perceived in a parallel fashion (Stokoe 1960). For instance, *handshape* is always articulated in a specific *location*. However, some degree of sequentially can also be found in sign language recognition. Following Grosjean (1981), Emmorey and Corina (1990) adapted the *gating* technique to sign language to investigate the processes of lexical access during sign recognition. Deaf ASL (American Sign Language) signers were presented repeatedly with a sign that increased in length of presentation every repetition and they had to guess what sign was presented and how confident they were about the responses given. Results showed that phonological parameters were recognized sequentially: *location* and *handshape* were recognized first, but word recognition only occurred when *movement* was recognized. These results suggest that both the visual and auditory input activated a cohort of potential lexical candidates (Marslen-Wilson 1987) that share the initial segments (location and handshape) and this cohort narrows the number of candidates as more information unfolds (movement) until a single lexical candidate remains (see also Orfanidou *et al.* 2009 for a similar argument).

These results were important because they provide evidence that sign language recognition is sequentially determined and therefore the process of sign recognition might rely on activation and competition processes among the phonological parameters. These results also supported the notion that phonological parameters do not contribute equally to sign language processing. However, there is still no agreement on the specific role of each of the parameters in sign language comprehension (nor in sign language production, as we will see below).

One of the most commonly-used tasks to explore how lexical access is accomplished in word and sign recognition has been the lexical decision task, in which participants are normally shown a video containing a real sign or a non-existing sign that violates the phonotactic constraints of one of the phonological parameters (for instance, by signing out of the signing space). As in spoken languages, the effect of *lexicality* has been observed in sign languages: deaf signers are faster and more accurate in recognizing signs than non-signs (Emmorey 1991, Corina & Emmorey 1993, Carreiras, Gutierrez-Sigut, Baquero & Corina 2008). Moreover, effects of *lexical frequency* have also been observed in sign language recognition: high-frequency signs are recognized faster than low-frequency signs. However, lexical frequency counts are not currently available in sign language research and rather subjective familiarity ratings by deaf signers are used. Importantly, studies have shown that high-familiarity signs are recognized faster than low-familiarity signs (e.g., Carreiras *et al.* 2008), supporting a strong correlation between lexical frequency measures obtained from corpus and subjective familiarity ratings (e.g., Balota 2001). Carreiras *et al.* (2008) showed that *neighbourhood density* of the phonological parameters also had an influence on sign recognition. In the same lexical decision task, the authors also manipulated neighbourhood density of the LSE signs (Lengua de Signos Española / Spanish Sign Language) by comparing signs with dense or sparse neighbours in relation to their *handshape* and *location*. For the low-familiarity condition, signs with a dense handshape neighbourhood (signs that contained a *handshape* shared by many other signs) were recognized faster than signs in sparse neighbourhoods (signs that contained a *handshape* shared by few other signs). The opposite was observed for location neighbourhood density: those signs with dense location neighbourhoods were recognized slower than sparse location neighbourhoods.

It is important to note that these results differ from the inhibitory neighbourhood density effect observed in spoken/written recognition. However, neighbourhood density in oral languages is defined differently from in the experiments that investigated neighbourhood in signed languages. In any case, other studies have also provided evidence that the phonological parameters *handshape, location* and *movement,* influence the recognition of the sign in a different manner.[15] Some of those studies have used the priming paradigm in which the phonological relation between the sign prime and the sign target was manipulated. For instance, Corina & Emmorey (1993) showed an inhibitory priming effect when prime and target shared location, but no priming effect when they shared handshape. Conversely, Carreiras *et al.* (2008) also found an inhibitory priming effect for location but a facilitatory effect for handshape. Moreover when the inter-stimulus interval between prime and target was manipulated (ISI), Corina and Hildebrandt (2002) found a trend for an inhibitory priming effect for location and movement with an ISI of 100 ms, but Dye and Shih

[15] Corina & Emmorey (1993), Corina & Hildebrandt (2002), Mayberry & Witcher (2005), Carreiras *et al.* (2008).

(2006) found a facilitatory priming effect for location and movement when the ISI was 50 ms.

In sum, these studies have provided limited but encouraging evidence that similar mechanisms may underlie sign and word recognition, but also that some effects may be modulated by the modality.

3.2. Sign language production

As mentioned earlier, in order to communicate our ideas from intention to the actual articulation of the sounds, at least three stages could be accessed in the spoken modality: semantic, lexical and phonological. One question asked has been whether the same levels of representation might be accessed during sign language production. This question is not trivial in sign language research, given that during decades, approaches such as the *semantic phonology theory* (Stokoe 1991, Amstrong, Stokoe, & Wilcox 1995) and others (e.g. Cuxac 2000) proposed that signs are units based on semantically transparent representations and therefore there is no difference between the meaning of the sign and its phonological parameters. Under this view, signs were essentially considered as *iconic* communication systems, that is, every sign resembles some aspect related to its meaning (for instance the sign TO DRIVE resembles the action driving by moving the steering wheel).[16] However, recent evidence adapting the most commonly-used paradigms in oral languages (see above) to sign language production, slips of the hands (slips of the tongue), tip of the fingers (TOF; tip of the tongue, TOT) and the picture naming interference paradigm, have suggested that the semantic and phonological levels of processing are distinct and separate.

Sign error corpora collected by Newkirk *et al.* (1980) and Hohenberger (2002) in two sign languages (ASL and DGS, American and German Sign Languages) has revealed that deaf signers while signing commit errors that involve exchange, preservation and anticipation of handshape, location or movement or even a sign exchange (although 1%). These results provide evidence for phonological encoding during sign production. More recently, Thompson, Emmorey and Gollan (2005) investigated whether the feeling of knowing the sign but being temporarily unable to retrieve it (tip of the tongue phenomenon) was also experienced by deaf signers. By means of asking deaf signer to translate English word into signs,

[16] Later we will focus on the present debate in sign language research about whether iconicity influences sign processing and to what extent.

Thompson *et al.* (2005) observed that ASL signers experienced TOF (tip of fingers) in which they could retrieve semantic information, but they could not access form, thus supporting a clear separation (as in oral languages) between retrieval of semantic and phonological information in sign production. These two studies revealed also the different role of the phonological parameters when making an error or in a TOF state. Most slips of the hand involved a sign exchange in handshape and only a few involved location or movement. In a TOF state, deaf signers were able to recover handshape and location of the sign but not movement.

To explore the dynamics of sign production in an online manner, sign language research has adapted recently the *picture-word interference paradigm* to sign language production (Baus *et al.* 2008, Corina & Knapp 2006). Baus *et al.* (2008) examined semantic and phonological effects by presenting deaf signers with pictures that they were required to sign while ignoring the video-sign produced in the background. Results showed that deaf signers from LSC (Catalan Sign Language) were slower signing the picture when presented with a semantically related sign than when it was presented with an unrelated sign, revealing that semantic information is accessed first during sign production and that lexical selection is accomplished by competition between the activated lexical entries, as in oral languages. In contrast, results from phonologically related signs were mixed. A facilitation effect was obtained when picture and distractor sign shared handshape or movement, but an inhibitory effect when they shared location, supporting the different roles that phonological parameters play during phonological encoding in sign production. Using the same paradigm but presenting pictures and sign-distractors at different SOAs (-130, 0, +130), Corina and Knapp (2006) observed that as in speech production, a semantic interference effect arose only when the sign-distractor was presented before or at the same time as the picture. In contrast, no facilitation of phonological overlap was observed no matter if the picture and distractor-signs shared one, two or three parameters.

Compared to research in sign language recognition, research in sign language production is still in its infancy and for instance no studies have so far addressed how lexical variables such as lexical frequency/familiarity, neighbourhood density or age of acquisition influence lexical access and selection during production. Importantly, sign production research has revealed interesting results about the similarity of some processes in spoken and sign languages, but also some differences in relation to the modality in which the language is produced.

3.3. Iconicity

Both spoken and sign languages contain iconic words/signs that resemble their referents (e.g., onomatopoetic words), but the visual/gestural modality in which sign languages are presented allows more concepts to be expressed iconically and so many more signs than words are iconic (e.g., Mandel 1977, Taub 2001, Pietrandrea 2002). Deaf adults may be aware of the iconic properties of signs, and in fact iconicity is very helpful in the first stages of learning a sign language (Campbell, Martin, & White 1992). However, it is unclear whether iconicity bestows a lexical processing or memory advantage for deaf signers (see Perniss, Thompson & Vigliocco, in press, for a review). Some researchers consider that iconicity is a property of signs that is not linguistically relevant for language processing (e.g., Klima & Bellugi 1979, Newport & Meier 1985). Other researchers, as we pointed out earlier, assume that sign languages are basically iconic, that is, the link between form and meaning is so prevalent that in fact there is no separation between them (e.g., Armstrong, Stokoe & Wilcox 1995). At present, evidence regarding whether iconicity plays an important role in sign language comprehension and production is mixed. On the one hand, some studies have reported that iconicity does not aid sign-language vocabulary acquisition in children (e.g., Anderson & Reilly 2002, Orlansky & Bonvillian 1984), that it has no effect on immediate short-term memory recall, that is, iconic signs are remembered as accurately as non-iconic signs (Poizner, Bellugi and Tweney 1981). Furthermore, iconic and non-iconic signs engage the same language-related neural regions (Emmorey *et al.* 2004), are processed equally in a semantic priming task (Bosworth and Emmorey, in press), are equally affected during TOF states (Thompson *et al.* 2005) and are equally impaired in cases of sign aphasia (Marshall *et al.* 2004). On the other hand, some recent studies have provided evidence of the role of iconicity in sign processing. For instance, Thompson, Vinson, and Vigliocco (2009) in a sign-picture matching task showed that iconicity aided response (see also, Grote & Linz 2003, Ormel, Hermans, Knoors & Verhoeven, 2009) when the task mapped onto semantic processing and in contrast, negatively influenced reaction times when the task tapped onto phonological processing, as in a sign-phonological decision task (Thompson, Vinson and Vigliocco; in press).

In sum, there is no clear answer so far as to whether iconicity influences sign language processing, and much work will be required in the future to provide a better understanding of the role of the iconic properties of signs during sign processing.

4. Final remarks

This chapter has provided a review of the most relevant literature regarding spoken and written word and sign language processing, both in comprehension and production. As has been described above, similar processes seem to underlie the recognition of written words, spoken words and signs. This also seems to be the case for speaking words or signing signs. However, while stressing the core common processes underlying word/sign recognition and production, there are some differences between these modalities and further research is required to establish to what extent these differences are peripheral or affect core processes of recognizing and producing the building-bricks of comprehension and production.

CHAPTER SIX

MORE THAN ONE LANGUAGE IN THE BRAIN

ITZIAR LAKA

1. Introduction

What difference does it make to have one language in the brain or to have more than one? This is an intriguing and currently much inquired question, which can help us unravel more than one mystery concerning language and the brain. At present, we only know bits and pieces of the answer. As research progresses and more pieces of this large and complex puzzle fit together, we discover some general outlines of the answer, and realize the intricacies of the detail. In this chapter, I will attempt to keep our eyes set in that general outline, occasionally dwelling into a detail or two, in the hope of giving you a glimpse of how research is conducted in this field of inquiry.

As we will see, there are indeed differences that relate to having more than one language in the brain. Some of these differences involve cognitive abilities that lie outside of the linguistic systems proper, such as the capacity to ignore irrelevant information when changing tasks, or a certain degree of resilience towards symptoms of neurodegeneration. Other differences between monolinguals and bilinguals involve the interplay of the two linguistic systems: their simultaneous activation and the need to select or inhibit one at a time, or the cost involved in having two lexicons and grammars instead of one. The developmental patterns of preverbal bilingual babies—who detect very early that there is more than one language in their environment—are also different from those of monolinguals. These differences eventually emanate from the intensive cognitive training undergone by bilinguals in their lifetimes, given the frequency and speed at which they switch from one language to another.

From a more narrowly linguistic point of view, humans that have more than one language in their brains can provide crucial evidence regarding the neurocognitive nature of the language faculty, with its innate/universal properties and its acquired/variable aspects. Thus, for instance, we will see that variable (parametric) aspects of the grammar appear to be sensitive to *when* they are acquired, and to *what* type of grammatical knowledge was there before, so that native speakers and non-native speakers do not process certain aspects of the grammar in the same way, even at high levels of language proficiency. In contrast, the lexicon seems to be insensitive to when it is acquired and what the words of a previously acquired language look like: at high proficiency, non-native speakers are native-like in lexical matters.

So, in broad terms, it matters a great deal to the brain whether it carries more than one language, and when there is more than one, which one developed first and which one later, and it also depends on the language component what is the length of time that matters (shorter lapse between native and the non-native language for phonology, longer lapse for syntax, insensitive for the lexicon).[1] It also matters to the brain how present these languages are, that is, what the proficiency achieved in each language is, and how frequently the brain uses them, or changes from one to the other, so that in some cases one language can be dominant over another, or not. There are also reasons to think that individuals vary in their abilities to acquire a second language, and that this abilities relate to intrinsic neurobiological differences (Díaz *et al.* 2008). The degree of morphosyntactic similarity of the languages coexisting in one brain might also make a significant difference in how we represent and process them, though currently, more is known about the lexicon and phonology in humans that know more than one language, than about the syntax. Perhaps this is partly due to the fact that the finer experimental studies exploring syntax in the brain often require the cooperative work of syntacticians, psycholinguists and neuroscientists, which hopefully books like this will encourage.

The literature covering various aspects of the topics to be discussed in this chapter is vast, and it is therefore not possible to cover it exhaustively. I will provide an overview of the main topics of research and the general outlook that emerges given the evidence found; and since it is not possible to discuss all issues on an equal footing, I will mainly concentrate on those that I find most significative and revealing for linguists with an interest in the neurobiology of language. I will review recent findings on cognitive

---

[1] Throughout this chapter, "the lexicon" refers to the open class lexical items, not to the closed class composed of functional elements such as inflectional morphemes, case markers, complementizers and the like. Functional-inflectional elements are taken to be part of syntax, which therefore includes inflectional morphology (morphosyntax).

advantages of bilingualism that are not directly related to the language faculty, and discuss the impact of age and language proficiency in the neural underpinnings of bilingualism. Within the components of language, I will concentrate more on the lexicon and the grammar. Bilingual phonology is a fascinating area of research that directly touches upon earliest stages of language acquisition, but I will not discuss it here. There are excellent overviews of bilingual language acquisition that focus on the early development of phonology and the lexicon, which address the issues that I will leave aside, such as Sebastián-Gallés and Bosch (2002), Sebastián-Gallés *et al.* (2005), and Sebastián-Gallés and Kroll (2003).

2. The bilingual is not two monolinguals in one person

François Grosjean, a pioneer researcher on bilingualism, warned in his 1989 paper that the bilingual is not two monolinguals in one person. In this chapter, we will review some discoveries on the neurobiology of bilingualism that show the extent to which this statement is true. We will also keep our focus on what these discoveries reveal about the human language faculty, and the new questions they pose to language research.

Bilinguals outnumber monolinguals in our species: according to some recent estimates, between 60 and 75 per cent of the world's population is bilingual. Although we do not have direct evidence, it has been argued that the capacity to learn more than one language is an adaptive trait in human evolution (Hirschfeld 2008); given what we know of interactions between human groups, is not unlikely that people throughout history have more often than not known more than one language.

For the purposes of this chapter, a person who knows more than one language and can use them to communicate efficiently qualifies as bilingual, even without reaching native-like command in both languages. We will thus adopt the view that "bilingualism is the regular use of two (or more) languages, and bilinguals are those people who need and use two (or more) languages in their everyday lives" (Grosjean 1992). We will not limit our attention to people who can speak two languages with equal mastery, often referred to as "balanced bilinguals"; we will also review studies of "unbalanced bilinguals", for whom one language is dominant over the other in some or another aspect. Finally, we will also consider adults in the process of learning a second o third language from the start. All these people are of interest to the neurobiology of language; we can learn a great deal about language in the brain by considering all kinds of different types of populations.

Psycholinguistic and neurolinguistic research initially tended to restrict itself to the study of monolinguals, and there was not much interest in the study of bilingualism, because it was generally (though tacitly) thought that the neural representation and processing of a given language was not affected by another language, whether acquired simultaneously or later in life. Recent findings challenge this assumption, and suggests instead that research beyond monolingualism holds a great potential for generating knowledge about the neurobiological nature of the human language faculty and the way in which language is organized in our brains.

Broadly speaking, neurocognitive studies of language and bilingualism reveal that the patterns of activation related to language processing are consistent across languages and native speakers; research shows that the processing of different languages occurs in much of the same brain tissue (Kim *et al.* 1997, Perani *et al.* 1998, Díaz *et al.* 2011). When differences between languages are found, they obtain in bilinguals and they correlate with differences in proficiency levels attained in each language, and differences in age of acquisition for each language. This is why the impact of language proficiency and of the age at which a language is acquired have been to date the factors that have received most research attention. The impact of the degree of similarity of the grammars located in one brain is a far less explored issue, as are the differences among different types of grammatical phenomena, though as our knowledge advances, grammatical specificity emerges as a likely relevant factor to be kept into account. As the volume and level of detail of the studies carried out increases, it also becomes increasingly clear that, although all these factors have often been studied separately, there are strong connections between them: proficiency in the language, age of acquisition, and grammatical similarity are likely to be intertwined rather than separate factors.

3. Cognitive advantages of bilingualism.

Perhaps the most striking findings from neurocognitive research on bilingualism relate to cognitive differences that result from using more than one language frequently during a lifetime. In fact, it is becoming increasing clear that speaking more than one language yields cognitive benefits that extend from childhood, through life, and into old age. Recent research into the neurobiology of bilingualism reveals that being fluent in two languages, particularly from early childhood, not only enhances a person's ability to concentrate and ignore irrelevant information, but also delays the onset of dementia and other age-related cognitive decline by an average of 4 to 5 years (Craik *et al.* 2010). Furthermore, early acquisition of more than one language has other consequences outside of the linguistic domain, such as an earlier development of theory of mind (Kovacs 2009).

The cognitive benefits associated to bilingualism were originally revealed by the research of Bialystok and collaborators (Bialystok 1999, Bialystok 2001, Bialystok and Martin 2004). In a series of experiments, they showed that bilingual children outperform monolinguals in tasks that required attentional control. Bialystok (1999) asked 4 to 5 year-old children to first sort a set of cards according to their colour and then to sort them according to shape, and found that bilinguals made less errors when changing tasks. This evidence strongly indicates that bilingualism contributes to the development and strengthening of the attentional control mechanisms that are involved in these changing task-rules. This group of researchers also showed that bilingualism has an impact on the attentional control abilities throughout the lifetime. Bialystok, Craik, Klein, and Viswanathan (2004) tested groups of bilinguals and monolinguals of different ages on the so-called "Simon task", that requires to pay exclusive attention to a specific property of the stimulus, like its colour or shape, while ignoring its position on the screen. People take longer to respond when objects that share the relevant property appear in different locations in the screen, the so-called "Simon effect". Bialystok and colleagues found that the magnitude of the Simon effect is smaller for bilinguals, and this difference between bilinguals and monolinguals was more evident in older participants (over 60 years old), which indicates that bilingualism not only facilitates the development of more efficient attentional control mechanisms in childhood, but it also delays cognitive decline (for a review see Craik and Bialystok 2006). Costa and collaborators (Costa *et al.* 2008, Hernandez *et al.* 2010) have further shown that this bilingual advantage is also present in adults, at ages at which individuals are at the peak of their attentional capabilities, and that this cognitive bilingual advantage specifically involves conflict monitoring and conflict resolution, two subcomponents of the executive control network that are neuroanatomically distinct.

Bilinguals extract cognitive benefits in the areas of the brain that constitute the *executive control system*, located in the dorsolateral prefrontal cortex; these areas have been found to be active during language switching in bilinguals (Hernandez *et al.* 2004). The cognitive capacities that are enhanced by bilingualism are not language-specific, because these mechanisms involve domain general cognitive control processes (Abutalebi and Green, 2008), though it is through the acquisition and frequent use of more than one language that they have been trained and strengthened in bilinguals.

The neural underpinnings of language control in bilinguals are still not understood in detail, though much progress has been made in identifying them during the last years. In a pioneering study on the neurocognition of language switch and translation in bilinguals, Price *et al.* (1999) used positron emission tomography, a technique that provides a high topographical resolution of brain activity, to study language processing in six German-English bilinguals who started learning English at nine years of age. They found that the most active brain areas during translation fell outside Broca's or Wernicke's areas, among them the anterior cingulate cortex that controls attention. More recently, Crinion *et al.* (2006) studied different groups of proficient bilinguals (German-English and Japanese-English) and suggested that the left caudate is strongly involved in language monitoring and control.

A recent study by Garbin *et al.* (2010) finds similar evidence in favour of the involvement of Broca's area in language switch in bilinguals, but moreover, these researchers find significant differences in the cortical networks involved in cognitive control between monolinguals and bilinguals: the most interesting difference is the involvement of the left IFG (Broca's area) also in non-linguistic switch tasks for bilinguals, whereas monolinguals activated the right IFG for the same switch tasks. The fact that the left IFG has been consistently related to bilingual language control indicates that there is a certain degree of overlap between the cortical network responsible for language control and general-purpose non-linguistic cognitive control in the case of bilinguals, but not in monolinguals. This means that early and proficient bilinguals recruit Broca's area not only for language switching tasks, but also for other non-linguistic switch tasks, whereas monolinguals do not. These results thus suggest that, although the attentional mechanisms that are better developed in bilinguals are not language-specific (they also monitor other types of tasks where language is not involved), and although in non-bilinguals these attentional mechanisms are controlled by bi-hemispheric or right-hemispheric regions of the brain, in bilinguals they are controlled and computed by Broca's area (Brodman 45/44), one of the areas crucially involved in linguistic computation.

4. How separate do bilinguals keep their languages?

How the brain organizes languages in bilingual individuals has been an intensely investigated question in the last years. Is each language located in separate areas of the brain or in overlapping regions? Studies of whole-brain functional neuroimaging show that highly proficient bilinguals activate the same brain regions when they use any of their two languages (Kim *et al.* 1997, Perani *et al.* 1998). Hernandez *et al.* (2001) run an fMRI study of six Spanish/English early bilinguals, all of whom had acquired both languages before the age of five, and found that the two languages were represented in overlapping regions of the brain. That is, given the degree of detail that current neuroimaging techniques allow for, it emerges that early and proficient bilinguals use the same neural circuits for the two languages they know.

However, when a bilingual has learned the second language later in life, linguistic tasks involving this second language activate broader areas of the brain, partially overlapping but distinct from the native language. Kim *et al.* (1997) run an fMRI study comparing early and late bilinguals while processing their two languages. Results revealed distinct physical loci for native and non-native languages along the periphery of Broca's and Wernicke's regions in the case of late learners, but not in the case of early learners.

Dehaene *et al.* (1997) run a study of French-English bilinguals, all of whom had acquired the second language after the age of seven. In listening tests, an fMRI revealed common areas of activation in the left temporal lobe for all subjects when the native language was used. When the non-native language was used in testing, researchers found highly variable areas of activation in both hemispheres. Several studies also find that Broca's area is generally more activated when listening to the native language than to other lesser known languages (Mazoyer *et al.* 1993, Perani *et al.* 1996, 1998). Some of these fMRI studies have suggested that there may be smaller-scale circuits specialized for each language (Dehaene *et al.* 1997, Kim *et al.* 1997). Halsband *et al.* (2002), for instance, studied ten Finnish–English adult bilinguals, all of whom had acquired the second language after the age of 10, using Pet scan. They found differential areas of activation for the two languages in both Broca's area and in the supramarginal gyrus, one of the convolutions lying between Broca's and Wernicke's Areas. Wartenburger *et al.* (2003) used fMRI testing to study 32 Italian-German bilinguals in three groups, (i) eleven subjects who acquired the German in early childhood and were fluent native speakers, (ii) twelve subjects who acquired German in adulthood but managed to attain a high level of proficiency, and (iii) nine subjects who had German late in life and had limited proficiency. They found that age of acquisition was a statistically significant variable in determining loci of grammatical processing in the brain, but less so in determining semantic processing.

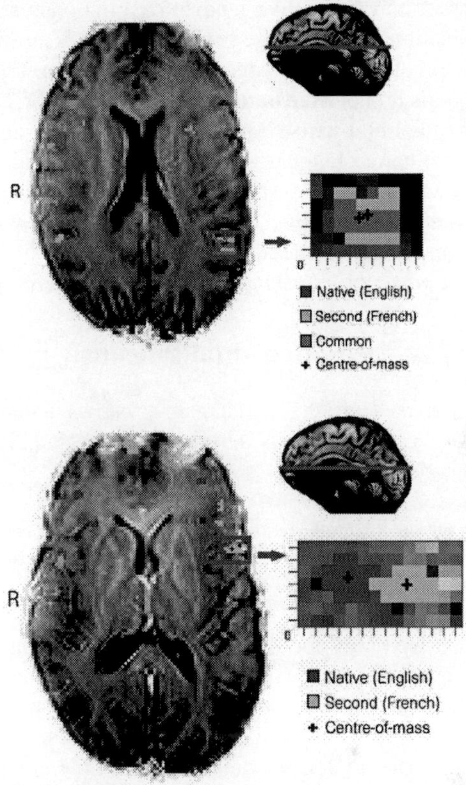

Fig. 6-1 Differences between early (supra) versus late (infra) bilinguals and brain activations during language processing in the two languages (from Kim et al. 1997)

Other findings strongly suggest that early and sustained bilingualism significantly alters the brain's structure: Mechelli *et al.* (2004) identified an increase in the density of grey matter in the left inferior parietal cortex of bilinguals relative to monolinguals, more pronounced in early bilinguals, and showed that the density in this region increases with second-language proficiency and decreases as the age of acquisition increases. This means that bilingual adults have denser gray matter (the brain tissue that is packed with information-processing nerve cells and fibers), especially in the left hemisphere, where Broca's area (Brodman 45/44) is found, and where linguistic computations are handled by the brain. This density effect is strongest in people who were bilingual before the age of five and are most proficient at both languages.

As we can see, neurocognitive studies on bilinguals strongly suggest that age of acquisition and language proficiency are determinant factors in the neural underpinnings of language and bilingualism, so that early and proficient bilinguals do not "separate" languages in the brain, but as age of acquisition of the non-native language increases and proficiency decreases, the non-native language tends to be located in more extended and individually variable areas: The brain does not represent and compute non-native grammars like native ones, though we still must understand more precisely, and in greater detail when and why this differences emerge (at what ages, for what components, given what conditions, etc).

5. Languages are activated simultaneously in the bilingual

The brain activates all the languages it knows when it has to use language. In particular, bilinguals activate both of their languages when they have to use one:

> "Interactions between languages have been observed at all representational levels of language, even when people were tested in purely monolingual language contexts." (Desmet and Duyck 2007: 168-69)

Much of what we know about how the brain copes with more than one language relates to the lexicon. The simple hypothesis that bilinguals have two separate lexicons, one for each language, so that when they use one language, only its lexicon is activated has been shown wrong by many studies: both lexicons are active whenever the bilingual speaks, either in one or the other language.

People who know more than one language must have knowledge of the two phonological and morphosyntactic systems, and also, inevitably, they must know that a given meaning has (at least) two different words attached to it; that is, for instance, a Spanish speaker who knows English must know that the lexical item *uña* in Spanish stands in correspondence to the word *nail* in English because both share the meaning "horny scale of plate of epidermis at the end of the fingers and toes of humans and many apes", and she must also know there is another homophonous *nail* in English that corresponds to Spanish *clavo*, because they both mean "a slender, pointed piece of metal, usually with a head, used for fastening pieces of wood or other material together, by being driven into or through them".[2]

[2] For the purposes of this chapter, the form of a word will be named "lexical item", and the meaning it expresses will be named "semantic content".

Different lexical items can share semantic content, either in case of synonymy or semantic similarity, but more to the point of our discussion, different lexical items in the brain of the bilingual can share semantic content, like in the case of pairs like *uña/nail* in our example above. The activation of semantic content in the brain is independent of the language the lexical item belongs to (Crinion *et al.* 2006), that is, the semantic/conceptual content associated with lexical items is neurocognitively distinct from the words themselves, so that different word-forms from the languages of the bilingual touch upon the same semantic/conceptual networks in the brain (Kovelman *et al.* 2008).

In a pioneering study, Van Heuven *et al.* (1998) found that the lexical items from a bilingual's native language are active while the bilingual is engaged in recognizing words from a non-native language. It was already known that, in monolinguals, the time it takes for a word to be read and recognized depends on the number of orthographic neighbors the word has. An orthographic neighbor is another word that results from changing one letter in the original one: for example, the Spanish word *una* "one (feminine)" is a neighbor of the word *uña* 'nail', because they only differ in one letter. Van Heuven *et al.* (1998) discovered that the time it takes Dutch–English bilinguals to recognize an English word like *farm* did not only depend on the number of English neighbors that lexical item has (e.g. *firm*, *fart*), but also on the number of *Dutch* neighbors it has (e.g. *darm* 'colon', *faam* 'fame'). This result show that lexical representations from the native language are active during a word recognition task in the second language. Following this discovery, other studies revealed the extent of this cross-linguistic activation of the lexicons of the bilinguals: Dijkstra *et al.* (2000) also found that both lexicons were active when reading cross-linguistic homographs, that is, different words from both languages that have the same orthographic form, like *room*, which is one lexical item in English (synonym of *chamber*), and another one in Dutch (meaning 'cream'), because these words took longer to read than those that had no cross-linguistic homographs.

Further studies have shown that this activation of the lexical items irrespective of the language for different types of bilinguals and language-pairs. For instance, further evidence that the native language is activated when using the non-native one had been uncovered by a large number of studies (among them Costa *et al.* 2000, Duyck, 2005, Duyck *et al.* 2004, Schwartz *et al.* 2007). It has also been repeatedly shown that the non-native language is active when the native one is used (see for instance Duyck 2005 and Van Hell and Dijkstra 2002, among others). Similar results have been obtained in studies where participants, instead of reading, heard the words they had to recognize (for instance Marian *et al.* 2008, Marian *et al.* 2003), and in studies where participants had to actually say the words (Costa *et al.* 2005, Kroll *et al.* 2006, Costa *et al.* 2008, Santesteban and Costa 2006).

Initially, classical psycholinguistic models of the lexicon did not consider the possibility of interactions between two languages in one brain; they were models that only contemplated one lexicon and its relation to semantics and grammar, like the word recognition model from McClelland and Rumelhart (1981) or the word production model by Levelt (1989) (see Chapter 5 of this volume). However, research on bilinguals reveals that people who know more than one language do not "turn off" the lexicon of the language they are not using, and therefore it calls for a revision of the "isolated lexicon" stance. "The Revised Hierarchical Model" (Kroll and Stewart 1994) became a very influential model of the lexicon in bilinguals; this model was mainly designed to account for successive bilingualism, where the non-native language is learned after the native language has been fully mastered, but could also be applied to simultaneous bilingualism. The conceptual level was assumed to be shared by the two languages, and a division was made between a lexicon for the native language and a lexicon for the second language.

The compelling evidence that linguistic systems are co-activated in neural events that involve the use of language has led current models of the lexicon to assume that the flow of activation from the conceptual/semantic system to the lexicon is not specific to a language (see among others Costa and Caramazza 1999, Green 1998; Hermans *et al.* 1998, Poulisse and Bongaerts 1994). That is, when we hear or say a word, the conceptual/semantic system activates all the corresponding lexical item of the languages we know regardless of the language in use. Ivanova and Costa (2008) have shown that this full activation of the lexicons has a cost in the process of lexical selection and retrieval when monolinguals and bilinguals are compared, such that bilinguals take longer in retrieving a word even when they do so in their dominant language, an effect that has been named "the bilingual cost".

This discovery naturally leads us to the question of how bilinguals manage to produce the words of the target language and prevents words from the non-target language from being uttered. If all the languages of the bilingual are active when language is processed, then there must be some further cognitive operation that controls what language is used at a time.

6. How do bilinguals control what language they speak in?

We have seen there is ample evidence that all languages are active when bilinguals or multilinguals use language. However, bilinguals easily control when to use one language or another, and in order to do this, they must employ some kind of neurocognitive control mechanism.

Proposals as to how bilinguals control their languages in order to produce the one they want to use generally agree that bilingual lexical access must involve some kind of attentional control mechanism (Costa 2005, Costa *et al.* 1999, Finkbeiner *et al.* 2006, Green 1998, Kroll *et al.* 2006, La Heij 2005). Some researchers argue that language control in bilinguals entails the active inhibition of the linguistic representations of the other language, which, despite being activated, is not intended to be used (Green 1998, Meuter and Allport 1999).

The most revealing evidence for inhibitory mechanisms in bilinguals has been provided by Costa and collaborators, in a series of experiments on language-switching, where participants are asked to name a picture in one language or another, depending on the colour of the picture. These studies, carried out with various types of bilinguals (of Basque, Catalan, English, French and Spanish), show that low-proficient bilinguals take longer to switch from their less dominant non-native language to their native one than the other way around (Costa and Santesteban 2004, Costa *et al.* 2006). This effect has been named the "asymmetrical switching cost"; at first sight this result may appear counterintuitive, because it entails that it is "harder" to change from the language you know worse to the language you know better than it is to change from the language you know better to the language you know worse. This is how the researchers explain it: when the bilingual has to speak in the weaker, non-native language, the native language is activated, and therefore it has to be very strongly inhibited. As a consequence of the strong inhibition applied to it, if later these low proficient bilinguals want to speak in the dominant native language, they need to undo the strong inhibition applied to words form their native language. In contrast to this, changing from the strong native language to the weaker non-native language does not require undoing such a strong inhibition, since the words of the weaker language need not be strongly inhibited.

Interestingly, this team of researchers have also discovered that early and highly-proficient bilinguals employ a different mechanism for language control. Given the same language-switching tasks, these bilinguals revealed symmetrical switching costs. That is, it took them the same time to switch from either of the two languages to the other one.

Surprisingly, the mechanism employed by these balanced bilinguals yielded a symmetrical pattern not only when switching between the two languages they had known and used throughout their lives, but, surprisingly, also when they had to switch to a third language they had learned much later and knew less.

So, while the switching performance of low-proficient bilinguals leads to an asymmetrical pattern, depending on language dominance, in early and proficient bilinguals it yields a symmetric pattern, which does not only apply to the dominant languages, but also to non-dominant, later learned languages. This neurocognitive difference between early and proficient bilinguals on the one hand, and late not very proficient bilinguals on the other, is reminiscent of the findings by Garbin *et al.* (2010) we have reported in section 4; this fMRI study reveals that early and proficient bilinguals recruit Broca's area for linguistic and non linguistic switching tasks, whereas monolinguals do not. It is thus very plausible that the symmetrical switch mechanism of balanced bilinguals reflects the involvement of Broca's area, whereas the asymmetrical mechanism reflects lack of involvement of Broca's area in the same tasks.

7. Bilingual Syntax: native and non-native grammars

There are less studies on syntactic processing in non-monolinguals in comparison to studies on lexical processing, but in the last years this area of research has experienced enormous growth. While evidence on nonnative syntactic processing is still sparse, "even so existing data clearly indicate that syntax is a phenomenon that deserves full consideration" (Kotz 2009).

One general finding that emerges is that the language proficiency of the bilingual has a direct impact in the neurocognitive representation and processing of syntactic phenomena. The impact of age is another factor under close scrutiny, though it is less understood and still subject to much debate. So, whereas there is widespread agreement that proficiency has a direct impact on the neurocognition of syntax, as it does in all domains of language, data concerning the precise impact on syntactic processing when a non-native language is learned and what the native language looks like do not yet provide a coherent picture that researchers can widely agree upon.

It is generally agreed that native versus non-native differences in language representation and processing are not found in the lexicon /semantic interface, so that as proficiency increases, native and non-native lexical processing are indistinguishable. It is also generally agreed that syntax is different in this respect, so that, within syntax, some aspects of it yield non-native effects, but others do not. What is currently missing is a clear and principled picture of what aspects of syntax fall in one or the other category, and why that is so.

Part of this disagreements relate also to the research methods utilized, and with the advent of experimental studies that employ techniques such as reaction-times, eye tracking, event related potentials and fMRI, we discover native versus non-native syntactic processing differences that could not be detected by off-line techniques such as error-production, grammaticality judgments, etc. In any event, and whichever the ultimate explanation for what in syntax differ in native versus non-native speakers, it is essentially neurobiological and must involve neural differences in language representation and processing between the first and the later acquired languages; the experimental methods of cognitive neuroscience are thus best suited to pursue this question in depth.

Similarity to the native language facilitates native-like processing of a second language, although it is not yet known whether this effect is due to transfer from the native language representation to non-native language representation (Sabourin *et al.* 2006), or to the use of shared neural networks for both languages (MacWhinney 2005). Studies on proficient bilinguals have shown that violations of morphosyntactic contrasts in the non-native language that are shared with the native one elicit stronger neural and behavioral responses as compared with violations of contrasts that are not shared by both languages (Sabourin and Stowe 2008).

When we consider research on syntactic processing that measures the electrophysiological activity of the brain by means of evoked response potentials (ERP), we also find a noisy landscape, with studies that do not find age-effects in non-native syntactic processing, and studies that do. A review of ERP studies in bilinguals indicate that the ERP components characteristic of language processing are modulated by age of acquisition and language proficiency of the bilingual person. That is, the "native-likeness" of a bilingual hinges upon the onset and extent of exposure to the languages (Moreno *et al.* 2008). Some indications that age impacts bilingualism have obtained in ERP studies on syntactic anomaly detection, where results suggest that late bilinguals fail to develop automatic short-latency syntactic processing mechanisms for the non-native language.

Some ERP studies report that syntax yields differences between natives and non-natives, because they detect different ERP signatures in non-native speakers for certain syntactic tasks, as compared to natives. But other studies report that very proficient speakers are indistinguishable from natives because they show the same electrophysiological components as natives (Steinhauer *et al.* 2009, see Kotz 2009 for a review).

The evidence is still scarce and the picture that emerges is incomplete, probably because all kinds of morphosyntactic phenomena have been considered on equal footing, without resorting to linguistically motivated criteria given current syntactic theories when the phenomena to test are selected. A review of the literature on ERP studies of native versus non-native syntactic processing suggests that neurocognitive differences obtain when the grammars of the bilingual differ in a linguistically significant way with respect to the syntactic property tested, and there has been a time delay in the acquisition of the non-native syntax; however, non-native speakers appear to process syntax in a native like fashion when the syntactic property tested in the second language has an equivalent correlate in the native language of the subjects investigated. This does not necessarily entail that all syntactic differences between the languages of the bilingual lead to detectable native versus non-native effects, nor does it entail either that all grammatically significant differences should be sensitive to non-native acquisition (Watenburger *et al.* 2003).

For example, and without intending to be exhaustive in this review of ERP studies, Weber-Fox and Neville (1996), Mueller *et al.* (2005), Ojima *et al.* (2005), and Chen *et al.* (2007), all found non-native effects that obtained when very proficient non-native speakers were processing syntactic phenomena that had no identical correlate in their native language: in the case of Weber-Fox and Neville (1996), non-native effects obtained when testing native Chinese speakers with subjacency effects in English Wh-questions. But Chinese is a Wh in-situ grammar that lacks overt Wh-movement while English is an overt Wh-movement language (see Cheng 1997). The syntactic phenomenon tested thus involved a property absent (or valued otherwise) in the native language of the participants. Mueller *et al.* (2005) tested Japanese classifier morphology, which German lacks, in Japanese natives versus German natives learners of Japanese. In Ojima *et al* (2005) and Chen *et al.* (2007), the phenomenon tested was verb agreement, in proficient bilinguals of English who were native speakers of grammars that lack verb-agreement morphology (Japanese, Chinese). This initial picture strongly suggests that it is in systematic and true grammatical phenomena that diverge in the two languages of the bilingual that we might find native versus non-native processing differences. Although, as we have seen, both age of acquisition and proficiency have been hypothesized and scrutinized as relevant factors conditioning non-native syntactic processing, less attention has been paid so far to the issue of what syntactic phenomena are tested, and why.

In the last decades, a rapidly growing body of studies using experimental methods and neuroimaging techniques has explored syntactic processing, and as a result, findings from linguistic theory and the neurosciences are progressively reaching increasing levels of convergence and reciprocal relevance (Marantz 2005, Moro 2008, Pullvermüller 2002). However, the vast majority of language processing and neuroimaging studies still focus on typologically similar languages (English, Spanish, Italian, French, German, or Dutch, for instance). With the exception of a few recently emerging studies on Japanese, Chinese and Korean, the languages most intensively studied share many parametric properties. In Linguistic Theory, a significant expansion of the language pool investigated, and systematic cross-linguistic inquiry was crucial to uncover the interplay between universal and variable aspects of the language faculty (Greenberg 1963, Chomsky 1981). Research on language representation and processing in the brain should similarly benefit from cross-linguistic studies pursued with criteria rooted in syntactic theory, so that we can differentiate language-particular, parameter-driven effects from universal, invariant properties, and understand the interplay between the two. In order to achieve this goal, it is necessary to conduct studies and gather evidence from a wide array of languages pertaining to different typological groups, and it is crucial to study bilinguals whose languages have opposite parametric specifications.

In a recent review of ERP studies on non-native language processing, Kotz (2009) concludes that "it is necessary to consider and investigate multiple structural subtleties at the linguistic and the neurophysiological level". In Generative Linguistics, one prevalent view of cross-linguistic variation, the Principles and Parameters (P&P) approach, holds that specific grammars result from combinations of a finite set of linguistic parameters. Thus, syntactic variation results from differences in the values of this combination of parameters (Chomsky, 1981, see Baker 2001, 2003 for overviews), and the acquisition of syntax would consist in determining the values of these syntactic parameters for the input language. Given the P&P model, native/non-native effects, if at all, should be expected only in computational components of language that are subject to specification via input (i.e. parameters), becausc in this approach language distance results from differences in the values of a finite number of discrete parameters. If the task of the language learner is setting the values of parameters, then non-native effects might arise with respect to specific parameters, when the native and non-native grammar differ in their value.

Within this approach, Zawiszewski *et al.* (2011) explored native and non-native syntactic processing in a group of native Basque speakers compared to a group of early and very proficient Spanish-Basque bilinguals, paying special attention to the parametric distance factor. To this end, the study compared how native speakers of Basque and early highly proficient non-natives whose native language is Spanish process certain core parameters of Basque syntax that either diverge from or converge with Spanish syntax. Natives and non-natives behaved alike in those tasks that involved the same parametric value for Basque and Spanish (such verb agreement), but differed in tasks that involved syntactic parameters of opposite values in the two languages such as the head parameter or the case system (nominative in Spanish versus ergative in Basque). The results hence suggest that divergent parameters have a deeper impact in non-native syntactic processing than other seemingly variable but superficially different aspects of language variability.

Clahsen and Felser (2006) propose that only non-local dependencies (such as antecedent-trace relations) yield non-native effects in syntax, whereas local dependencies (such as verb agreement) are processed native-like by non-natives at high degrees of language proficiency and use, regardless of the specifications of the native language. This claim has been contested by Steinhauer *et al.* (2009), who discuss evidence that syntactic processing becomes native-like at high proficiency, some of which has also been discussed in the previous section, and argue that language proficiency is the ultimate determinant of differences between native speakers and second language learners. There is wide agreement that second language learners tend to rely more on lexical-semantic information and less on syntactic structure to process various morphosyntactic phenomena, a difference that is reflected in the generation of the N400 ERP component in learners for those morphosyntactic violations that generate either a LAN-P600 biphasic pattern or a P600 in natives. This difference between natives and second language learners is repeatedly encountered across different studies on non-native syntactic processing, even at the earliest stages of language learning in adults (see Steinhaur *et al* 2009 and MacLaughlin *et al.* 2010 for reviews).

8. Is there a critical period for second languages?

Ever since Lenneberg (1967) suggested that there is a critical period for language acquisition, the impact of age of early linguistic experience for adult neural representation and processing has been a much debated issue. Concerning first language acquisition, there is wide agreement among language researchers that there is a critical window for language, which closes around puberty. For instance, Mayberry *et al.* (2002) showed that deaf adults that had had early language experience achieved native like proficiency in American Sign Language, while deaf adults that had no early language experience did not. See Hagen (2008) for an argument that the critical period for language acquisition is an adaptive trait in human evolution, tightly tied to the biological roots of the language faculty, but crucially involving only native languages learned during childhood.

However, the status of the critical period for language acquisition has been and still is intensely debated in bilingualism, and very especially in the field of Second Language Learning (SLA), that focuses on late childhood, pubescent or post-pubescent language learning once a native language has been fully acquired. Whether a second language learner's linguistic knowledge is represented and processed like the native one has been intensely debated for almost two decades (Gregg, 2003, White 2003); researchers report similarities in the production of second language learners and children acquiring their native language, which cannot be accounted for in terms of transfer from the learners' first language (White 2003). These production data favour hypotheses that place great emphasis on the similarity between first and second language grammar-building (Leung 2007). Recently, however, cognitive neuroscience methods have started to be used in second language learning research, and have provided new evidence on the similarities and differences between first and second languages (Abutalebi, Cappa and Perani 2001, Mueller *et al.* 2005, Diaz *et al.* 2008, McLaughin *et al.* 2010). Also, as discussed in previous sections, both electrophysiological studies (ERPs) and functional brain imaging studies (fMRI) reveal the general picture that both proficiency and age of acquisition has an impact on the representation of language in the bilingual, with an increase relevance of age as the time lapse between the native and non-native languages increases. This general picture has generated growing agreement in the field of SLA that the distinctiveness of second language learners do not involve lexico-semantic aspects of the second language, but do involve some aspects of syntax, although their extent and nature are still poorly understood. Given the uncontroversial status of proficiency as a relevant factor in the representation of language in the brain, it is worth underlying the finding that age of acquisition also makes a difference, for this is the point of debate in the field of SLA, within applied linguistics.

It must also be kept in mind that proficiency and age of acquisition are likely to be correlated factors. Hence, the possibility exists that second language learners achieve higher proficiency levels at earliest ages of onset of learning. The impact of individual variability is also a potentially relevant, though still obscure factor in the neurobiology of language. For instance, Golestani *et al.* (2006) studied syntactic processing in second language learners tested at 20-28 years of age, who started to learn English at the ages of 10–12 and studied it in school for a total of 5 to 7 years. The proficiency levels varied across individuals. During syntactic production, activation in Broca's area of the second language increased as proficiency increased. As the authors discuss, this result could be due to individual differences, so that more grammatically proficient bilinguals can use cortex that is more "tuned" for native-like processing, due to either their architectonics and pattern of connectivity, or to differences in the critical period window involving degree to which individuals recruit more "optimized" neural representations or processes for a second language. This evidence points to the individual variability in non-native syntactic processing.

Thus, the age at which language, whether native o non-native is acquired has significant effects in the domain of phonology and syntax, the two computational components of language.[3] This variable affects not only the acquisition of phonological or syntactic representations, but also the processing mechanisms. Wartenburger *et al.* (2003) argued that different linguistic components are affected by age of acquisition to different extents, without affecting semantic information. Hence, there is a growing consensus that, given a relatively high degree of proficiency and use, age acquisition does not impact the lexical-semantic component, it impacts phonology, and it also impacts syntax in ways that still need to be better understood.

In comparison to child/early learners, young-adult/late learners display selective problems in phonology, that are most obviously manifest in foreign accent: Oyama (1976) did a correlational study of foreign accents among immigrants to the United States; the variable age at arrival was a strong predictor of degree of accent but length of stay in the USA was not (see also Flege *et al.* 1999). This effect emerges also in morphosyntax (Flege *et al.* 1999, White 2003). Johnson and Newport (1989) tested of knowledge of English grammar among Korean and Chinese immigrants to the United States, and found that proficiency correlated negatively with age of arrival; those who arrived at an earlier age were more likely to have

[3] Birdsong (1999), Birdsong and Molis (2001), Flege (1999), Flege, Yeni-Komshiam and Liu (1999), Newport (1990, 1991), Sebastián-Gallés and Bosch (2002), Sebastián-Gallés, Echeverría and Bosch (2005), Sebastián-Gallés and Soto-Faraco (1999), Weber-Fox and Neville (1996, 1999); for a review, see Sebastián-Gallés and Kroll (2003).

a full mastery of English grammar, and this command decreased as age of arrival increased. In sharp contrast lexical-conceptual processing (Weber-Fox and Neville, 1996; Hahne, 2001; Wartenburger *et al.*, 2003).

Age effects in language have been explained in a variety of ways (Birdsong, 1999), including the loss of language-specific learning mechanisms (Bley-Vroman 1989, Pinker and Prince 1994), the advantage of small working-memory capacities in childhood, such as the "less is more" hypothesis (Newport 1993), and "neural commitment" or "entrenchment" and consequent interference of second language by earlier learned knowledge (Marchman, 1993).

Among the neurobiological proposals to account for age-related differences in language, an influential ones is the model advocated by Ullman (2001b, 2004), in terms of procedural versus declarative memory. In Ullman's DP model, the brain systems underlying two well-known memory capacities, declarative and procedural memory, also subserve aspects of the mental lexicon and the mental grammar. The grammar is subserved mainly by procedural memory, whereas lexical-semantic knowledge is mostly subserved by declarative memory. These two systems have a maturational pattern such that the brain can incorporate new knowledge to the procedural memory system easily in childhood, but this capacity is diminished in favour of declarative memory after puberty. In this view, the difficulties experienced with syntax by second language learners are interpreted as a consequence of the fact that, especially after puberty, the grammatical/procedural system is less available than lexical/declarative memory (Ullman 2001b).

This dual system for language has been incorporated to models of bilingualism like Paradis (2004), who joins Ullman's claim that the ability to incorporate knowledge into procedural memory decays in adulthood. Post-pubescent language learners rely more on the declarative memory system to compensate for the decline in procedural memory. The DP model could account for the different loci of language activation in early versus late bilinguals, since declarative and procedural memory involve distinct brain regions, and also for why the acquisition of semantic knowledge—subserved by declarative knowledge—remains unaffected by age, and is only sensitive to proficiency.

The impact of age in language remains somewhat controversial, particularly in educational and social studies, although it is much more accepted within the neurocognitive community. Whether the impact of age in language is due to a critical period for language acquisition is still a matter for debate, particularly in the case of a second language, where some researchers accept it as an effect of the critical period for language acquisition, but others interpret it as reflecting less exposure and practice (and less proficiency).

9. Conclusion

The curse that allegedly dispersed humankind in the biblical story of Babel could never work, because humans are not necessarily monolingual. In fact, humans are distinctively and characteristically capable of learning, knowing and using more than one language, a trait that is unheard of in other species with communicative systems. Given this inherent feature of the human language faculty, linguistic diversity does not constitute an insurmountable obstacle in the path of human cooperation, and Babel's curse can be easily overcome.

Despite the fact that bilingualism is pervasive in our species, not much is known about its neural underpinnings, though its study is a fast growing research area, of which we can only provide a partial view. Bilinguals and multilinguals are particularly relevant subjects of study to understand the interaction between aspects of language that are independent from experience, part of the human genetic endowment, and those that depend on experience, and yield language diversity. In their review on native versus non-native language processing, Clahsen and Felser (2006) stated that "non-native language processing has long been the subject of much speculation and little empirical investigation." The advent of experimental methods, the increasing number of researchers that take part in the goal of understanding bilingualism and multilingualism, and the ever growing variety of languages and types of bilinguals studied have radically changed this picture. Bilingual and non-native language processing is a thriving area of discovery today, as reflected in these words from neuroscientists that study bilingualism:

> Possibly, in an unexpected twist, it is the study of bilinguals that may reveal the language processing potential not fully recruited in monolinguals and lead us to the biological extent of the neural tissue underlying all human language. (Kovelman *et al.* 2008:1468).

As we have seen, a brain that holds more than one language does not keep the two languages separate, isolated one from the other. Instead, when the two languages are acquired simultaneously or nearly so, the brain hosts them together, in the same neural tissue, as a monolingual brain would do with a single language. As age of acquisition increases and proficiency decreases, the representation of the language is less specific to language areas and more widespread, indicating less computational efficiency. Sustained bilingualism enforces specific neural networks for the executive control system that differ from those in monolinguals; as a result, bilinguals have enhanced cognitive capacities in these domains, which can detectably delay symptoms of neurodegeneration, given the compensatory resources available to the lifetime bilingual. Languages are simultaneously activated in the brain and inhibited if not required; again, early, sustained and balanced bilingualism yields distinct and more efficient neural mechanisms for language control in language areas, while other types of bilingualism do not. Proficiency and frequency of use are significant factors in neurobilingualism, in all language components. Although there is much we do not know at present about syntactic processing in bilinguals, what we know indicates that variable (parametric) aspects of syntax may be sensitive to age of acquisition (similar to what is found in phonology), whereas other invariable aspects of syntax are not, provided a native language has been normally acquired. Lexical-semantic aspects of language are impervious to age, only sensitive to proficiency.

These findings suggest that the brain is not very much interested in separating one language from another. Rather, the brain appears to compute and store "language" as one single cognitive function, with phonology, syntax and semantics as somewhat distinct components. Only when a language is learned late, and given less exposure and use, does the brain start to signal a difference, and treat this second language distinctively, locating it more broadly, less specifically, and hence computing it less efficiently.

Any thriving research field is full of disagreements and debates, and the neurocognition of language is no exception. However, there is ample agreement that research on bilingualism and multilingualism are privileged windows into brain plasticity, critical learning periods, and the degree to which language is constituted of specific neurocognitive substrates. Linguistics has undoubtedly much to contribute in this challenging and fascinating endeavour. For those linguists interested in the study of the neurobiological foundations of language, bilingualism and multilingualism offer a unique opportunity to explore the nature of those properties shared by human languages, and those that vary and are patently dependent on experience. As we discover this, we will learn how grammar is shaped by cognitive forces, and how the human brain evolved its computational capabilities to beget words, and knit them into sentences of limitless expressive power that allow us to wonder what it is like to have more than one language in the brain.

PART III

LANGUAGE AND THE SPECIES

INTRODUCTION TO PART III

UP THE CUDGELS FOR LENNEBERG!

SERGIO BALARI*

He tried to become original [...],
—original, to steal Our Lord's authority,
to command his own destiny, to bear his own light!
—William Gaddis, 1955

What did you expect?
—Uma Thurman (in a TV commercial)

Georges Canguilhem once wrote that "L'histoire pure de la botanique au XVIIIe siècle ne peut comprendre sous le nom botanique rien de plus que ce que les botanistes de l'époque se sont assigné comme leur domaine d'exploration" (Canguilhem 1977: 14).[1] Contrary to what it may seem at first, Canguilhem's is not a statement in defence of scientific relativism—far from it. Indeed, few philosophers of science would maintain that a necessary and sufficient condition for scientificity is the claim by the practitioners of some discipline that it is a science.[2] The question Canguilhem is raising is orthogonal to criteria of scientificity and rather

* This paper is a refurbished and highly revised version of Balari (2011). I wish to thank Guillermo Lorenzo for his comments to that earlier version and to Cedric Boeckx, María del Carmen Horno and José Luis Mendívil for their kind invitation to participate in this book. This work has been partially supported by two research grants from the Spanish Ministerio de Ciencia e Innovación to the Universidad de Oviedo (FFI2010-14955, with FEDER funds) and the Universitat Autònoma de Barcelona (FFI2010-20634), and by research grant 2009SGR1079 from the Generalitat de Catalunya to the Centre de Lingüística Teòrica of the UAB.

[1] "The pure history of botany in the 18th century can comprise under the name botany just what botanists of the time considered to be their domain of inquiry." (My translation; SB).

[2] With the exception, perhaps, of Paul Feyerabend in some of his moods (e.g. Feyerabend, 1975).

concerns the right of a community to proclaim themselves as such and to define what is their object of inquiry, a right that Feyerabend (1970a) defended most vehemently by quoting Bakunin, but which none of his contemporaries in the philosophy of science ever actually objected to (see, e.g. Kuhn 1970a; Kuhn 1970b; Popper 1963, among others).

I produced a fair number of false starts for this brief introductory note before I eventually made the decision that this was the most appropriate one. The reasons for this settlement are that, in my opinion, there is an upward trend nowadays within the language sciences (in the broadest sense of the term) to deny this right to biolinguistics and to those who claim to be its practitioners. I won't get into a detailed analysis of these acts of intellectual imperialism—but, for a taste of it, the reader is referred to Koster (2009) and Postal (2009) for two recent, and quite different, examples. This is not the time or the place to enter into such debates, which, I'm afraid, in the end just boil down to terminological questions concerning what "language" is or is not.

I'd rather do something else.

Chance wanted that, while doing some preliminary reading for the preparation of this text and while dusting my personal copy of Eric Lenneberg's *Biological Foundations of Language*, my eyes fell on the following lines from the Preface:

> We may pursue investigations that concentrate on what man has done with or to specific languages; or we may regard language as a natural phenomenon—an aspect of his biological nature, to be studied in the same manner as, for instance, his anatomy. Which of these approaches is to be chosen is entirely a matter of personal curiosity (Lenneberg 1967: vii).

So, Lenneberg seems to have anticipated the attacks to his "naturalization program" of the kind I mentioned above and he guarded himself against them from the outset by proclaiming his right to define his object of inquiry. These words prompted me to re-read Lenneberg—now with the mind of someone who sees himself as part of the biolinguistic community co-founded by him—and try to gauge the reach of his foresight.

The results are impressing, and for the purpose of illustration I compiled a number of quotations from Lenneberg's book, which I reproduce below:

> a. "A major objective of this monograph is [...] to show that reason, discovery, and intelligence are concepts that are as irrelevant for an explanation of the existence of language as for the existence of bird songs or the dance of bees." (p. 1)

b. I believe there is evidence that behavior has the same history and the same origin as form and physiological processes; in fact, the division between physiological function and behavioral function is an artifact of our mode of looking at animals, and these functions shade into each other and are, thus, objectively indistinguishable. (p. 3)

c. [T]he systematics of behavior do not have the same hierarchical relationships. Discontinuities and unique traits are common: specializations of behavior seem to deviate more markedly from general patterns, and in many cases the specializations are so pronounced that the abstraction of general behavior types is impossible or hazardous. (p. 27)

d. In general, it is not possible to assign any specific neuro-anatomic structure to the capacity for language. However, this capacity may be due to structural innovations on a molecular level. Language is probably due to the peculiar way in which the various parts of the brain work together or, in other words, to its peculiar function. (p. 72)

e. In order to reconcile the experimental findings regarding the relationship between the cerebral cortex and behavior in animals, and to relate those findings to our clinical experience with man, it appears reasonable to assume that complex, species-specific behavior patterns, such as language, result partly through subcortical, highly centralized integrating mechanisms, and partly through interaction of activities of the most rostral levels. (p. 222)

f. A study of design features may give us insight into some of the biases that enter into the process of natural selection, into the biological usefulness of certain features of animal communication but it is not relevant to the reconstruction of phylogenetic history. For the latter we are only interested in the relation of types of anatomical structure (including molecular structure) and physiological function (including motor coordination and sensory acuity), but we disregard *usefulness* or *efficiency* of these features to the contemporary form. (p. 234; emphasis in the original)

g. [I]t is possible to talk about language in connection with genetics without having to make shaky assumptions about "genes for language". (p. 244)

h. Our present capacity for language may well go back to species-specific alterations in genetic material (intracellular changes) which, however, affected certain rates and directions of growth during ontogeny (temporo-spatial gradients), producing a peculiar ontogenetic phase of an optimal confluence of various abilities. (p. 265)

i. In the mechanisms of language we find a natural extension of very general principles of organization of behavior which are biologically adapted to a highly specific ethological function. (p. 324)

This selection just lists each quotation as it appears in the book, ordered by page number; for the sake of brevity, I mutilated some of them, although the reader may easily verify through the page references the general context in which they appear.

I believe that from this collection of quotations one may distil a research methodology for biolinguistics that is entirely up-to-date and valid today, one that might perhaps be summarized in two words—*interdisciplinarity* and *development*, but that I'd rather try to break up in a series of basic principles:

a. Functional/behavioural considerations are of little use unless they come coupled with detailed structural analyses at all biological levels. [from (a–c) and (f)]
b. In particular, phylogenetic inferences established on the basis of functional/behavioural analyses are useless because it is impossible to define robust functional/behavioural categories—the comparative method must be applied at the different levels of structure (from genes to anatomy) and activity (development, physiology, etc.). [from (c) and (f)]
c. Naïve notions like "genes for language" and "language brain areas" must be eschewed, since little specialization both at the genetic and neuroanatomical levels is expected. [from (d–e) and (g)]
d. It may well be the case that the structural and organisational principles underlying language are not unique to humans or to language and that homologues exist in other animals. [from (1e) and (1i)]
e. The emergence of language may be the result of heterochronic perturbations that affected an ancestral developmental system, thus creating a new developmental pathway favouring the structural changes that made language possible. [from (h)]

It is sad to see, almost 45 years after the publication of *Biological Foundations of Language*, how little influence Lenneberg's recommendations have had on the language sciences. Indeed, a quick review of most of the work done since then shows that research either proceeded along the lines of what Boeckx and Grohmann (2007) characterized as "business as usual" (with, perhaps, the appropriate buzzwords—like for example "genotype" or "phenotype"—being dropped here and there) or, when evolutionary considerations have been on focus, more often than not the paths taken ran in the opposite directions from the ones suggested by Lenneberg. I'll spare a list of references to the reader since the majority of the most relevant ones already appear in the bibliographies of the three papers that make up this part.

And, thus, we come to the papers, which, interestingly, have the common feature of not following the highly consolidated (but not necessarily well-grounded) trend of the past years and, rather, come closer to the kind of methodology Lenneberg proposed for the biolinguistic enterprise in 1967. In the following paragraphs I'll offer a brief summary of each paper, where I'll try to highlight their main points and how these relate to Lenneberg's principles.

Antonio Benítez-Burraco (in Chapter 7 *The "Language Genes"*) strikes what is perhaps a deadly blow to such naïve or folk notions like "language genes" and "language brain areas". In his detailed, accurate and well-documented study, Benítez-Burraco shows us how little sense it makes to assume the existence of genes specific to language. At the same time, however, and taking a developmental perspective, he demonstrates that genetics is an invaluable source of information to understand how the brain areas that participate in the computational processes underlying language develop and organise themselves. In the final section, the author offers an insightful reflection on how the biolinguistic perspective may change our way of understanding some notions, like "innateness" and "modularity," that have played a key role in the language sciences for the past fifty years. Without actually eschewing them, Benítez-Burraco maintains that they must be substantially revised even to the extent to make their past and present meanings actually incommensurable. This point, to which I'll come back below, is, I think, another important common feature of the three papers.

Guillermo Lorenzo (Chapter 8, *The Evolution of the Language Faculty*) presents a fresh and renewed perspective on how to approach the problem of the "Origins of Language". By expounding a series of arguments—very close to the ones already developed by Lenneberg, although also extending them—against the application of the comparative method at the functional/behavioural level, and turning to such venerable notions as the pre-Darwinian concept of homology, he shows how a number of new avenues open up for evolutionary linguistics. Lorenzo demonstrates that, despite the fact that such notions as "functional" or "behavioural homology" make little sense, homology is perfectly applicable at the dynamic level of organic activity, physiological or, and here lies the core of Lorenzo's proposal, computational, where the latter is the organic activity associated to nervous systems. The author's conclusions are striking and, putting again strong emphasis on the importance of development, he contends that several homologues of "language" may exist in nature, even within evolutionarily very distant species.

Finally, Bridget Samuels (in Chapter 9 *Animal Minds and the Roots of Human Language*) runs a project that in many senses may be seen as parallel to Guillermo Lorenzo's. She identifies a number of "design features" of human phonology, defined as specific computational tasks that all phonological systems exhibit. She then sets to determine whether other animals are also capable of performing these tasks by examining a variety of behaviours, not necessarily of the communicative kind. Her conclusions, as the reader may already have anticipated, are that all these abilities, without exception, are found in some species or other, although it is perhaps the human species the only one in which all of them are present together. In other words, homologues of all characteristic features of human phonological systems exist in nature, although, of course, with other, different functions and associated to different, not always comparable behaviours at a superficial level.

From these necessarily brief summaries I hope nonetheless that readers will be able to appreciate, as I already advanced above, that these papers, in spite of their apparent disparity, share much more than just being direct heirs of Lenneberg's methodological principles. I am tempted to use the words "intellectual honesty" and leave it there, but I feel that I should expand on what I mean by that.

The point I want to stress is that the three authors start off with a more or less common set of assumptions, a basic frame of reference that most researchers working in the Chomskyan tradition would accept without (much) hesitation, and where words like "module," "language" or "faculty of language" have a fairly stable meaning. As their analysis proceeds, however, the authors find themselves in the need to revise and refurbish this original framework, often seeking new meanings to the original concepts, even to the extent of eventually denying what Canguilhem (1977) would have called their *scientific ideology*. This I take to be a very important feature of the biolinguistic methodology established by Lenneberg and followed by our three authors here. It is where, in my opinion, resides the originality of biolinguistics—in its commitment to taking biology seriously, without fear of instigating a crisis capable of shaking the current foundations of the language sciences.[3]

---

[3] As an aside, I surmise that this explains some (if not most) of the hostile reactions against biolinguistics, for its being perceived as a discipline with suicidal tendencies. This cannot be further from the truth, however, and, as pointed out by Canguilhem (1977) this readiness to renounce to its original scientific ideology is a common feature of the biological and life sciences, where change is fast and new discoveries are often challenging. It also refutes Jackendoff's (2011) contention

To conclude, let me, for once, quote Lenin:

> Two very important practical conclusions follow from this: first, that in order to accomplish its task the revolutionary class must be able to master all forms or aspects of social activity without exception [...]; second, that the revolutionary class must be prepared for the most rapid and brusque replacement of one form by another (Lenin 1920: p. 95).

Now I propose the readers the exercise of substituting the phrase "scientific activity" for all instances of the phrase "social activity" occurring in the quotation and the word "biolinguist" for all instances of the words "revolutionary class," and they will have a rather faithful description of the methodology of biolinguistics; to this, as we have just seen, we should add something Lenin would perhaps not have approved of, namely the readiness of applying this methodology *at any cost*, even at the risk of eventually denying its original conceptual underpinnings in order to reconstruct them anew.

"*But, hey!*," some will protest. "*This is not linguistics anymore*." "*Why, of course!*," shall we retort. "*It's biolinguistics! What did you expect?*"

---

that involvement in the biolinguistic enterprise somehow implies some form of commitment with Chomsky's Minimalist Program. Hopefully, these papers will show that the biolinguistic agenda substantially transcends that of the MP and that it is in fact quite independent from it.

CHAPTER SEVEN

THE "LANGUAGE GENES"

ANTONIO BENÍTEZ-BURRACO

1. Introduction

In the last fifty years, the hypothesis that language acquisition by the child cannot be properly and fully achieved by just inductive learning, i.e., in the absence of some aprioristic grammatical knowledge, has progressively emerged and become strengthened (Chomsky, 1959a; 1980: 34; Pullum and Scholz, 2002, among many others). A hot topic of discussion regarding this possibility has been the nature and the source (ontogenetically but also phylogenetically) of this aprioristic knowledge we are allegedly endowed with at birth. For many years, Chomsky has conceptualized it as "a common human attribute, genetically determined, one component of the human mind", (Chomsky, 1977: 164), famously labelled by him as the Universal Grammar (henceforth, UG). "Through interaction with the environment", Chomsky states, "this faculty of mind becomes articulated and refined, emerging in the mature person as a system of knowledge of language [i.e. a grammar of his native language]" (Chomsky, 1977: 164). Therefore, it seems, at the outset, that we essentially face a biological enterprise when trying to resolve what Chomsky has also claimed as being the five key questions concerning the nature of language, i.e, what type of knowledge language is, how it is acquired, and how it is used, how it is implemented at brain level and how it has evolved (Chomsky and Lasnik, 1993). In fact, Chomsky has frequently pointed to his conviction that Linguistics should be a branch of Biology (Chomsky 1975b: 123; 1986: 27). But the former quote crucially exemplifies, as well, the particular position which Chomsky has traditionally adopted regarding the somewhat slippery—but at the same time crucial—dichotomy that pervades the whole debate about language acquisition (and also many other similar issues in cognitive science): the nature vs. nurture controversy, or putting it into more conventional terms, the innate vs. acquired dispute. Although we must be cautious here, since Chomsky's claims on innateness can be biologically interpreted in different ways (see Bateson and Mameli, 2007; more on this below), it seems that we are confronted with a particular (and restrictive) construal of what can be considered as innate: in essence, innate will be what is genetically determined. Consequently, to bring about our enterprise (i.e. to resolve the former central questions regarding the nature of language) would not only entail formulating "a restrictive theory of UG [, which] is nothing other than a commitment to discover the biological endowment that makes language acquisition possible and to determine its particular manifestations" (Chomsky, 1977: 164), but more importantly to assume in particular that "linguistic competence [...] necessarily results from the interplay of the genetically-determined language faculty, specified by UG, and the person's (accidental) experience" (Chomsky 1977:164). As a result of what became known as the second main conceptual turn in his own theory about the nature of language (Chomsky, 1986: 6), by the eighties Chomsky had equated such a theory to "a specific theory of genetic factors in the acquisition of this knowledge [of one's own language] or the development of these systems of mental representation [which ultimately constitute such a knowledge]" (Matthews, 2001: 97).

Consequently, Anderson and Lightfoot (1999) postulated the existence of both a linguistic genotype and a linguistic phenotype. The former would be "that part of our genetic endowment that is relevant to our linguistic development", and it is assumed that this "linguistic genotype (what linguists call UG) is uniform across the species (in the absence of a fairly severe and specific pathology)" (Anderson and Lightfoot, 1999: 702). Conversely, the linguistic phenotype would be the linguistic competence finally acquired by the child:

> A grammar represents what a speaker comes to know, subconsciously for the most part, about his or her native language. It represents the fully developed linguistic capacity and is therefore part of an individual's phenotype (Anderson and Lightfoot, 1999: 711).

In essence, language is acquired because "the triggering experience [i.e. Primary Linguistic Data] causes the genotype to develop into a phenotype" (Anderson and Lightfoot, 1999: 712). In sections 3.2, 4.1, and 4.3 we will critically discuss such a conceptualization of the Faculty of Language (henceforth, FL) and also such a theory of language acquisition, both from a linguistic perspective (by briefly presenting non-maximalist characterizations of both phenomena) and from a biological point of view (by critically discussing the gene-centric view of development). But for the moment, we will go on by assuming that the hypothesis of the existence of genes that, when mutated, impair linguistic competence is of maximum interest for linguistic theory. Such a piece of evidence has ultimately come to light as a consequence of the discovery of different cognitive disorders in which only language seems to be impaired and at the same time are characterised by an inherited nature. Among others, dyslexia deserves to be mentioned, as well as SLI (an acronym for *specific language disorder*), SSD (an acronym for *speech-sound disorder)*, and some other rare (i.e. of low prevalence) conditions, such as the Landau-Kleffner syndrome, the rolandic (or sylvian) epilepsy and speech dyspraxia, or the chromosome 22q13 deletion syndrome (for a review see Benítez-Burraco, 2009: 83-227). After several decades of intensive (and sometimes controversial) analyses in phenotypic (i.e symptomatic) terms, neurobiological analyses are currently contributing to a more accurate characterization of these conditions from a clinical point of view, but, of vital importance, also to disentangle their genuine aetiology. At present, it seems that such disorders are actually caused by the mutation of particular genes, which gives rise to different structural and/or functional anomalies in diverse brain areas, which in turn originate in the underlying deficit(s) traditionally linked to these conditions. Therefore, a widespread (and at first glance, fairly attractive) working hypothesis in the field has commonly been the following: (inherited) specific language impairments or disorders are the consequence of the dysfunction of diverse genes, which, affecting (a specific component of) linguistic competence, leaves nevertheless unaffected the remaining cognitive capacities/abilities (even those involved in performance). Consequently, such genes would be those ideally comprising the linguistic genotype. Note that we are (wrongly?) assuming that such specifically linguistic disorders do really exist; and that a univocal and causal relationship does always exist between them and the mutation of the identified genes (but see section 3.2). What is more, we are also (even more wrongly?) inferring that the wild versions of such genes are (univocally and specifically) involved in the regulation of the development (and functioning) of what we conventionally call "language areas" (which would ultimately represent the implementation of the linguistic competence at the brain level), and even that neural structures exclusively devoted to the processing of linguistic stimuli do actually exist. These are controversial issues, which will be briefly, but critically, discussed in sections 2.1, 3.2, 4.1, and 4.2. But before continuing by exploring where the genetic analysis of these kinds of conditions can lead us, we will present a brief sketch of the structure of the present contribution.

After this introductory remark, in section 2 we review major methodological issues regarding the way in which genes involved in the regulation of the development and functioning of the neural substrate of the FL are identified, cloned, and structurally and functionally characterised. In section 2.1 we begin by critically reassessing the fact that to achieve this we have usually circumscribed to inherited phenotypes in which only language seems to be impaired (i.e. the aforementioned specific language disorders), even though such specificity is actually elusive at clinical, cognitive, and even neurobiological levels, in part due to methodological caveats (but not only, as we discuss later in section 3.2). Section 2.2 examines the very case for heritability, pointing to the crucial concern that genes are not the only foundation for such inherited traits. In section 2.3 we specifically address most relevant cloning strategies, highlighting their strong points, but also their caveats. Section 2.4 discusses in turn how cloned genes are structurally and functionally characterized in order to disentangle the real biological functions they contribute to.

In section 3, we (briefly) depict the most noteworthy genes presently related to inherited language disorders. In section 3.1 we particularly consider genes related to the (allegedly) foremost specific language disorders. In section 3.2 we review genetic evidence which seems to further put into question their specificity, thus reinforcing the clinical, cognitive and neurobiological evidence previously discussed in section 2.1. Consequently, in this subsection we will finally wonder if such a specificity is either a required feature of language (at all levels of biological complexity), or merely demanded by just particular models and conceptualizations of the FL. We will thus proceed by both critically discussing the key concept of *modularity* regarding the FL (which will be reviewed in depth in section 4.2), and also by introducing fresh Chomskyan ideas about its nature and (biological) properties, which heavily minimize the claim for a grammatical machinery specifically derived from genetic instructions. In accordance, we will finish by arguing for the convenience (and the necessity) of considering broad (i.e. nonspecifically linguistic) disorders when trying to identify genes involved in the regulation of the development and functioning of the neural substrate of language, which we (briefly) examine in section 3.3.

In section 4, we scrutinize the so-called "black box" of development, by discussing the real contribution of genes to the regulation of developmental processes, and the actual nature of the neural structures which emerge at the term of growth. In section 4.1 we specifically look over the diverse biological factors which channel developmental processes and make a claim for a non gene-centric view of development. In section 4.2 we reassess the whole case for modularity, by trying to reconcile both the extensive inspecificity of genes and brain structures related to the FL (as characterised in sections 2.1 and 3.2), the phenotypic idiosyncrasy of language as "one component of the human mind", and crucially, the pervasiveness of modularity as a fundamental property of biological structures at different levels of complexity. In section 4.3 we conclude by discussing why Minimalism is the model of the FL which seems to fit most accurately the way in which developmental processes take place and biological components organize and interact during growth. Moreover, we will claim that the whole scenario depicted in this section is still fully compatible with a (revised) nativist account of language development. The contribution concludes with a reflection (section 5) on the reasons why (improved) genetic/molecular data will still heavily contribute to a better understanding of the real nature of language in subsequent years.

2. How to identify and characterize genes related to language

2.1 Dealing (fiercely) with the phenotype

Because of ethical concerns we cannot deliberately mutate human DNA sequences of interest in order to subsequently establish and evaluate the consequences for language of such mutations at the phenotypic level. Hence, the search for genes related to language must necessarily come as discussed in section 1, from people imbued with a dysfunctional or afunctional linguistic competence. Crucially, any experimental characterization of the competence must necessarily come from an assessment of the performance, i.e., the way in which people use their knowledge of language in particular contexts (*vid.* Chomsky, 1965: 4). In the traditional Chomskyan account of the FL, performance is the output of the interaction (i.e. the interface) between competence and other cognitive devices (Chomsky, 1981: 18). Consequently, a non-trivial problem arises at this very initial stage: which precise phenotype(s) should we consider for our genetic analyses—and which other(s) should we keep aside, and particularly, in which precise way are we going to assess linguistic performance (and inferring linguistic competence) of our probands? Suppose that we find some children who are unable to correctly resolve the type of tasks conventionally employed to evaluate the achievement of the usual milestones during language acquisition, but who are also unable to resolve simple mathematical problems or even draw elementary geometrical figures, these deficits being suggestive of an impairment of other cognitive domains besides the FL. Suppose, as well, that we discover that this is a hereditary condition. Are we taking this disorder into account when trying to characterize the linguistic genotype? Plausibly not, since we could argue that in these people linguistic deficits are secondary to their mental retardation. Similarly, a hearing deficit can affect the normal acquisition of language, but we would not probably regard the resulting anomalous performance as a consequence of any alteration of the linguistic genotype, to the extent of not including it in the diverse genes plausibly involved, for instance, in the development of the inner ear. Hence, and paraphrasing Chomsky's quote in section 1, we seem to be impelled to only search for people whose impaired UG, after becoming somewhat articulated and refined by linguistic stimuli, emerges in the mature person as a damaged system of the knowledge of language, which ultimately gives rise to different errors when confronted by linguistic tasks, but at the same time, who are not impaired in any other domains of cognition. In other words, we seem interested in what are clinically labelled as (inherited) specific language impairments and disorders, i.e., conditions that are present in subjects who exhibit an abnormal process of language acquisition in the absence of nonlinguistic causes, such as acquired neurological dysfunctions, cognitive (in a broad sense) or mental delay, auditory impairments, or an inadequate or insufficient exposure to linguistic stimuli during development, plausibly as a consequence of the socio-educational peculiarities of the milieu in which they have been reared.

Defining and characterizing at the phenotypic level these kinds of clinical categories, and accurately examining the linguistic abilities of a particular experimental subject to correctly assign any of them to her are not trivial concerns. An inappropriate delimitation, characterization and/or categorization of language impairments and disorders at this level will crucially condition the relevance and the meaning of the results ultimately derived from their analysis at the remaining levels of biological complexity, particularly, at the genetic level (more below). Hence, we must be cautious here, and be aware of the different problems and caveats we routinely face when trying to do so.[1]

To begin with, since the experimental analysis of competence must necessarily come from an assessment of performance, a main objective at this stage should be a maximization of the linguistic nature of the

[1] For further and perhaps unexpected implications of these concerns, regarding, in particular, the implementation of the FL at the brain level, see sections 3.2 and 4.2.

experimental tests designed to evaluate the linguistic capacities of the subjects, in order to minimize the effect exerted on performance by the cognitive or pragmatic factors necessarily involved in communication. Note that it has been suggested that an experimental differentiation between competence and performance might be impossible, due to the obliged implication of other cognitive systems (perhaps not yet characterised) in passing from the former to the latter (Newmeyer, 1997); in fact, we will return to this scenario in section 3.2, when discussing an alternative depiction of the FL indebted to some of the recent papers by Chomsky. Incidentally, such tests should be designed, if possible, to allow us to discriminate as a minimum among what we usually categorize as the diverse functions or functional components of competence (phonological processing, syntactic processing, etc.), but ideally also among diverse linguistic categories and phenomena of interest (nouns, verbs, agreement, etc.). Obviously, when tests employed for the diagnosis cannot suitably differentiate competence from the remaining cognitive abilities involved in performance, we can wrongly conclude that the nature of a particular disorder is cognitive (in a broad sense) and/or that the condition is comorbid with other language disorders and/or other cognitive disorders. Another common problem at this level is that there are usually different tests routinely used for the diagnosis of the same disorder: the possibility that such tests are based on dissimilar criteria can lead one again to erroneously conclude that several subtypes of the conditions exist, and/or the disorder is caused by two (or more) underlying deficits, and/or the disorder is comorbid with other language impairments (and/or other cognitive disorders).

Even in ideal circumstances (let us assume that the perfect test is available and there is only one such perfect test) physicians usually face a somewhat difficult scenario when trying to diagnose these types of language disorders. To start with, notable variations concerning the symptomatic profiles of the affected individuals are characteristically observed. Ultimately, this circumstance has usually led them to postulate the existence of different subtypes of each disorder, allegedly characterised by a more marked dysfunction of one (or some) of the different aspects of competence. Moreover, in most clinical cases the inferred dysfunction of the linguistic competence affects quite general aspects of language, but not necessarily or sharply any of the linguistic functions, entities of phenomena which are nuclear to the different theoretical models of language developed by linguists (Newmeyer, 1997). For instance, in the case of SLI clinical practitioners usually distinguish the following different fundamental categories or basic subtypes: (i) phonological, (ii) expressive o (iii) expressive-receptive (*sic*) (Rapin and Allen, 1993; American Psychiatric Association, 1994). Conversely, notice that the very definition of clinical categories always involves some sort of homogenization of the results obtained in the tests used to diagnose and evaluate them. Although in doing this we habitually follow properly normalised statistical procedures, we must assess if the boundaries which separate the different groups that represent the diverse clinical categories (or even the different subtypes inside each category) are of biological significance, and/or if we are rejecting a variability of real biological relevance, ultimately simplifying the neurobiological (and hence, genetic) nature of the disorder (see Shaywitz *et al*, 2008 on dyslexia for a discussion). Whatever the case may be, there exists a growing body of evidence which does point to the convenience of describing language impairments as concrete intervals within a continuum of which those individuals who exhibit a linguistic competence regarded as normal would also be part of, (to some extent competence is also variable in the normal population!). Such a depiction leaves aside an old-fashioned conceptualization of competence as a cognitive capacity with a bimodal distribution (affected/non-affected), in turn characterising it (but also the diverse clinical conditions in which language is impaired) as continuous (i.e. non bimodal) variables. For diverse reasons, such a characterization is of relevance in the context of the present piece of work (and for any genetic analysis of language). Firstly, it does not significantly simplify their genuine genetic (and molecular) nature, suggesting that only a few principal genes are involved, and that these, when mutated, always give rise to the condition. As we will see in section 3, the number of candidate genes for the diverse language impairments and disorders, of genes which can be considered risk factors for their appearance, and of *loci* (i.e. physical places on the chromosome) which can be linked or associated to them, have exponentially increased over time (Smith, 2007; Benítez-Burraco, 2009: 240-281), each one subtly influencing what is in fact a (genetically) complex trait. Indeed, the characterization of language disorders (and of linguistic competence itself) as continuous variables is at the core of the use of the so-called QTLs (quantitative trait *loci*) for their genetic analysis, which comprises a noteworthy implementation of the genetic analysis of the FL (see below). Secondly, such a characterization allows for different genes to be involved in the emergence of the same disorder in different populations (see section 3.2). Thirdly, it suggests that the variance observed in the normal and affected populations regarding their linguistic competence depends on common genetic factors (in impaired people some of these common genes will be dysfunctional, but the remaining will not). Lastly, it does not minimize the foremost role played by other non genetic factors in the regulation of the development of the neural substrate of the FL (see section 4.1).

On the other hand, another widespread but disturbing outcome of clinical practice is the fact that, even in such ideal circumstances, results obtained by the affected subjects in diverse experimental tasks are still interpretable in terms of simultaneous occurrence (*comorbidity*) of different specific language impairments and of diverse disorders in which language and other cognitive abilities seem to be simultaneously impaired. Some of the disorders listed in section 1 are actually comorbid in a significant percentage of cases. For instance, SLI is frequently comorbid with dyslexia (Smith *et al.*, 1996; Catts *et al.*, 2005) and SSD (Shriberg *et al.*, 1999), but also with autism (Norbury, 1995; Tager-Flusberg, 2006).

Trying to reconcile (and explain) the former symptomatic variability with the latter observed comorbidity, two non-excluding aetiological hypotheses have frequently been pointed out: (i) purportedly specific language disorders are rooted in broader (even non compulsory specifically linguistic) underlying deficits (see Nobre and Plunkett, 1997); such a deficit can manifest itself as a non linguistic disorder in other populations and/or circumstances (hence the alleged heterogeneity and/or comorbidity); and (ii) different—even non compulsory specifically linguistic—underlying deficits contribute to the emergence of a particular disorder (i.e, clinical categories are heterogeneous conglomerates of diverse cognitive deficits with similar symptomatic profiles), some such deficits can manifest themselves as a non linguistic disorder in other populations and/or circumstances (hence the purported heterogeneity and/or comorbidity). For example, an impairment of the phonological memory seems to be the core deficit of SLI (Bishop, 2002), dyslexia (Shaywitz *et al.*, 1998), and SSD (Shriberg *et al.*, 1999). All the same, a deficit in the capacity for processing (and discriminating between) sensorial impulses of an acoustic (linguistic or nonlinguistic) nature which are brief and/or quickly follow one another has allegedly contributed to the emergence of both dyslexia (Temple *et al.*, 2000), and SLI (Tallal and Pierce, 1973; Tallal *et al.*, 1991), but since the deficit is seen as affecting all acoustic modalities, it gives rise to other diverse (non-linguistic) perceptive hearing impairments. Additionally, other non-specifically linguistic deficits could contribute to the emergence of dyslexia, such as a cerebellar dysfunction (Nicolson and Fawcett, 2006), a deficit in the capacity for visual processing (Lovegrove *et al.*, 1980) or a dysfunction of the magnocellular pathway (Livingstone *et al.*, 1991; Stein and Walsh, 1997).

This entangled state of affairs has not been unravelled by recent neuroanatomical and neurophysiological evaluations of affected people. An increasing body of evidence does suggest that diverse brain regions in them are structurally or functionally impaired and the impaired regions normally match the standard "language areas". Such regions are also frequently involved in computational tasks not directly related to language processing, while to some extent their precise identity, extension and boundaries are variable among different individuals (the same holding for "language areas" in the normal population!). The impairment of these purportedly "crucial for language" areas commonly gives rise in other subjects to different deficits and/or different (including non-linguistic) disorders. An illustrative example here could be the ventral portion of the occipito-temporal region, a classic area for dyslexia (Horwitz *et al.*, 1998; Shaywitz *et al.*, 1998; Paulesu *et al.*, 2001), but whose dysfunction gives rise as well to a non-linguistic disorder known as prosopagnosia (Sorger *et al.*, 2007; Dricot *et al.*, 2008). It could be alleged that these shortcomings are just a consequence of the limited resolution of most current non-invasive neuroimaging techniques (usually they do not go beyond 0.1 mm) (Koizumi, 2004); in fact, it could actually be the case that some kind of histological and/or functional dissociation exists among different (but closely located) neural populations in such (apparently) multifunctional areas. It may be the case, but we will continue under the hypothesis that this kind of evidence is linked to crucial aspects of the way in which the brain develops, functions and is structured (see also sections 3.2 and 4.1). Anyway, a related and non trivial concern is also the convenience of (re)assessing the way in which we tend to correlate the structural anomalies detected by such neuroimaging analyses (and especially the abnormal visual outcome of the functional/clinical studies) with the diverse linguistic dysfunctions or dysfunctional components of language. We should be aware that we are systematically interpreting such visual representations as impaired homogeneous linguistic functions/categories and isolating them from other functions/categories of a dissimilar nature, when in fact neural systems are not (so) discrete in functional terms (see Kosik, 2003 and Poeppel and Embick, 2005 for interesting discussions on these topics).

This has been a long digression, just when we were about to grasp the very genes. But in fact, it has not been a pointless detour, but an important insight into very relevant concerns, particularly if we bear in mind that most productive strategies for gene cloning substantially proceed by establishing statistically significant correlations between certain chromosomal fragments and certain phenotypes independently of how (and how precisely) such phenotypes are defined, categorised, and delimited. An imprecise phenotypic definition, categorization or delimitation of one particular disorder can give rise to results that could be interpreted as indicative of the involvement of the same gene in the emergence of more than one different disorder, while in fact this common aetiology would be false. Even more crucially, an incorrect assignment of one particular proband to a particular clinical category (i.e., an inaccurate diagnosis), could be incorrectly interpreted as indicative of the involvement of a particular gene in the emergence of such a condition, while this would not be the case. Let us continue by assuming that specific language disorders really do exist and see how we should identify the involved genes and where we arrive (for a critical reappraisal of such an endpoint, see section 4.1).

2.2. The case for heritability

A second fundamental assumption that we have made so far is that language disorders to be used as the starting point for our genetic analyses are those allegedly transmitted to the probands by their ancestors, with the worthy aim of discarding those impairments due to traumatic brain injuries or tumoral processes (among many other causes) occurring during the individual's life. Nevertheless, we would be mistaken if we only assume that all the observed resemblance at the phenotypic level between parents and offspring is inherited and inheritable (in the sense of emerging during development and relying on transmitted information via the germ line), or that in all inherited traits the transmitted pieces of information are genes. On the contrary, at the cognitive level the connection between heritability, innateness, genes, environment, experience and learning is not as straightforward as a first glance might suggest (Bateson and Mameli, 2007). Notice, for instance, that some acquired traits (namely, reading ability) could be regarded as inherited (in the naïve sense of regularly emerging at concrete stages during child development), while conversely different inborn disabilities (caused by some sort of chromosomal rearrangements, for instance) can be ultimately reverted by therapy. Moreover, most cognitive faculties (paradigmatically, the FL) rely on external stimuli to properly develop, suggesting that the boundaries between the organism and its environment, or between the internal and the external information needed for development could be blurred (more on this in section 4.1).

Conventional tools used to evaluate the heritability of cognitive traits rely on the assumption that this is a statistical concept that applies to the variation of a phenotypic trait existing in a given population at a given time (Falconer and Mackay, 1996; Mameli, 2004). Among such tools, we could mention studies of familial aggregation, analyses of identical and non-identical twins, studies of adopted individuals or studies of kinship (Bishop, 2001; Stromswold, 2001). Notice also the somewhat reductive way these tools proceed. Studies of familial aggregation, for instance, determine the incidence of the impairment among the affected subject's relatives, though some type of the observed aggregation can be of cultural origin. Moreover, analyses of identical and non-identical twins (adopted or not) try to evaluate the incidence of the environment on the emergence of the trait, though this is a changing variable along with ontogeny, as is development itself (more on these issues in section 4.1).

2.3. Cloning strategies

Whatever the case may be, heritability measurements (along with analyses of performance) emerge as a productive tool for a gross delimitation of the traits to be considered and later analysed. We can now make use of diverse cloning strategies to identify and physically isolate the gene (or genes) presumably mutated in the affected probands. The most fruitful of such approaches is the one known as *positional cloning*. It allows us to correlate, in the absence of significant evidence of the aetiology of a particular disorder, the anomalous phenotype with a specific chromosomal fragment, by just measuring the co-heritability of the trait with a suitable number (in statistical terms) of known polymorphic genetic markers. Depending on whether the kin relationships among the experimental subjects are known or unknown (a circumstance that crucially conditions the number of markers to be used), the correlation test is known as linkage or association analysis, respectively (Cardon and Bell, 2001; Francks *et al.*, 2002); association studies provide a further advantage in the sense that the identified genomic regions are much narrower than in linkage studies. Moreover, the analytical yield of positional cloning has been implemented by the recent development of the so-called *genome-wide association studies* (GWASs), which make use of the whole genome and consequently not only make linkage analyses unnecessary, but also allow us to simultaneously establish the presence and the localization of multiple *loci* of susceptibility to a certain disorder (Zondervan and Cardon, 2007; Potkin *et al.*, 2009). Moreover, the categorization of language disorders as continuous variables (thus going beyond the traditional but simplifying dichotomy of *affected* vs. *non-affected*), which we previously discussed (see section 2.1), has allowed us to identify diverse QTLs (quantitative trait *loci*) related to them (Lander and Kruglyak, 1995; Risch and Merikangas, 1996), and ultimately, to detect multiple genes which exert a relatively small effect on a particular trait (Bishop, 2002). QTLs more properly apprehend the variability detected in (and characteristic of) groups regarded either as affected or normal. Finally, the employment of endophenotypes—i.e. cognitive, neuroanatomical, neurophysiological, endocrine or biochemical quantifiable components of the different levels of biological complexity which can be identified and described between the linguistic phenotype and the genes (Gould and Gottesman, 2006)—as the starting point for linkage and association analyses has allowed us to obtain more direct evidence of the links which plausibly exist between certain genes and certain cognitive (dys)functions, as they refer to more concrete (and more physiological) aspects of brain activity (Gottesman and Gould, 2003). Notice however that to an important extent we are implicitly assuming that there is a direct relationship between the genotype and the phenotype, while this is clearly not the case—in section 4.1 we will reject the transparency of such a relationship and will move into the "black box" of development (Walsh, 2007: 179, 184). Conversely, as these endophenotypes count on uncontroversial homologs in other species, they seem to justify the use of animal models when trying to clone genes related to language (see below), but also (and crucially) when analysing the structural and functional features of the cloned genes (see section 2.4). It is worth remembering at this point that language has been regarded—by Chomsky (1968; 1980), but also by many others—as a trait exhibiting a fundamental discontinuity at the phenotypic level with the communicative systems employed by the remaining species. The possibility that genuine language homologs ultimately arise when considering the genetic level (but also other in-between levels of biological complexity of the FL, such as neural structures) emerges as a promising field of research if we are seriously intending to disentangle the evolutionary history of the FL(see also Lorenzo in the present volume).

Anyway, all those otherwise productive techniques cannot be regarded as a panacea either. To begin with, positional cloning merely establishes statistical correlations, but not compulsory causal relationships, between certain genes and certain phenotypes, whose validity is restricted as well to concrete populations and particular environmental conditions (Hofmann, 2003; Fisher, 2006). Once again, there seems to be much more between genes and traits than a direct link. On the other hand, the search for QTLs cannot either properly detect highly polymorphic *loci*, and evidently leaves as well unidentified other non-genetic factors plausibly involved in the emergence of these disorders, such as epigenetic, maternal, or ontogenetic factors (more on this in section 4.1.). Furthermore, since GWAs examine such a vast number of SNPs (an acronym for *single nucleotide polymorphisms*) for association with diagnosis simultaneously, they must incorporate substantial statistical correction, thus losing power to detect genes of small effect—on the contrary, candidate gene studies, which assess only a small number of polymorphisms, while exhibiting a greater statistical power, necessarily demand functional studies in order to test the biological plausibility of the results (Cannon, 2010) (see also section 2.4).

Conversely, the identification of genes related to language (and whose mutation gives rise to different language disorders) becomes simplified when there are evidences of chromosomal rearrangements in the karyotype of the impaired individuals, plausibly affecting some gene sequences at the breaking points. These anomalies can easily be detected by routine techniques such as fluorescence *in situ* hybridization (FISH) (Volpi and Bridger, 2008). Moreover, these kind of genes seem not to be randomly localized in the genome, plausibly because of being coordinately (co)expressed (Ramsay, 2000; Benítez-Burraco, 2009: 197-205). Likewise, when the impaired linguistic phenotype specifically entails the atypical presence, accumulation or degradation of particular biochemical products, it is possible to identify the affected gene by *functional cloning*, if the identity and the biological activity of the enzyme involved in its biosynthesis or catabolism is previously known (generally in other organisms) (Brzustowicz, 1998). Additionally, *comparative cloning* also simplifies the identification of the affected genes (Brzustowicz, 1998). In this case we know the sequences (and the functional properties) of genes that, when mutated, give rise to akin disorders in other species, i.e., disorders characterized by similar structural and/or functional brain anomalies and/or whose phenotypic profiles resemble those observed in our own species at the cognitive level. This leads us again to the question of the homology of the FL across diverse levels of biological organization (see above).

Finally, remember that while

> the discovery that the presence of a specific genetic mutation is correlated in a small population with a specific case of linguistic impairment might suggest that the mutation (or some nearby region of the genome) is causally involved in the impairment the correlation in itself does not tell us whether (and in what way) that region of the genome plays a role in normal linguistic development" (Bateson and Mameli, 2007: 821; see also Balaban, 2000).

We cannot make such (key!) inferences unless we can also assess (as a minimum) that (i) there exists an obligate relationship between the mutation of the gene and the dysfunction of the FL (i.e. the development of the FL is always impaired in all subjects bearing such mutation); (ii) the anomalous phenotype appears as a consequence of the loss of a crucial component (i.e. the functional product encoded by the gene), but not because of the appearance of an interfering component (i.e. the dysfunctional product encoded by the mutated version of the gene); (iii) some of the subsets of language impairment are not an indirect consequence of the mutation of the gene (i.e. all the observed linguistic deficits must share a common aetiology); and (iv) the anomalous phenotype (presumably) indebted to the mutation of the gene is reversed (ideally at every level of biological organization) after the supplying of the wild copy of the gene (but we are not allowed to do such an experiment on humans, only on animal models! (see section 2.4) (Balaban, 2006). As we discuss below (section 3.1) this could not in fact be the case for even those genes typically regarded as "language genes". We must be cautious again in order to avoid that "too little information about biological development is getting converted into too much cognitive explanation" (Balaban, 2006: 299). Consequently, additional studies of examining the identified genes in detail (both structurally and functionally) must be developed in order to unravel (at the minimum) these crucial questions. We turn to this issue in the next section.

2.4. Structural and functional analysis of the cloned genes

Once the relevant chromosomal fragments are cloned, they should be sequenced to settle on the identity and nature of the gene (or genes) they comprise (Brzustowicz, 1998). DNA sequences are routinely subject to intensive computational analyses, in order to obtain as much relevant information as possible about the structural features of the putative genes, and to infer as well the nature and function of the biochemical products presumably encoded by them and even the phylogenetic relationships they exhibit with homologous genes from other (closely and distantly related) species. Candidate genes are subsequently subject to functional analyses (*in vitro* and *in vivo*), to accurately characterise their transcriptional and translational profiles, the biochemical properties of the products they encode and ultimately the physiological role carried out by such products, and (crucially!) the manner in which their mutation contributes to the emergence of the disorder and the way in which the wild allele is involved in the development and functioning of the neural substrate of the FL (see

Gibson and Gruen, 2008 for a comprehensive exemplification concerning dyslexia).

Nowadays functional analyses are commonly implemented by recurring to animal models (typically the rat and the mouse, but also different songbirds). As was previously pointed out (see section 2.3), continuity—i.e. (deep) homology—regarding the FL quickly arises when molecular, neurobiological, and even cognitive levels are considered, thus reasonably justifying the use of such models but also allowing for (deep) comparative studies (including cloning strategies). In such model organisms homologous genes are deliberately disturbed—inactivating them (*knockout*) or altering their expression levels (*knockdown*)—to help to clarify the physiological role of the products encoded by the candidate genes. We are particularly interested in finding neural anomalies and/or articulatory, perceptive, or cognitive deficits which resemble those detected in humans. Dyslexia notably exemplifies how the manipulation of such animal models and the analysis of the resulting phenotypes can contribute to unravel the link existing between the mutation of certain genes (usually identified by positional cloning) and the neurobiological and phenotypic profiles of diverse language disorders, and ultimately between certain genes and the FL. In rats and mice an induced decrease of mRNA levels of the candidate genes for dyslexia *DYX1C1* and *DCDC2* (see section 3.1) gives rise to structural changes in the brain (plausibly as a consequence of the disruption of the normal pattern of neural migration and interconnection) which are similar to those observed in dyslexic people (Paracchini *et al.*, 2006; Rosen *et al.*, 2007; Burbridge *et al.*, 2008), but significantly to auditory and cognitive deficits as well which resemble those detected in dyslexics (Galaburda *et al.*, 2006; Threlkeld *et al.*, 2007). Eventually it is possible to replace diverse genes of interest with the wild or mutated copies of the corresponding human homologues in order to assess the consequences at different levels and ultimately to infer the functional value of the human versions (in section 3.1 we discuss one experiment of this kind involving *FOXP2*, a well-known "language gene").

Last but not least, special attention is deserved by the elucidation of the functional value of the different variants of a particular candidate gene naturally arising in different human populations. These variants can indistinctly be functional, dysfunctional or afunctional compared to the wild type. Consequently, different alleles and protein polymorphisms can be ideally correlated to different linguistic deficits (this test is known as allelic association analysis) (Wahlsten, 1999).

3. Genes related to language disorders

3.1. Genes related to specific language impairments

In section 1 we listed most relevant cognitive disorders with a phenotypic profile alleged to be specifically linguistic (though see also section 2.1). In recent years, different genes that can be regarded as causal or risk factors for their emergence have been cloned and characterised. A well-known case is *FOXP2*, popularly known as "the language gene" *par excellence*. This gene encodes a transcriptional repressor (Lai *et al.*, 2001; Shu *et al.*, 2001; Vernes *et al.*, 2006) and its mutation gives rise to a plausible subtype of SLI, though this is still a controversial claim (for a symptomatic profile of the affected people, see Gopnik, 1990; Vargha-Khadem *et al.*, 1995; Watkins *et al.*, 2002a; Vargha-Khadem *et al.*, 2005; Shriberg *et al.*, 2006). In any case, there is ample evidence supporting a relevant role of this gene in modulating the development of brain areas implicated in language processing. Firstly, the primary pathology of the disorder associated to the mutation of the gene is located in the caudate nucleus, a subcortical structure which seems to play a key role in the computation of the sequential tasks involved in phonation and syntax (Ullman, 2001a; Lieberman, 2002; 2006; see also below and section 4.2), while at the same time other brain regions also relevant for language processing (Broca's area, Wernicke's area, the angular gyrus, the ventral portion of the cerebellum) exhibit morphological and/or functional anomalies to diverse degrees (Vargha-Khadem *et al.*, 1998; Watkins *et al.*, 2002b; Belton *et al.*, 2003; Liégeois *et al.*, 2003). Secondly, *FOXP2* is expressed (during development and/or in the adult stage) in various brain areas important for language, including the basal ganglia, the cerebral cortex (particularly the frontal and occipital lobes, and crucially the perisylvian region of both hemispheres) and the cerebellar cortex (Ferland *et al.*, 2003; Takahashi *et al.*, 2003). Thirdly, FOXP2 protein seems to be involved in the regulation of the neural differentiation (but plausibly also in establishing the cellular identity and/or function) needed for the correct organization and/or development of certain cortico-thalamic-striatal circuits associated to motor planning, sequential tasks, and procedural learning (for a review, see Marcus and Fisher, 2003; Vargha-Khadem *et al.*, 2005; Fisher and Scharff, 2009), and ultimately to language (see also Lieberman, 2002; 2006, below, and section 4.2). Fourthly, the mutation of some of the gene's physiological targets seems to give rise to (different) language disorders, as *CNTNAP2* regarding SLI (and autism!) nicely shows (Vernes *et al.*, 2008). Finally, in the mouse the mutation of *Foxp2* reduces the long-term plasticity related to learning tasks in which basal ganglia play a relevant role (Groszer *et al.*, 2008; see Yin *et al.*, 2006 for a review of the molecular mechanism involved in striatal long-term depression), but also gives rise to characteristic structural anomalies, predominantly in the cerebellum (French *et al.*, 2007; Fujita *et al.*, 2008; Groszer *et al.*, 2008)[2]. Simultaneously, mice carrying the human (non-mutated) sequence of the gene exhibit physiological effects that are substantially opposed to the former (Enard *et al.*, 2009). Moreover, at the phenotypic level the *knockout* of *Foxp2* typically leads to a decrease in the frequency of the ultrasonic vocalizations in the pups (Shu *et al.*, 2005; Fujita *et al.*, 2007), conceivably resembling the deficit in the capacity for discriminating brief or very close auditory stimuli (and plausibly certain sound frequencies) which is found in SLI (McArthur and Bishop, 2001). On the other hand, it is noteworthy as well that the *knockdown* of *FoxP2* in zebra finch mainly affects neurons of the X-area of the song circuit, a structure which is homologous to the basal ganglia. As a consequence, a significant shortening of the critical period for song learning is observed, but also a reduction of the accuracy of the song itself (Haesler *et al.*, 2007), reasonably resembling as well the reduced capacity for repeating words and pseudowords, and even sentences, which is characteristic of affected people carrying a mutation of the *FOXP2* gene (Watkins *et al.*, 2002a).

Notice that the controversy around the phenotypic profile (and the aetiology) of the disorder linked to the mutation of *FOXP2* is neither trivial, nor an academic debate perhaps restricted to a medical audience. On the contrary, we are confronting three possible aetiological scenarios with diverse implications for our endeavour of biologically characterising the FL: (i) a motor deficit which impairs the normal programming of articulatory movements; (ii) a strictly linguistic deficit (which would ultimately resemble those deficits observed in people affected by "standard" variants of SLI), or (iii) a unique (and broader) deficit affecting both motor and linguistic computational tasks. The first scenario would

---

[2] Remember that this structure plays a key role during language processing as a component of the verbal working memory (Gathercole and Baddeley 1993), plausibly facilitating as well the interface between language and other cognitive domains (Desmond and Fiez, 1998); consider also that this is an area where the gene is expressed and which is structurally and functionally impaired in people carrying a mutated version of *FOXP2*.

confine this impairment to the realm of speech disorders[3]. The second scenario is perhaps the expected one, in as much a putative specific language disorder would be the outcome of a specifically linguistic underlying deficit (for a critical reassessment of the unavoidability of such an assumption, see section 3.2). The third scenario would probably imply the existence of a computational device relying on a pattern generator (or sequencer) capable of indistinctly processing both motor and cognitive elements (a plausible neural substrate for such a device being the basal ganglia; see Ullman 2001a, Lieberman 2000, 2002, 2006; see also sections 4.2 and 4.3).[4] Whatever the case may be, it seems that *FOXP2* "is not the gene for language or even grammar", but represents at the best "a fortuitous entry point of discovery into molecular networks that, in the human brain, contribute to the language phenotype and, in other brains, may contribute to language subsystems" (White, 2010: 26). We will return to this issue in section 4.1.

Mutations of *FOXP2* itself are rare, to the extent that gene sequences are normal in most people affected by SLI. In fact, while the mutation of the gene is linked to a monogenic disorder, SLI is usually regarded as a complex trait, hence being the result of the accumulative effect of diverse (mutated) genes of lesser importance, but not of the effect caused by the mutation of a small group of principal genes (*cf.* Bishop, 2001). Moreover, if the nuclear pathology associated to *FOXP2* mutations ultimately corresponds to a "sensorimotor impairment" (Watkins *et al.*, 2002a; Shriberg *et al.*, 2006), this circumstance would contravene the common clinical criteria for diagnosing SLI—nonlinguistic symptoms or deficits are avoided! See Bishop and Leonard (2001), Leonard (2002). Hence the interest in finding QTLs linked or associated to "standard" variants of SLI (Bartlett *et al.*, 2002; SLI Consortium, 2002; Fisher *et al.*, 2003; SLI Consortium, 2004). One of such candidate genes plausibly corresponds to

---

[3] And we are supposed to be aware of the convenience of distinguishing between the FL, as a computational device, and speech as a peripheral device for exteriorising the output of the former (Hauser *et al.*, 2002); but more on this hypothesis in sections 3.2 and 4.

[4] The latter scenario is as well of maximal interest in evolutionary terms (Lieberman, 2000; 2002; 2006; Balari and Lorenzo, 2009; Balari *et al.*, 2011; Balari *et al.*, forthcoming; see also Lorenzo in the present volume), suggesting the feasibility of central components of the computational system of the FL having been evolved as a result of the tinkering (and partial modification) of regulatory networks controlling the plasticity of neural circuits mediating motor-skill learning (Fisher and Scharff, 2009; White, 2010).

CNTNAP2, which encodes a neurexin presumably involved in the stabilization of the location of certain K^+ channels (Poliak *et al.*, 1999), but also in the regulation of synaptogenesis (Dean *et al.*, 2003). The mutation of this gene has also been associated to autism (Alarcón *et al.*, 2008), thus suggesting that the CNTNAP2 protein could be a shared regulatory mechanism mediating language-related deficits in distinct disorders (Fisher and Scharff, 2009). Crucially, and as we noted above, CNTNAP2 is as well a direct neural target of FOXP2 (Vernes *et al.*, 2008). Maybe the same holds for some of the genes that encompass the regulatory network in which *FOXP2* is presumably integrated. The recent identification of several dozen of its physiological targets (Spiteri *et al.*, 2007; Vernes *et al.*, 2007; Konopka *et al.*, 2009) represents a crucial step in our effort for unravelling the molecular mechanism involved in the modulating of the development and the functioning of the brain substrate of the FL (Spiteri *et al.*, 2007; Vernes *et al.*, 2007; Konopka *et al.*, 2009). We still do not know if the mutation of (some of) such targets is also a causal factor for the emergence of (specific) language disorders. Interestingly, two other promising candidate genes for SLI (not yet related to the FOXP2 network), *CMIP* and *ATP2C2,* have been associated to the phonological component of the working memory (Newbury *et al.*, 2009), whose impairment is a core deficit in SLI, dyslexia, and SSD (see section 2.1). While the former encodes one of the molecular devices which anchor the cellular membrane to the cytoskeleton, and seems to modulate neural migration and/or the assembly of synaptic complexes (Grimbert *et al.*, 2003), the latter encodes a cation transporter, which promotes the interchange of calcium and manganese between the cytosol and the Golgi apparatus (Missiaen *et al.*, 2007). A last putative risk factor for SLI (also not apparently belonging to the FOXP2 network) is *ATP13A4*, which encodes a cation-transporting P_5-type ATPase (Kwasnicka-Crawford *et al.*, 2005).

The case of dyslexia represents another promising insight in the entangled relationships that genes and language seem to maintain. Up to nine different *loci* for this condition have been identified at present (DYX1 to DYX9) (Williams and O'Donovan, 2006; Gibson and Gruen, 2008). Nevertheless, there seem to be many additional *loci* which confer susceptibility to the disorder and which would correspond, consequently, to genes that can be regarded as risk factors for reading disability (Smith, 2007). From three of the former *loci* (DYX1, DYX2 and DYX5) a total of four different genes have been cloned. *ROBO1*, which corresponds to *locus* DYX5, encodes a protein that is apparently involved in the regulation of axonal growth (Hannula-Jouppi *et al.*, 2005; McGrath *et al.*,

2006), possibly of those nerve fibres which conform the so-called thalamo-cortical projections, as has been attested to in mice (Bagri *et al.*, 2002). The remaining three genes[5] encode proteins that contribute to the regulation of the radial migration of cortical neurons (Taipale *et al.*, 2003; Meng *et al.*, 2005; Paracchini *et al.*, 2006; Velayos-Baeza *et al.*, 2008). The analysis of chromosomal rearrangements observed in different dyslexic individuals has made feasible the identification of other candidate genes for this condition, or at least of genes which can be considered as risk factors for the disorder in certain subjects or certain populations. One of the most promising is *DIP2A* (Poelmans *et al.*, 2009), which encodes a protein which plays a key role in the regulation of synaptic plasticity (Yu *et al.*, 2001; Collingridge and Isaac, 2003), and whose mutation impairs certain cognitive processes which depend on hippocampal activity, as is the case for learning and memory (Collingridge and Isaac, 2003), and which are also impaired in dyslexics (Swanson *et al.*, 2006).

Finally, speech-sound disorder (SSD) is another purportedly specific language impairment with a genetic basis. Significantly, one of the *loci* related to this disorder (3p12-q13) corresponds to *ROBO1*, which has been also associated to dyslexia (Nopola-Hemmi *et al.*, 2001; see above). As in the case of *FOXP2* and *CNTNAP2*, this is clearly an interesting analytical outcome, since it suggests that the comorbidity observed between two different disorders can ultimately root in the mutation of the same gene (we will explore deeper implications of such a possibility in sections 3.2 and 4.1). The existence of a remarkable linkage has been documented as well between SSD and the 15q14 region (Stein *et al.*, 2006). It is worth mentioning that the duplication of this chromosomal fragment has been associated to autism (Cook *et al.*, 1997; Schroer *et al.*, 1998; Filipek *et al.*, 2003; Shao *et al.*, 2003), while its deletion gives rise to both Angelman and Prader-Willi syndromes (Magenis *et al.*, 1990; Kishino *et al.*, 1997).[6]

3.2. Dealing (fiercely) with the genotype (but not only)

At this stage we apparently own a pretty nice bunch of putative "language genes", suggesting that the achievement of our initial project of characterising the linguistic genotype merely depends on the completion

[5] *DYX1C1*, which corresponds to *locus* DYX1, and *DCDC2* and a close counterpart, *KIAA0319*, which both correspond to *locus* DYX2

[6] For a review of the genetic basis of other rare specific language disorders (like those mentioned in section 1), see Benítez-Burraco (2008) or Benítez-Burraco (2009: 83-227).

of the list we commenced in section 3.1. (Un)fortunately, the real scenario is much more complex, appearing at first glance as astonishing or even disappointing. To begin with we should address ourselves to the fact that (i) genes which are mutated in individuals who exhibit this kind of specific language disorders are also expressed in brain regions not related to language processing in healthy people, but even in diverse tissues outside the nervous system;[7] (ii) it occasionally occurs that these supposed "language genes" are mutated in people affected by (a) language disorders unlike the one from which it was originally cloned or (b) cognitive (i.e., nonspecifically linguistic) impairments, while it is also the case that (c) sometimes they are simultaneously linked or associated to diverse linguistic (and cognitive) impairments;[8] (iii) it frequently happens as well that in some of the individuals affected by a particular language disorder the sequence of such "language genes" is normal (*phenocopy*), while (iv) the competence (or more properly, the FL) of some of the individuals who are endowed with an anomalous variant of one of these "language genes" seems not to be impaired at all (*null penetrance*) or is just mildly impaired (*reduced penetrance*); moreover (v) the identity of such genes differs (to a certain extent) from one population to another and/or depending on the subtype of the disorder: it is clear that for each language disorder there exist numerous candidate genes and multiple genes that can be considered as risk factors for its emergence. As was for the case of the problematic (i.e. suggestive of inspecificity) results observed at the clinical, cognitive, and neural levels (see section 2.1, but also below), at the genetic level we cannot either rely on just methodological caveats to explain evidence like the former.[9] On the contrary, this significant corpus of facts seems to suggest, as a minimum, two crucial things. Firstly, there exists a non univocal relationship between genes and the linguistic phenotype. In fact, *pleiotropy* (that is, the same gene playing different roles—i.e. contributing to different physiological functions—in diverse moments and body tissues during development) is not the exception, but the norm, and (partially) explains the former scenario (from now, *FOXP2* should not only be regarded as "the gene for language", but also as "the gene for breathing"

---

[7] For instance, during the embryonic development *FOXP2* itself is expressed in the lung, the intestine and the heart, besides the brain! (Shu *et al.*, 2001)

[8] Remember that *ROBO1* has been associated both to dyslexia (Hannula- Jouppi *et al.*, 2005) and SSD (Nopola-Hemmi *et al.*, 2001).

[9] Remember that in section 2.3 we considered some of the shortcomings concerning the conventional tools employed for the genetic analysis of language disorders.

or the "gene for eating"!). Secondly, and bearing in mind that the same genotype can give rise to diverse phenotypes (and ultimately to dissimilar disorders), it seems that other (nongenetic) factors must contribute to the emergence of the latter, with environment plausibly playing some sort of key role in the process. It is evident that the "black box" of development is going to crucially condition the effect of the involved genes upon the growth of the FL (more on this in section 4.1). Incidentally, such things as "the gene for tense" or the "gene for agreement" cannot properly exist!

On the other hand, we must also address the disturbing circumstance that there exist a vaster group of genes that when mutated simultaneously impair (partially or totally) both the FL and other cognitive abilities, ultimately causing what we describe in clinical terms as (inherited) nonspecific language disorders or (inherited) cognitive disorder in which language is impaired (for a review, see Benítez-Burraco, 2009: 88-94, 168-172, and 177-227). We could thrust aside this huge set of genes by merely stating that they are plausibly involved in the regulation of the development of the neural substrate of "general learning devices", "broad cognition", "intelligence", or even "cognitive devices involved in performance" (i.e. implying that they do not specifically regulate the development of the FL or of the linguistic competence, and hence implicitly suggesting that they are not part of the linguistic genotype). But this would both be an inaccurate and a poor depiction of the real scenario (see below, sections 4.2 and 4.3).

Adjoin to such a problematic circumstance all that we have previously discussed not only on the plausible inspecificity of specific language disorders at the symptomatic level, but also at the cognitive and neural levels (see section 2.1). If we recall the evidence we then pointed to, what we actually observe regarding this issue could be summarized as follows. At the symptomatic level, (i) there frequently exists a (certain) comorbidity among different language (and cognitive) disorders, while at the same time (ii) these disorders do not show an unvarying symptomatic profile, but usually represent heterogeneous clinical categories. At the cognitive level, (i) these disorders usually entail other (cognitive) dysfunctions besides those which can be regarded as specifically linguistic, and/or (ii) they seem to arise as a result of a broader cognitive dysfunction; consequently, (iii) the same disorder can be caused by (or its emergence can contribute to) several different underlying deficits, while (iv) the same underlying deficit can give rise to diverse linguistic (and cognitive) disorders. At the neural level, (i) brain regions which seem to be structurally or functionally disturbed in affected people are also frequently involved in computational tasks not directly related to language processing, while (ii) the identity and extension of such areas is variable (as it is also in the normal population); hence, (iii) the impairment of these purportedly "language areas" commonly gives rise in other subjects to different deficits and/or different (including nonlinguistic) disorders.

On the whole, the former evidence hints at the existence of diverse (and complexly interconnected) layers of biological complexity regarding the FL, with genes certainly playing a causal role in the emergence of such a device during ontogeny, but with many other factors also contributing to the phenomenon, and with functional labels ("language gene", "language area", and the like) not only being quite inexact (and simplifying), but crucially impeding a proper apprehension of the intricacy of the real scenario (remember section 2.1; more on this in section 4.1).

Finally, it is perhaps worth considering here some of the recent Chomskyan intuitions about nature and the (biological) properties of the FL, posited in the context of the so-called Minimalist Program (Chomsky, 1995; Chomsky, 2000c). Such intuitions have substantially implied a detachment from the original hypothesis of the existence of a UG (i.e. an autonomous system of knowledge based on idiosyncratic principles and categories which would not be shared with other cognitive systems, endowed with the properties we discussed in section 1) and the subsequent attachment to a conception of the FL as an interface device between the cognitive systems responsible for thought and the sensorimotor systems involved in perception and motricity (usually known as "external"). In a famous paper (Hauser *et al.*, 2002) Chomsky furthermore posits a crucial distinction between a Faculty of Language in a Narrow sense (henceforth, FLN) (i.e., a computation system capable of recursive processing), and a Faculty of Language in a Broad sense (henceforth, FLB) (i.e., all the aspects related to contents to be expressed and interpreted, and also to signals employed in their transmission). Moreover, according to the strongest minimalist thesis (Chomsky 2000a: 96), the optimal version of the FL would imply a minimum of grammatical machinery, the FL mandatorily emerging during development from the direct coupling between the conceptual system and the sensorimotor systems whenever growth takes place in presence of a threshold amount of linguistic stimuli (Hauser *et al.*, 2002).[10]

[10] To a certain extent the development and the functioning of the FL would rely as well on the general laws which regulate the organization of biological systems, which are labelled as "the third factor" [Chomsky 2001: 1-2; 2005]; more on this in section 4.1.

If we are really suggesting that the structure of the FL is to be minimal and lack specific principles, its mechanism responding to the exigencies of the peripheral modules or to the simplest solutions that "come for free", then grammatical principles derived from genetic instructions (in the sense discussed in section 1) could be ultimately unnecessary. As Chomsky himself has conjectured, the Minimalist Program implies "shifting the burden of explanation from the first factor, the genetic endowment, to the third factor, language-independent principles of data processing, structural architecture, and computational efficiency" (Chomsky 2005: 9). Consequently, we could speculate about the possibility that "language genes" do not actually exist, or that they are unnecessary, or more ultimately (and more accurately), that they really match those genes involved in the development and functioning of the external systems. In such a case, it seems that any "genetics of language" should also necessarily consider that those genes when mutated give rise to broad cognitive disorders in which language impairment is a prominent symptom. Of course, not every gene of this kind will be of interest. But we can always select among them actual candidate genes after the pertinent structural and functional analyses (see section 2.4).

In fact, we could examine this key question from an opposite perspective, by hypothesising that the pertinacious difficulties which seem to universally arise when trying to completely discriminate among the different language disorders at all levels of biological complexity (phenotypic/clinical, cognitive, neuroanatomical, neurophysiological and genetic), and to distinctively separate them (at all those levels) from other cognitive disorders, are problematic just for particular models of the anatomical and functional organization of the brain, and for concrete hypotheses regarding the biological nature of both cognition and language. We can plausibly label such views as strictly modular. Although *modular* is a pretty slippery concept (see Barrett and Kurzban, 2006 for a review), in cognitive sciences there seems to exist an ample consensus regarding the conceptualization of *module* as an encapsulated and autonomous computational device, innately specified, which exhibits a domain specificity and which evolved independently to satisfy particular functions (*cf.* Fodor, 1983; Coltheart, 1999; see also Barret and Kurzban, 2006, and Griffiths, 2007 for critical reviews). It is true that at first only peripheral devices were characterised in this way (see Fodor, 1983). However, soon after diverse (if not all) higher cognitive capacities were also judged to be modular in that precise sense (see Pinker, 1994; Spelke, 1998; Sperber, 2001, among many others). Consequently, the hypothetical modular nature of the FL has been an outstanding theme of controversy for many decades

(Fodor, 1983; Chomsky, 1986; Karmiloff-Smith, 1992; Smith and Tsimpli, 1995; Anderson and Lightfoot, 1999; Coltheart, 1999; among many others). Indeed, we also went into the topic in section 1, when we considered the hypothesis about the existence of the UG, but we have also gone into the issue in the last two paragraphs, when suggesting that the Minimalist Program actually relativizes both the condition of language as a distinct module of the mind and the idea of its modular internal organization (see Longa and Lorenzo, 2008 for details). But what concerns us at this precise moment is the fact that under such a strictly modular conception of the mind, what we would expect is precisely what we were looking for and what we have not found almost anywhere (see section 2.1 and above), to be precise, that (i) specific language disorders are univocally linked to strictly linguistic symptoms (and underlying deficits); (ii) specific language disorders are univocally linked to structural alterations and/or dysfunctions of particular brain regions (exclusively dedicated, in principle, to the processing of linguistic stimuli = "language areas"); and (iii) specific language disorders are linked or associated to the mutation of specific genes (= non mutated in other cognitive impairments), which are exclusively involved in the development and functioning of "language areas". The problem with this kind of strictly modular hypotheses is that they cannot properly explain, not only (i) the aforementioned unexpected difficulties for achieving an effective separation (at all levels of biological complexity) between the diverse afunctional or dysfunctional phenotypes of the competence and those corresponding to the impairment of other cognitive capacities, but (ii) the dynamics followed by brain development during ontogeny either (see Karmiloff-Smith, 1998) or, consequently, (iii) the effect exerted by experience on the structuring and functioning of the FL (experience is worth!)

In section 4.2 we will discuss a slightly different (but perhaps more productive) conceptualization of *module*, this time defined in developmental terms, and will also review the whole case for modularity, with the appealing aim of trying to link the developmental itinerary of the neural substrate of the FL, with its structural organization in the adult brain and the kind of processes it contributes to, and ultimately with the cognitive function it encompasses. Before doing that, we shall conclude this section by appealing to less strictly modular models of the mind as a way of beginning to disentangle the former puzzling *cul-de-sac*. In doing so, we will quote Marcus (2006), in particular, who has pertinently pointed out that two cognitive modules, being functionally distinct (in fact, they have to be if they are to be considered modules), are never completely independent in genetic, neurobiological and evolutionary terms. This would imply that most genes which plausibly contribute to regulate the development (and to some extent, the functioning) of neural structures involved in language processing can actually be shared with other cognitive capacities (or modules) (as would also be the case for most of such neural structures), whilst at the same time rendering as idiosyncratic the whole of set of (functionally related) genes, ultimately suggesting that a functional autonomy at the cognitive/phenotypic level is not incompatible with an (extended) overlapping at other deeper biological levels. For the moment, let us avoid labelling such a set of genes a "genetic program", a "developmental module" or something of that ilk (we shall deal with this issue in sections 4.1 and 4.2). Anyhow, what we have concluded in the present section seems to allow us to satisfactorily explain many of the most remarkable and apparently paradoxical results of the analysis of (specific) language impairments and disorders at the phenotypic/clinical, cognitive, neuroanatomical, neurophysiological and genetic levels (see sections 2.1 and 3.1, and also above), which clearly cannot be solely imputed to just methodological caveats (see sections 2.1 and 2.3). Thus, in a pleiotropic context the mutation of a particular gene can affect the normal development (and functioning) of two (or more) different brain areas, giving rise to structural and functional anomalies which will in turn originate in two (or more) diverse deficits; these deficits will subsequently give rise to different symptoms, susceptible of being clinically categorised as two (or more) dissimilar disorders (sometimes heterogeneous, sometimes comorbid). As the context is simultaneously polygenic, it frequently occurs that the mutation of two (or more) functionally related genes can give rise to similar structural and functional anomalies in the same brain area(s), and consequently to an equal deficit, and ultimately to akin symptoms susceptible of being clinically categorised as a unique (and the same) language disorder (which will sometimes be heterogeneous). In other cases, the mutation of two (or more) of such functionally related genes will conversely give rise to diverse structural and functional anomalies in two (or more) brain areas, which will originate in two (or more) different deficits; these deficits will in turn give rise to diverse symptoms, susceptible of being clinically categorised as two (or more) dissimilar disorders (sometimes heterogeneous, sometimes comorbid). What is more, as the contribution of each dysfunctional or afunctional product to the anomalous phenotype will always be subtlety conditioned by the effect exerted by the remaining involved genes, and crucially, by the remaining involved modulatory factors (epigenetic, maternal, ontogenetic, environmental, etc.) (see section 4.1), it also frequently happens that (i) the mutation of the same gene gives rise, in different people or populations, to diverse levels of affectedness regarding the structural and functional integrity of (the same) brain area, and the cognitive and symptomatic profiles of the individuals, which will in turn be clinically categorised as two (or more) subtypes of the same disorder or even as two (or more) different disorders (which sometimes will be comorbid); or (ii) the mutation of two different genes gives rise, in different people or populations, to similar structural and functional anomalies in the same or different brain areas, which will in turn originate in a common deficit; this deficit will give rise to akin symptoms, susceptible of being clinically categorised as the same disorder or as different subtypes of a common condition. Consequently, a noteworthy corollary, of particular interest for clinical linguists, is that the actual contribution of genes to a final dysfunctional or afunctional phenotype will, in general, be limited, difficult to predict, and substantially conditioned to the effects exerted by the remaining involved factors.[11]

Whatever the case may be, and after this again long digression (never a pointless detour, we hope yet again!) we must seriously consider the (reasonable) convenience of taking into account general cognitive (i.e. nonspecifically linguistic) disorders if genes involved in the regulation of the development and functioning of the neural structures implicated in language processing are to be identified and characterised. At the present moment, a reasonable hypothesis seems to be that these kinds of genes will be principally related to what we previously labelled as the FLB. But in section 4.1 we will discuss in depth the possibility of such genes being part of the linguistic genotype, ultimately putting into question the concept itself.

3.3. Genes related to broad language impairments

When also considering cognitive disorders which cannot be regarded as specifically linguistic, but which simultaneously exhibit a characteristic impairment of the linguistic competence, the number of genes related to language substantially increases (for a review see Benítez-Burraco, 2009:

[11] Nonetheless, and according to the consequences caused by their mutation, some of these genes are customarily regarded as principal (i.e. their mutation constitutes a main causal factor in the emergence of a particular disorder in most affected people), while others are considered as secondary (i.e., their mutation just represents a risk factor for the appearance of the condition in some individuals) (see also Winterer and Goldman, 2003).

88-94, 168-172, and 177-227). Consequently, the biochemical nature and function of the products encoded by them are diverse, as they include (i) enzymes involved in basic brain metabolic reactions, (ii) membrane transporters (or proteins associated with them), (iii) enzymes involved in basic cellular metabolic reactions or essential cellular structural proteins (including, significantly, those related to the functioning of the cellular cytoskeleton); (iv) proteins implicated in cell-to-cell interactions (including, notably, those responsible for adhesion and recognition processes between neurons); (v) extracellular signalling proteins, (vi) membrane receptors and proteins integrated in signal transduction pathways, and (vii) transcriptional and translational factors, and other regulators of gene expression (including, ncRNAs and proteins which interact with the DNA).

At the same time, these products mediate diverse physiological processes at the brain level. While some of them are (i) regulators of basic brain metabolism, others regulate (ii) fundamental cellular processes (such as cell-to-cell interactions and/or cellular adhesion; inwards and outwards cell vesicle trafficking; organelle morphology, location, and interaction; DNA replication and reparation; transcriptional/translational activity and mRNA processing; or cell cycle related processes, including the stabilization and remodelling of the cellular cytoskeleton and the cell size, shape, and movement); (iii) neuron specific cellular processes (including the diverse events related to the generation of action potentials, the nerve impulse transmission, and the different steps encompassed by the synapses); (iv) neural proliferation and migration; (v) synaptogenesis and axonogenesis; (vi) neural identity and/or functionality (including the establishment of brain basic patterning and regionalization, and the—embryonic, perinatal, and postnatal—maturation of diverse cerebral circuits), and (vi) essential brain processes, such as memory, long-term potentiation (LTP), neural plasticity and/or critical periods for synaptic development in response to experience.

Notice however that a high percentage of genes related to language play an eminently modulatory role, since their products regulate the expression of other genes, or else belong to cellular pathways involved in signal transduction. Crucially, and from a physiological perspective, these genes are mostly related, at the brain level, to the regulation of neural proliferation, migration, and specialization, or else to the establishment of initial contacts among differentiated neurons (synaptogenesis, and axonogenesis). This circumstance will be of particular relevance, ontogenetically (see also section 4.1), but also phylogenetically. For instance, the mutation of a single master gene can simultaneously impair many complex biological processes depending on its direct targets, ultimately rendering diverse (and often unpredictable) phenotypic outcomes. Moreover, the modification of a regulatory gene involved in brain wiring can eventually give rise to novel patterns of neuronal interconnection and even to the emergence of unexpected cognitive capacities. Hence the hypothesis that the evolutionary modification of the sequences of a few of these regulatory genes could have contributed to achieve in a brief time span the remodelling of neural circuits needed to attain the computational capacity which characterizes the human FL (see Balari *et al.*, In press, and also Lorenzo, this volume).

4. From genotype to phenotype

4.1. How genes actually contribute to the development of the Faculty of Language

What we have discussed in the previous sections would seem to have undermined the role of genes in the emergence of the FL and the primacy of the (linguistic) genotype as a key causal factor in explaining language acquisition, perhaps ultimately suggesting that a "genetics of language" is pointless. We certainly began by positing that "the genotypical principles responsible for language acquisition can be viewed as a theory of grammar, sometimes called Universal Grammar" and furthermore suggesting that such "genetic equipment [....] makes language growth possible" (Lightfoot, 1982: 22) (see also section 1). Nonetheless, we have finished by suggesting that "the ultimate foundations of FL are certainly not linguistic, the interaction of the external systems during development being responsible for the emergence of a linguistic capacity and with genes perhaps only in charge of the development of those non-linguistic systems" (Longa and Lorenzo, 2008: 548). However it is now, after such a long journey, when we have actually reached the necessary vantage point to rigorously discuss this kind of hypotheses. We will bring to a close the present section by concluding that genes do actually play an important role in the emergence of a FL as the end point of development, and crucially that "there is no reason to abandon the thesis of innateness" (Longa and Lorenzo, 2008: 548), which will basically mean that the FL robustly emerges at the end of development if (but only if) growth takes place in the presence of a sufficient amount of linguistic stimuli, regardless of the role ultimately assigned to the genes. But we must consider in detail other relevant issues before finishing in such a way.

From a neurobiological perspective, what we actually observe during the first stages of development is that brains tend to achieve a substantial level of internal organization in advance of experience.[12] Such a process fundamentally entails the establishment of basic interconnection patterns among the diverse types of differentiated neurons involved and hence the setting up of the plain histological organization of the main anatomic macrostructures which conform the neural substrate of language, although (crucially) without generating fully operative computational devices (Ramus, 2006). That's why this fundamental brain pre-wiring, frequently conceptualized as *developmental anticipation* (Balaban, 2006: 317), must be ultimately (and compulsorily) implemented by the feedback effect exerted by neural activity during language processing.[13] The latter is an eminently physiological phenomenon, encompassing structural and functional changes in neurons as the result of the interactions which take place among different brain regions, but also (and significantly) between them and the environment; nevertheless, many genes must eventually be involved, as is suggested by the recent analyses of the synaptic proteome (Grant, 2003) or of the neural plasticity (Kaufmann and Worley, 1999) (more below). Only in such a way, the definitive cytoarchitecture of the neural substrate of the FL is ultimately achieved and fully operative neural structures are eventually generated.

In any case, it could be tentative to posit the existence of some kind of "genetic toolkit" in charge of such pre-wiring processes—in essence, a set of coordinated genes that would ultimately regulate every step of development (True and Carroll, 2002; Carroll, 2005); collaterally, this has been a widespread hypothesis in the field of evolutionary developmental biology (Goodman and Coughlin 2000, True and Carroll 2002, Baguñá and García-Fernández 2003, Carroll 2005), with external stimuli simply acting as a sort of triggering device for such a toolkit, and perhaps only further affecting the latter stages of development (see above). But we must be very cautious here. To begin with, genes are not blueprints, nor sets of instructions belonging to any kind of engineering book on how to build an organism (Kay, 2000). Genes merely contribute to the synthesis of certain biochemical products, which will be subsequently engaged in particular physiological functions. But even the assumption that there is a direct connection between genes and (functional) proteins, or as a minimum

[12] To be exact, before starting processing external stimuli and in the absence of synaptic transmission (Verhage *et al.*, 2000; Bouwman *et al.*, 2004).
[13] Note however that precisely determining when experience begins is a complex challenge (Gottlieb, 1997); more below.

between genes and certain (functional) biochemical products—since we must seriously "call into question the long-standing assumption that most genes encode proteins" (Mattick *et al.*, 2009: 25)—must be appropriately reassessed, in view of the fact that many genes are subjected to post-transcriptional modifications, ultimately giving rise to alternative proteins from a same primary transcript (for a review, see Sholtis and Weiss, 2005: 505-510, among many others) or even to hybrid proteins from chimeric transcripts (Gingeras 2009). Moreover, many proteins are post-translationally modified to obtain (diverse) functional products and/or to be suitably integrated into multiproteinic complexes, and/or be correctly transported to their appropriate target *loci* inside or outside the cell (for a review, see also Sholtis and Weiss, 2005: 505-510 among many others). The very case of *FOXP2* nicely exemplifies this intricate state of affairs, since its RNA is post-transcriptionally spliced in all species (Bruce and Margolis 2002, Haesler *et al.* 2004), diverse protein isoforms seem actually to play different roles *in vivo* (Vernes *et al.* 2006), and the FOXP2 protein conceivably interacts with other related proteins (FOXP1 and FOXP4, certainly) to properly (and functionally) join to the correct regulatory *cis*-regions of its target genes (Li *et al.* 2004). By the way, such a complex scenario seems to put into question the very definition of genes, which has recently been reformulated as "fuzzy transcription clusters with multiple products" (Hashimoto *et al.* 2009) or even as the "union of genomic sequences encoding a coherent set of potentially overlapping functional products" (Mattick and Makunin 2006).

Moreover, the extent to which a particular gene ultimately contributes to a certain biological process heavily depends on the precise time, place and amount in which it is expressed and in which its product(s) become(s) synthesized. The known fact that ordinarily the same gene plays different roles (i.e. contributes to different physiological functions) in diverse moments and body tissues during ontogeny (*pleiotropy*; see also section 3.2) ultimately makes a one-to-one correlation between genes and traits unrealistic. It has progressively become evident that the development and the emergence of any phenotypic feature much more crucially depend on the transcriptional status of the cell (i.e. which RNAs are present, in what amount, and what functions they contribute to), with genetic sequences remaining in a secondary place (Mattick *et al.* 2009). Consequently, the way in which gene expression is regulated emerges as the key phenomenon during (and for) development (more below). This is also an extremely complex process, specially after the recent discovery that not only *cis*-DNA sequences and regulatory proteins (in the form of transcriptional and translational factors) are involved, but that there exist additional layers of (pre- and post-transcriptional) regulatory complexity also implicated, mainly in the form of non-coding RNAs (ncRNAs), which can themselves act as regulatory factors, or confer specificity to the protein factors also concerned (Mattick, *et al.* 2009).

In addition, genes do not influence such particular biological processes or traits in an isolated way; on the contrary, many genes contribute (each to a different extent) to the development and functioning of the neural substrate of the FL (*polygenism*), as plausibly suggests the fact that the mutation of so many numerous genes impairs linguistic competence (see sections 3.1 and 3.3). Beyond the plain existence of multiprotein complexes, gene products routinely interact in the form of complex regulatory networks (Geschwind, and Konopka 2009), the relevance of the particular roles played by each of the involved genes being clearly surpassed by the joint effect due to the precise balance kept, in a particular moment and place, by the biochemical products encoded by the whole set of involved genes.[14]

Likewise, as the cell is not an isolated system either, multiple internal and external factors ultimately modulate gene expression, adding extra factors, besides genes themselves and their direct products, to the regulatory elements which contribute to the wiring of the neural substrate of the FL. On the one hand, these regulatory elements derive from the remaining levels of biological complexity which exist between cells and the FL, as in the case of nervous tissues, neural circuits, or brain areas, structures and macrostructures, but also from other body tissues (hormones, for instance, play a key role in brain development and functioning: MacLusky *et al.* 2006, Tournell *et al.* 2006, Bernal 2007, Brann *et al.* 2007). But, as we previously discussed, diverse environmental elements ultimately affect gene expression via diverse signal transduction and regulatory pathways, which eventually influence the homeostasis of transcriptional (and translational) regulators.[15] But what it is actually worth considering here is the fact that those internal and external factors not only modify gene expression in such a way, but also give rise to

---

[14] Notice that these products are typically arranged in the form of gradients or specific combinations of signaling molecules, which are endowed by physicochemical properties of particular relevance for developmental phenomena (more below).

[15] Incidentally, remember that in section 3.3 we highlighted that a significant percentage of genes related to language encode transcriptional or translational factors, or proteins belonging to pathways involved in signal transduction towards the cell.

epigenetic modifications of the DNA, i.e., to structural changes in the molecule that, not affecting the nucleotide sequences, do crucially condition gene expression by restricting access to them of regulatory factors acting in *trans*. Significantly, such modifications are inheritable (Isles and Wilkinson, 2000). Moreover, they have been linked to key neural processes (throughout both the development of the FL and language processing in the adult stage), such as neural proliferation and differentiation, and particularly, neural plasticity, which is at the bottom of essential cognitive abilities such as learning and memory (Levenson and Sweatt 2006, Gräff and Mansuy 2008, Mehler 2008), a consequence being that epigenetics has ultimately emerged as a widespread and strategic mechanism in the regulation of cognitive functions and behaviour (Franklin and Mansuy 2010).

Finally, other additional factors ultimately (and crucially) channel organisms' developmental trajectories. For instance, protein gradients inherited via the egg cytoplasm usually play a key role during the first stages of development (Davidson 1986). A special attention is also deserved by the hypothesis that part of the information which determines the features and functional properties of any biological structure is generated by developmental processes themselves (Oyama 2000a, Oyama *et al.* 2001), but especially the possibility that, to some extent, development relies as well on general laws which regulate the self-organization of biological systems (Kauffman 1995, 2000). In fact, as noted above, diverse physicochemical parameters and properties crucially condition developmental processes, as is the case for viscoelasticity, differential biochemical diffusion and oscillation, the dynamics followed by sedimentation and diffusion gradients, mechanochemical excitability or the very dimensions of the space in which chemical reactions take place, these factors acting in combination with basic properties of cells, such as polarity and differential adhesion (Newman and Comper 1990, Newman *et al.* 2006). These kinds of "generic" factors (Newman and Comper 1990) ultimately determine the way in which the remaining involved elements (proteins, ncRNAs, hormones, etc.) behave and function, yet explaining basic dimensions of the manner in which developing tissues organize, such as the appearance of regionalization patterns or the emergence of morphological regularities (Goodwin 1994, Sholtis and Weiss 2005: 513-515). Notice also that developmental processes are, to some extent, stochastic phenomena—hence two identical developmental processes can actually lead to different phenotypic outputs (Balaban, 2006: 320), and that, significantly, these stochastic effects are particularly relevant with regards to brain development (Balaban, 2006: 321- 325). In fact, such general laws regulating the self-organization of biological systems and playing such a relevant role in development, are precisely what we labelled in section 3.2 (when we introduced Chomsky’s Minimalist Program) the “third factor”; notably, they are independent from the genome and the environment (Chomsky, 2001a: 1-2; Hauser *et al.*, 2002; Chomsky, 2005). Finally, notice as well that the whole set of these additional factors which also contribute to the initial wiring of the neural substrate of the FL (epigenetic elements, maternal factors, regulatory elements belonging to all levels of biological complexity located between genes, ontogenetic information, general laws which regulate the self-organization of biological systems), while robustly appearing and acting during (certain stages of) development can be legitimately regarded as innate (see Bateson and Mameli, 2007, for a critical view of such a conceptualization of innateness). Consequently, what can be regarded as “innate” clearly and necessarily transcends what can be considered “genetic” (see also Wimsatt, 1999). In section 4.3 we will re-examine such a depiction of innateness under the prism of the Minimalist Program and also consider the derived accounts of language acquisition (see Longa and Lorenzo 2008, this volume), but perhaps it would be already worth bearing in mind that “canalization offers yet another way [besides genes themselves] of interpreting Chomksy’s claims about the development of linguistic abilities” (Bateson and Mameli, 2007: 823).

Overall, as gene activity (and the activity of the products encoded by genes) is necessarily (and decisively) conditioned by (while simultaneously conditioning) such epigenetic, maternal, and ontogenetic factors, by the physiological factors derived from the remaining levels of biological complexity of the FL, and by environmental factors, we cannot still regard genes as a primary cause regarding the development of the FL. On the contrary, they just represent one more among diverse regulatory devices involved in the modulation of such a process, with the particularity that each of them regulates (and is regulated by) the activity of the rest. Genes are not “first among equals” (Schaffner 1998), nor is environment still a triggering factor which switches on a developmental program residing in the genome (Rosenberg 1997; Maynard Smith 2000). In fact, development is the outcome of a non-additive but synergistic interaction among the whole set of involved factors (Robert 2008: 397), which are equally necessary, hence responding to the so-called “thesis of parity”: “Parity is the idea that genes and other material causes are on a par” (Griffiths and Gray 1998: 254). Ultimately “a gene initiates a sequence of events only if one chooses to begin analysis at that point” (Oyama, 2000a: 40). On the whole, this is the intricate suite of factors which ultimately explains why the same genotype can give rise to different phenotypes—the whole range of possible phenotypes is commonly characterised as its *reaction norm* (Pigliucci *et al.*, 1996)—when confronted with diverse environmental conditions. We generally refer to such a property as *phenotypic plasticity* (West-Eberhard, 2003; see also section 4.3). But notice that, as we will (more o less explicitly) discuss in sections 4.2 and 4.3, the environment and the very dynamics followed by developmental systems can also promote that the same phenotype robustly emerges at the term of growth even when the corresponding genotypes are (slightly) different.[16]

4.2. The case for modularity

From the genetic (and the neurobiological) analyses of inherited language disorders (but also from the FL itself) succinctly sketched in this chapter we have ultimately concluded that genes do really contribute to regulate the development (and to some extent, the functioning) of certain neural structures which in turn compute the tasks involved in language processing. At the same time, we have rejected the existence of both genes exclusively related to the emergence (and activity) of linguistic competence, and a genetic program "for language" envisaged as a set of instructions capable of guiding and performing by itself the development of the adult FL—but note that this is quite a common assumption in developmental biology: "development is genetically programmed" (Raff 2000: 74). Ultimately it seems that the role actually played by genes during development is much better apprehended by alternative models of development, an outstanding instance being the one rooted in the so-called DST (an acronym for *Dynamic Systems Theory*).[17]

Furthermore, in the last section we also briefly depicted the way in which the neural substrate of language grows and wires during development. It is actually out of our scope to in depth characterise such a substrate in the adult individual (but see the contributions by Cuetos and Laka in this volume). Nevertheless, it is worth remembering that in section 3.2 (and implicitly in section 2.1) we came across the concept of *module* as a feasible label for (the output of) such a neural device, though we then concluded that a strictly modular view of the mind could be incompatible

[16] As pointed out in the last paragraph, this condition is usually referred to as *canalization* (Waddington 1957).
[17] See Oyama (2000a, 2000c) Oyama *et al.* (2001); see also Lorenzo and Longa, (2009) for an implementation of minimalism from a DST perspective.

with the results of the genetic and neurobiological analyses of inherited language disorders (as it could be, as well, with other diverse evidence of a similar nature, like dissociation analyses in acquired language disorders). At that moment we consequently argued for a more "functionalist" conceptualization of module, tentatively suggesting that to understand language in neural terms can plausibly mean dissecting the particular way in which certain neural structures are interconnected in order to allow them to compute the tasks involved in language processing, while simultaneously assuming as improbable the finding of portions of brain tissue exclusively related to linguistic competence. In fact, we had implicitly sketched in section 3.1 a plausible model of language processing in accordance with such a hypothesis—it was actually the one usually known as the "grammar of the basal ganglia" (Lieberman, 2000; 2002; 2006; more on this below). And the same holds, of course, for genes, in the sense that to understand language in genetic terms can fundamentally mean to precisely establish the particular way in which a set of functionally related genes contributes to regulate the development (and to some extent, the functioning) of such neural structures, while simultaneously assuming as improbable the finding of genes exclusively related to linguistic competence.

Anyhow, a reasonable question necessarily arises at this very stage: if both the particular instructions (= genes) which contribute to regulate the development of the neural substrate of the FL and the particular pieces (= brain circuits/structures) which conform such substrate are substantially shared with other cognitive capacities, does it make sense (or even is it possible) to still conceptualize the FL as a discrete cognitive entity? Putting it into other words, while admitting the substantial unspecificity of genes and brain structures related to language, we must simultaneously acknowledge the existence of a set of genes that co-ordinately contribute to the emergence, at the term of development, of a set of functionally interconnected brain devices whose output is what we phenotypically describe as *competence*, the *FL* or even merely *language*, in essence (and putting aside the fact that these labels identify substantially different entities) "one component of the human mind" (Chomsky, 1977: 164), which we customarily consider as an idiosyncratic cognitive capacity/entity/ability. Consequently, are there any level(s) of biological complexity at which *modularity* can (still) be considered an appropriate concept for accurately apprehending what we actually observe in them? Conversely, at which other level(s) does *modularity* cease to be a useful or even acceptable concept or hypothesis? Notice that when wondering about the existence (and the extent) of potential (functional) links among diverse genes, different regulatory elements, certain neuronal structures, or even various cognitive abilities, we are ultimately harmonizing with a growing corpus of evidence which suggests that the brain is substantially structured at the anatomical level, but particularly so at the functional level.[18]

In section 3.2 we partially answered (positively) this key question of to which extent (if any) it is still licit to persist in talking about language as a module of the human mind. We did so by pointing out—following Marcus (2006)—that two cognitive modules can be functionally distinct (and in fact, they have to be if they are to be considered modules), while not needing to be completely independent in genetic, neurobiological or evolutionary terms (and in fact, they never are). Moreover, when quoting Chomsky's claims about the minimal requirements for implementing the FL at the brain level, we were also hypothesising about the latter being merely the outcome of the interface between two pre-existing modules, but simultaneously an autonomous entity at the cognitive level. As we then pointed out, we shall now try to contribute to answering the former question by appealing to (and exploring the implications of) the conceptualization of *module* which Evo-devo (i.e. evolutionary developmental biology) usually makes use of. In such a paradigm—since "no unified theory of Evo-devo exists" (Hall and Olson 2003: xv)—a *module* is merely a set of related elements which maintain a strong connectivity within, but a weak connectivity among other equivalent sets of elements (i.e. other modules) (Wagner 1996, Wagner and Altenberg 1996, Kirschner and Gerhart 1998). Consequently, our set of functionally interrelated genes can be legitimately categorized as a (genetic) module, as well as our interrelated set of neuronal structures that can also be categorized as a (neurobiological) module. In fact, modularity emerges as a fundamental property of living things at every level of organization, conferring both robustness (i.e. insensibility to external perturbations) (Kitano, 2004) and flexibility (i.e. the capacity to respond to such perturbations) (Bergman and Siegal, 2002) to living organisms, and therefore plausibly (and crucially) proving indispensable for understanding the structure of the mind as well (Barrett and Kurzban 2006).

Crucially, Evo-devo posits a basic dichotomy between developmental modules (i.e. a set of coordinated interactions among regulatory elements

---

[18] The persistence and repeatability of certain neuroanatomical features, the existence of substantially similar patterns of brain activation in response to similar tasks, the typical discrepancy which exists between the ontogeny of diverse cognitive capacities (including linguistic competence!) during growth, etc. (Crain 1991; Felleman and Van Essen 1991, among many others).

responsible for the development of particular biological structures) and functional modules (integrated units of traits, i.e. biological structures, serving common functions) (Breuker *et al.*, 2006). As also discussed by Breuker *et al.* (2006: 490), a crucial concern regarding this dichotomy is the diverse evidence plausibly putting into question the universal validity (and the mandatory character) of the "matching hypothesis", i.e. the possibility that "developmental modules evolve adaptively to match functional modules" (Wagner and Altenberg, 1996). Accordingly, certain functional modules can emerge as the result of the co-ordinated activities performed by two (or more) biological structures which are in turn the output of two (or more) developmental modules, the latter acting as ontogenetic restrictions (i.e. favouring or disfavouring certain developmental trajectories) to ulterior evolutionary changes (Raff, 1996). For instance, the scutellum and the wing of a true bug are (the outcome of) two distinct development modules, with the former and the basal part of the latter comprising one unique functional module (involved in protection) (Breuker *et al.*, 2006).[19] Moreover, and concerning cognition, Griffiths (2007) discusses the convenience of distinguishing among three different kind of modules: developmental, (neuro)functional, and mental (or even virtual), with the former corresponding to separable neural structures performing specific activities (Griffiths, 2007: 198) and the latter being "a pattern of dissociability between aspects of the systems of performance that does not correspond to the existence of separate neural systems" (Griffiths, 2007: 201), thus apprehending those

> aspects of an organism's psychological performance profile that can be developmentally and functionally dissociated from one another in such a way as to allow performance in one domain to be optimized independently of performance in the other (Griffiths, 2007: 201).

It is out of our scope to contribute to any conceivable terminological dispute (Breuker *et al*'s *function* and Griffiths' *function* could actually address the two alternative senses of the term, as discussed by Love 2007), or to consider the (evolutionary) rationales for mental modules. Appealing

---

[19] Notice however that *function* is a somewhat slippery concept in Biology, as it is indistinctively used to denote the *activity* performed by a particular organic structure and the *use* given to such structure as a consequence of its connections with other structures, but also as a result of the relationships existing between the organism and the environment (Love, 2007; see also Lorenzo in the present volume for a comprehensive discussion on this issue).

hypotheses in the context of this work are the proposals that mental modules do not necessarily correspond to separate adaptative problems (Sterelny and Griffiths 1999: 328-332), as well as that "phenotypic novelty is largely reorganizational rather than a product of innovative genes" (West-Eberhard 2005: 6547), ultimately suggesting that innovations could arise in neutral conditions as a consequence of the dynamics and the generative properties of development systems (Müller and Newman 2005), later being (or not) sanctioned by the "boundary condition" of natural selection (Müller 2007: 947; see also Lorenzo in this volume), or even to confront a hypothetical depiction of the FL according to the principles of Evolutionary Biology, with Chomsky's fresh characterization of the FL (but see section 3.2). Remember from section 3.2 that what actually concerns us is how to relate the developmental itinerary of the neural structures involved in language processing, with their structural organization and with the kind of process they perform at the end of development, and eventually with the ultimate cognitive function they contribute to (always bearing in mind that our main goal is to correctly place genes somewhere onto this entangled scenario). With the notions of developmental and functional (or even mental) modules in hand, we can actually hypothesise that when growing, the embryonic brain is a mosaic of different developmental modules—with genes being just one among the diverse types of developmental factors involved (see section 4.1)—ultimately giving rise to diverse neuronal structures which perform concrete activities, and which can become interconnected (during the brain pre-wiring) with other structures, thus anticipating possible functional devices (with some genes probably contributing to this process as well). Plenty of operative devices further develop under the influence of (additional) external stimuli which come from the environment in which growth takes place (this process encompassing different and crucial structural and functional changes in neurons!), ultimately emerging as dissociable entities from other similar entities in terms of performance, while simultaneously being not necessarily dissociable from them in genetic or neurobiological terms. As Griffiths (2007: 202) points out:

> a modularity concept that simply identified a module with a neural system that produces double dissociations would not be useful for neuropsychology because it would serve only to blur distinctions between different neural architectures.

Notice that it is modularity itself that eventually favours certain trajectories of development (and also of evolutionary change), a consequence being that phenotypic variation in a particular population follows certain tendencies ultimately determined by the properties of the developmental systems as a whole (*generative bias*) (Müller 2007: 945, 946).

Our "grammar of the basal ganglia" model fits plausibly and nicely to such a scenario. According to Lieberman (2000, 2002, 2006) the FL is basically equalled to a computational device capable of processing symbolic elements (and ultimately externalising and internalising the output of such computations). The required computational system is conceivably the outcome of the interaction between a sequencer (the activity performed by the basal ganglia) and a working memory (the activity executed by diverse cortical structures), although such a computational device can in fact be applied to the processing of inputs of a diverse nature, one kind being the symbolic elements stored in the lexicon (see also Balari and Lorenzo 2008, 2009b; see Balari *et al.*, 2011, Balari *et al.*, In press, and Lorenzo in the present volume, particularly for a discussion in evolutionary terms). Consequently, one "component of the human mind" believably results from the interface between different functional modules (*à la Griffiths*) that in turn are the output of diverse developmental modules, there not being a compulsorily existing one-to-one relationship between them. This scenario is actually pretty close to the one we have already discussed in section 3.2 in the context of the Minimalist Program. In the last section of the present contribution we will examine whether minimalism is compatible with the genetic (and neurobiological) account of language that emerges from the topics we have considered (or better if among the diverse hypothesised models of the FL minimalism is the one which best matches such an account).

4.3. Minimalism revisited

Different biolinguistic approaches attempt to achieve a biological depiction of the FL (both ontogenetically and phylogenetically) which is in accordance with current linguistic insights about the nature and structure of language, obviously facing in their progress diverse obstacles (Boeckx and Grohmann 2007, Hauser and Bever 2008, Fitch 2009b). At present it seems that if any, minimalism (Chomsky 1995, Hauser *et al.*, 2002; see section 3.2) is the paradigm that is boosting the most productive biolinguistic inquiries (Boeckx and Grohmann 2007). We will finish by somehow reinforcing such prominent status, arguing why minimalism is the model which most accurately fits the biological characterization of the FL emerging from its genetic—and in general, (neuro)biological—analysis, as outlined in the present paper, but also from the remaining topics (crucially, how development takes place) which we have also addressed.

Three aspects at the core or, or derived from, such a program deserved to be highlighted. On the one hand, the distinction established between a FLN and a FLB (Hauser *et al.*, 2002), with the FLN emerging as a computational system capable of recursive processing, but simultaneously susceptible of being coupled to different cognitive or sensorimotor systems, thus potentially rendering pretty diverse kinds of outputs. In the last section we outlined a plausible neurobiological substrate for such a computational device. Besides we argued that the very developmental dynamics of the organism, heavily rooted in the modular organization of all biological systems, plainly favours phenotypic variability and the emergence of such functional (or even mental) modules, which are not straightforwardly the outcome of developmental modules *stricto sensu*, but the result of the functional coupling during growth between quasi-independently developed structures. Crucially, this is also in accordance with recent models of brain evolution in vertebrates, which suggest that brains tend to reorganize themselves internally in complex ways (even promoting such kinds of functional interconnections) as they increase in size (Deacon 1990b, Striedter 2005). The latter scenario is clearly of maximum interest for any inquiry regarding the origin and the evolutionary trajectory of language (for a detailed proposal, see Balari *et al.*, In press, and particularly, Lorenzo in the present volume). Whatever the case may be, the essential point is that, from an ontogenetic perspective, we can ultimately arrive at a distinctive mental entity (hence susceptible of been conceptualized as a mental or a virtual module, a dissociable component of the human mind, etc.) whose biological idiosyncrasy yet resides, as discussed in sections 3.2, 4.1, and 4.2, in the particular way in which the diverse biological components related to it (genes and neural structures, but also many other relevant factors) interact at the different levels of biological complexity during development (and in the adult stage), these components otherwise being involved in other diverse developmental processes and hence related to other capacities.

Moreover, it is also according to our genetic depiction of the FL that the hypothesis of the FLN is the main evolutionary innovation for human language in biological terms, while the FLB simultaneously has a long-lasting evolutionary history (Hauser *et al.* 2002). If it was actually the case, an overwhelming part of the genetic information involved in the emergence of the FL should plausibly correspond to genes needed for the development and functioning of the external systems. This would imply that most "language genes" would ultimately be related to the FLB, and hence, that their mutation, while not affecting the FLN itself, would still impair the FL as a whole, plausibly giving rise to other diverse disorders (not specifically linguistic) in certain populations and/or environments. And this is in fact what we actually observe (remember sections 2.1, 3.1, and 3.2), since nearly no (if any) gene mutation impairs language (but only language) in all the affected people and in all environments (of course, pleiotropy also contributes to explain such an outcome). Clearly, if we subscribe to the strongest minimalist thesis (see section 3.2), there would not exist "language genes" *stricto sensu* at all, for the reason that all genes related to language would be really involved in the regulation of the development of the external systems, with the FL just emerging during development (as the result of the coupling between such external systems) whenever growth takes place in the presence of the required amount of linguistic stimuli (Hauser *et al.*, 2002). But even if we contend that some genes can actually be related to the development of the FLN (conceivably helping to connect the basal ganglia with different regions of the cortex—in fact, this could be the case with *FOXP2*, the circumstance where such a computational system can be functionally coupled to different mind modules/devices/components (see above) would also allow the mutation of one of such putative genes to be rendered non specifically linguistic dysfunctional phenotypes.[20]

Finally, we can term specially productive the very hypothesis that the coupling between the conceptual system and the sensorimotor systems would be mandatory during development (in the presence of a threshold amount of linguistic stimuli), with the development and the functioning of the FL relying to a certain extent on the general laws which regulate the organization of biological systems (Chomsky 2001a: 1-2; 2005). This would eventually imply that a considerable amount of the innate (in essence, not derivable from experience) information involved in the development of the FL is not genetic, but either epigenetic, or biologically determined by the features of the ontogenetic environment in which growth takes place, or even dependent on generic factors of a physicochemical nature. As we discussed in section 4.1, such a possibility is clearly in accordance with the role actually played by genes in regulating developmental processes and the very way in which developmental systems behave and development itself occurs. Moreover, and concerning the particular topic of this contribution (in essence, the

[20] Remember the contentious debate about the precise phenotypic profile of people bearing a mutated copy of *FOXP2* (see section 3.1).

nature of the putative "language genes" and their ultimate role in language acquisition), we must conclude by highlighting the pertinence and the accuracy of Lorenzo and Longa's (2003) claim about the very noteworthy reduction of the genetic charge necessary for the ontogenetic (but also for the phylogenetic) development of the FL which minimalism implies (the two key points of their argumentation being that [i] an overwhelming fraction of the genetic information needed for achieving the task would correspond to the genetic information needed for the development and functioning of the "external" systems; and [ii] a considerable amount of the overall information needed for the development and the functioning of the FL would be epigenetic or would depend on the general laws which regulate the organization of biological systems). Such a claim certainly matches both the outcome of the genetic and the neurobiological analyses of the linguistic phenotype (both functional and dysfunctional), as we have depicted to quite a detailed extent in this chapter, and the way in which the FL (both functional and dysfunctional) actually develops, as we have also sketched in the present contribution. At the same time, the latter pieces of evidence confirm as well the convenience of not abandoning the thesis of innateness regarding the way in which language is acquired (Longa and Lorenzo 2008: 548), while simultaneously the necessity of reformulating the innate nature of the FL, particularly by abandoning previous gene-centric conceptualizations of it.

On the whole, the minimalist conception of the FL fits more accurately than others (even the preminimalist Chomskyan conception itself) the recent findings concerning the way in which developmental (and evolutionary) processes take place in living beings (with regard to which the FL should not represent any exception, notwithstanding its significant idiosyncrasy at the phenotypic level). Notice that Chomsky (2010b: 1) has recently posited the existence of "some analogies between what some biologists call 'the Evo-devo revolution' and ideas that have been lurking in the background of biolinguistics since its origins". However, it is our contention (Benítez-Burraco and Longa, 2010) that Chomsky seems to adhere to the gene-centric current of Evo-devo, according to which genes would actually be *primi inter pares*, with such a primacy reflected in the central role conferred to genetic toolkits during (and for) development (see Carroll, 2005). Clearly, this conception neither accurately reflects the actual role played by genes during growth (see section 4.1), nor adequately considers the effect exerted by the remaining developmental factors also involved (see section 4.1 and above), nor ultimately even properly explains (some of) the phenotypic output of the mutation of the former (see section 3.2). Moreover, we also contend (Benítez-Burraco and Longa

2010) that such a gene-centric conception of development could be eventually incompatible with the very biolinguistic approach based on the Minimalist Program (in spite of plausibly being compatible with other approaches rooted in previous models of language and theories of language acquisition, such as Government and Binding, and the Theory of Principles and Parameters; see also Lorenzo and Longa 2009). To begin with, because it clearly contradicts the reduction of the genetic information needed for the development of the FL endorsed by the Minimalist Program (Lorenzo and Longa 2003, Chomsky 2005, see also above), which ultimately rejects the very ideas of a genetic program for language acquisition or a linguistic genotype (see section 1), this rejection being a consequence of its vindication for a highly unspecific architecture for language (Chomsky 2000a, 2000c; Longa and Lorenzo 2008; see also section 3.2).[21] Moreover, because according to a gene-centric (Evo-devo) view of development, architectural constraints to growth are fundamentally of a genetic nature, in the form of "deeply conserved [regulatory] gene expression patterns" (Raff 2000: 76), which clearly contradicts the ultimate aim of the Minimalist approach in looking for "the burden of explanation [...] from genetic endowment".

Notice that this contention on the whole does not mean that minimalism should be divested of its position as the chief booster of biolinguistic inquiries, but just that it should try to explore the possibility of building bridges to alternative biological approaches to development (see section 4.2) which as a minimum are prompted to (i) reduce the importance conferred to genes in regulating development; (ii) increase the relevance bestowed to the remaining non-genetic factors which also contribute to regulate developmental processes; (iii) consider, in particular, the significant causative role exerted on brain development by the very dynamics followed by the developmental systems implied in the structural and functional organization of the brain, and ultimately (iv) take into account hot topics in Evo-devo, such as canalization, generative bias, or epigenetic integration, and eventually essential properties of development such as its modular architecture (at all levels of complexity) or phenotypic plasticity (Walsh 2005: 193-194, Müller 2007), seeing that "a very large body of evidence shows that phenotypic novelty is largely reorganizational

---

[21] Remember the quote by Chomsky himself (2005: 9) we referred to in section 3.2: the Minimalist Program implies "shifting the burden of explanation from the first factor, the genetic endowment, to the third factor, language-independent principles of data processing, structural architecture, and computational efficiency".

rather than a product of innovative genes" (West-Eberhard, 2005: 6547). The recent paper by Lorenzo and Longa (2009) stands as one noteworthy effort in such a direction, in as much it represents an implementation of the minimalist theses from the DST perspective (see also section 4.2).

5. Conclusions

Nowadays judging language as a biological object and thus considering the possibility of fractioning it into different biological components appear as natural demands of most (consolidated) theoretical models of language and linguistic paradigms, which generally regard the FL as a component of the human mind and language as a product of brain activity. Such an approach to language can be deemed a viable aspiration if we take into account the diverse technical advances currently accomplished in the fields of genetics, molecular biology, and neurobiology. While merely identifying and structurally characterising such biological components of the FL is in essence a technical (but never trivial or easy to attain!) task, to subsequently explore and assess the way in which those components interact at different levels of biological complexity and jointly contribute to regulate the emergence of the FL at the term of growth (whenever growth takes place in a suitable environment) clearly represents a much more appealing endeavour, although simultaneously an urgent "must" if we actually intend to achieve a thoroughly accurate depiction of human language. Ultimately, our challenge is to relate such a hypothetical model of linguistic ontogeny in biological terms with existing theories on language acquisition by the child, and eventually, to develop bridging models capable of conciliating them (which parenthetically can entail a—substantial—modification of such pre-existing theories of language acquisition).

Genetic/molecular data (in essence, biological information regarding genes and their direct products) clearly represent a suitable and quite unproblematic (clearly from a methodological point of view, but to some extent even from a linguistic perspective, see sections 3.2, 4.2, and 4.3) entrance to such a biological network. Starting out from the diverse conditions in which language is impaired, in the last decades we have managed to identify and functionally characterize several dozens of genes putatively related to the development of the neural substrate of language. Recent insights into the developmental dynamics followed by biological systems are helping to properly apprehend the actual role played by genes in regulating the emergence of the FL, which has been clearly overemphasised by certain (and very successful) models of the FL, and subsequently by theories of language acquisition heavily relying on the existence (and the necessity for properly explaining language acquisition) of a substantial amount of genetically encoded grammatical information. Epigenetic factors, maternal gradients, ontogenetic constraints or "generic" factors ultimately determining the ways in which the remaining involved elements interact and biological systems self-organize, progressively emerge as being highly relevant elements crucially contributing to the robust development of an FL at the term of growth, thus minimizing the foremost role previously assigned to genes, while at the same time still making the case for innateness as an accurate (and necessary) way for explaining language acquisition. Incidentally, this reassessment of the manner in which developmental processes take place is also contributing to a better understanding of the (sometimes unexpected or fuzzy) phenotypic profiles of people bearing mutated versions of such genes, thus contributing to an improved setting up of the real aetiology of these kinds of conditions.

Moreover, since many genes related to language encode regulatory factors (see section 3.3), a reasonable hypothesis seems to be that some such genes could have been selected in order to accomplish the necessary remodelling of certain neural circuits so as to gain more computational power (eventually going beyond strictly lineal computational regimes and accomplishing context sensitive regimes) and/or to achieve the coupling between conceptual systems and sensorimotor systems previously evolved (incidentally, the long-lasting evolutionary trajectory of these systems makes the suggestion that most genes related to language are ultimately devoted to the regulation of the development and functioning of such "external systems" realistic, a hypothesis which satisfactorily matches the outcome of the genetic and neurobiological analyses of language disorders). Such remodelling processes would depend as well (both ontogenetically and phylogenetically) on the very dynamics followed by the nervous system during growth and on the external stimuli to which the individual is exposed to.

In the present contribution we have intended to depict, as accurately as possible, a plausible characterization of the genetic foundations of language, which is also in accordance with the recent (and appealing) achievements of (evolutionary) developmental biology. While certainly diminishing the central causal role previously assigned to genes in explaining language acquisition, we think that such a characterization emerges as a more realistic description of how the FL actually develops, but also as one conceivably amenable to being productively coupled with the most up-to-the-minute (and most engaging) conceptualizations and models of human language appearing in the field of Linguistics.

CHAPTER EIGHT

THE EVOLUTION OF THE FACULTY OF LANGUAGE*

GUILLERMO LORENZO

1. Introduction

A main challenge for the study of language evolution is customarily referred to as the "problem of continuity" (Chomsky 1968, Langer 1960): The behavioral repertoires of species with a close evolutionary link with humans (such as other primates) do not show anything that can really be classified as language. Actually, if comparisons are focused on the communicative behaviors of those species, the impression that nothing resembles the symbolic and computational richness of language becomes even more intense (Bickerton 1990, Hauser *et al.* 2002). An when the contention is made that forms of communication more akin to language are to be found within some species of birds and insects (Bickerton 2009, Chomsky 2000c), then we have entered the realm of confusion. From all these observations, the conclusion is reached that language defies the Darwinian formula of "descent with modification" from an ancestral form of communication.

Different theoretical strategies have been put forward in order to overcome the problem of continuity. Bickerton (1990), for example, argues that the problem disappears if language is disconnected from the evolution of communication and is instead related with other facets of animal cognition reflected in other behavioral areas. More recently,

* This work has been carried out through the project *Biolingüística: evolución, desarrollo y fósiles del lenguaje* (FFI2010-14955) from the *Ministerio de Ciencia e Innovación* (Spain) and partially co-funded by FEDER funds (EU). I would like to express my gratitude to Sergio Balari, who is almost as responsible as the author for the ideas put forward in this chapter—except for errors.

however, Bickerton (2009) has contended that language is a form of animal communication and that its discontinuity relative to other instances of this natural kind is to be explained as a consequence of the special environmental conditions that explain its origins. Furthermore, he argues that the fact that forms of communication similar to language are found within ants and bees should not surprise us, because these species have evolved under environmental pressures also similar to those that explain the recent evolution of the human species. The radically different stances that Bickerton has defended over time are actually illustrative of the two main positions currently held on the continuity question: (1) language is a form of communication that, at least in part, descends from forms of non-communicative behaviors; and (2) language is a descendant, among others, of an ancestral form of communication, modified under the exceptional circumstances of human evolution relatively to that of other closely related species.

The ideas that I will put forward in this chapter depart from both these positions: I will defend that language is not a modified version of any form of behavior, communicative or otherwise. There are many conceptual reasons underlying this position. Firstly, "communication" is not the name of a natural kind and, consequently, is not in itself a phenomenon subjected to evolutionary modifications. Furthermore, "behavior" is not a natural dimension of organisms exposed *per se* to evolutionary modification and descent; the true targets of evolutionary change are organic structures and, more specifically, the systems underlying their development in the individual. These developmental systems, by the way, are highly conservative, much more so than the highly variable structures in which they are expressed might lead us to think. This purports to say that the degree of the observed divergence between organic structures and the behaviors they serve is not a reliable index of the degree of divergence between their respective systems of development. Finally, another consequence of all this is that species are closer at this developmental level of biological analysis than they are at the anatomical or other higher levels of analysis, and that small changes in the systems of development can give as outcomes, affinities, or departures between species that apparently contradict their distance or proximity in terms of evolutionary kinship.

In this chapter I will explain how this collection of ideas must reflect on a biologically coherent explanation of language evolution. Because these ideas rely on considerations that defy some strong intuitions and current theoretical assumptions concerning the nature of language (Balari & Lorenzo 2009a), as well as on some technicalities of current Evolutionary Developmental Biology (Carroll 2005, Hall 1999, Minelli

2007), I will devote some space to justify these considerations before translating them into an articulate picture of how such a peculiar biological structure as language could possibly evolve.

2. What is it that makes language so special?

Let us start by observing the songs of Bengalese finches (*Lonchura striata domestica*), of which we have a detailed formal analysis by Okanoya (2002) (see also Berwick *et al.* 2011). These songs basically consist of a series of "notes", represented by lower case letters in Fig. 8-1, subject to the following constraints:

1) Some notes can only be followed by some other specific note (for example, "b" must be followed by "c" and "c" must be followed by "d"), which means that these notes compose fixed motives within the song template;

2) Some other notes, however, can be followed, in different songs, by one or another different note (for example, "d" can be followed by either "i" or "e"; after "e" or the motive "im", however, only "f" can appear), what means that at points like this the song is open to probabilistic transitions;

3) Finally, there is a single note ("b") which can be either the starting point of a song ("S") or the note following some other notes (namely, "p" or "q"; "ε" represents the possibility of omitting a certain note, such as "q", or a certain motive, such as "hnfo"), which means that this note allows the introduction of a new sequence obeying the constraints of the pattern so far described.

We thus conclude that, even if probabilistically at certain points, the possible transitions within the sequence of a particular song are always predictable: Each note always follows one or another note (two at the most). For this reason, this type of birdsong can be formalized by means of a "finite state automaton" (FSA) and we can assign it a language of "Type 3" (T3) within the scale of complexity of the Chomsky Hierarchy (Chomsky 1956, 1959b). This amounts to saying that, however relatively complex within their type, the execution of these songs requires a relatively simple type of working memory, from which a note can be deleted as soon as the following note is introduced.

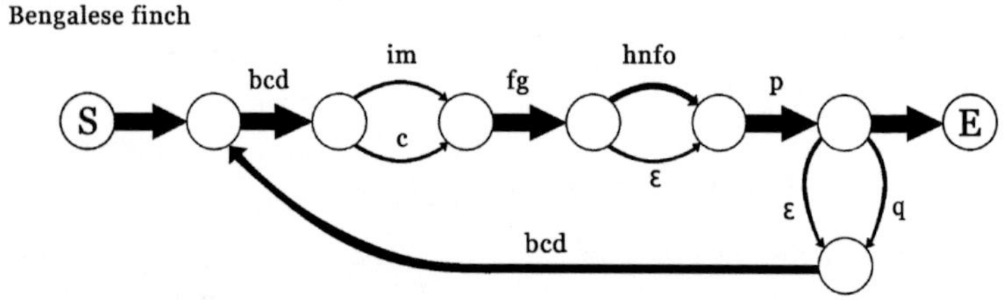

Fig. 8-1 Possible transitions between the notes (lower case letters) composing the songs of Bengalese finches by means of a "finite state automaton" (FSA). The introduction of each new note is only constrained by the previously introduced note. From Okanoya (2002: 49).

Let us move now to another apparently more complex form of birdsong, that of nightingales (*Luscinia megarhynchos*). According to Todt and Hulstch's (1998) analysis, the superficial patterns of the songs of this bird are certainly more complex than those of finches; anyway, their execution does not presuppose anything more complex than an FSA. The main peculiarities of this form of birdsong are the following:

1) Songs are composed of series of stereotyped sequences of notes (motives or strophes; Latin characters in Fig. 8-2);

2) Songs divide into three different sections (Greek characters in Fig. 8-2), each section showing a particular intensity and its own collection of motives;

3) Some motives can be followed by more than one different motive (alternative vertical boxes in Fig. 8-2), considerably opening the number of alternative sequences. A single motive can also be repeated (loops around a single box in Fig. 8-2) or it can be followed by a repetition of the previous motive (loops around two boxes in Fig. 8-2) at certain points near the end of the song.

These features obviously add a certain degree of complexity to the songs of nightingales relatively to that of other birds. However, the property is maintained that transitions from one motive to another are always predictable, as well as transitions from one section to another (the first motive of a new section is predictable from the last motive of the previous section). This means that sequences representing these songs can be formalized by an FSA flowchart. Accordingly, we conclude that the working memory system required for their execution only needs to maintain in active memory a single motive at each stage of the performance, and that it is deleted as soon as the next motive enters the computation. Therefore, these birdsong sequences also belong to a T3 language. The higher degree of complexity relative to other birdsongs is just quantitative.[1]

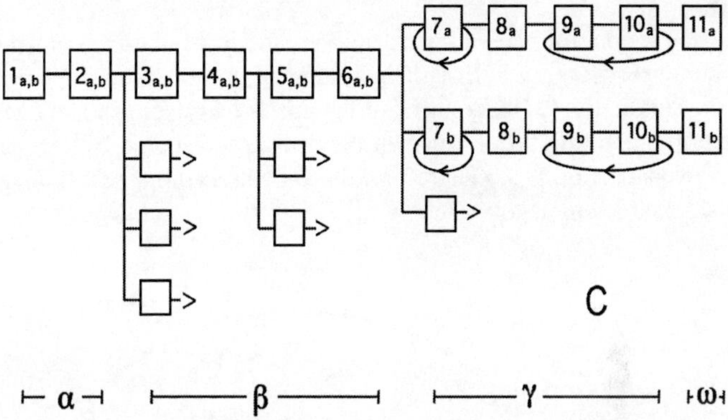

Fig. 8-2 Nightingale's songs are relatively complex, but transitions between motives (Latin characters) and even sections (Greek characters) are predictable in a way describable by an FSA flowchart. From Todt & Hultsch (1998).

Enough about birds, let us move now to non-human primates. Thus far, the most complex inventory of vocalizations known consisting in combinations of minor call units is that of Campell's monkeys (*Cercophitecus campbelli*), described by Ouattara *et al.* (2009). These monkeys make use of six different call units (B, K, H, W_+, K_+, H_+; the "+" symbol represents a certain enlargement of the unit in question), whose combinations correlate with some kind of predatory or non-predatory danger. The "meaning" of combinations modifies or, sometimes, completely cancels that of the units by themselves.[2] Some units are exclusively used within combinations.

If we focus our attention on vocalizations with a non-predatory meaning, we can observe, for example, that one repetition of B (BB) correlates with the loss of eye contact with the rest of the group, but if this sequence is followed by a certain number of repetitions of K_+

[1] See Slater (2000) for a catalogue of different types of birdsong, and Marler & Slabbekoorn (2004) for a complete overview of this form of animal behavior.

[2] Similar effects have been described by Arnold & Zuberbühler (2006a, 2006b) in the complex vocalizations of other species of *Cercopithecidae*.

(BBK_+K_+...), then it correlates with the sudden fall of a tree or a branch. And if the same BB sequence is followed by repetitions of H_+ followed by repetitions of K_+ (BBH_+H_+... K_+K_+...), then the sequence is normally used as an answer to calls coming from individuals of neighbor or strange groups.

Putting aside for the moment the semantics of these vocalizations,[3] what we observe again is that their system of composition does not offer anything qualitatively different from the case of birdsongs: The execution of the corresponding sequences can be easily represented by means of an FSA flowchart (Fig. 8-3). Therefore, these vocalizations belong again to a Chomskyan T3 language.

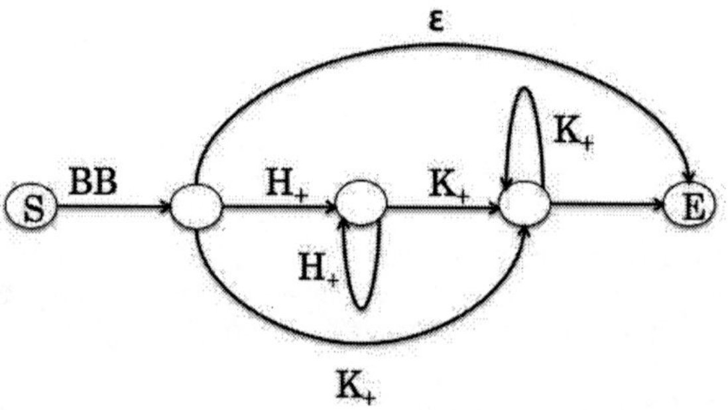

Fig. 8-3 Possible transitions between the different call units (capital letters) composing the complex non-predatory vocalizations of Campbell's monkeys, represented as an FSA flowchart. Each one of the three alternative transitions from BB (obligatory) introduces a different meaning. Based on data from Ouattara *et al.* (2009).

The meaning of this is that, from a combinatorial or computational point of view, performing this type of vocalization does not require more complex cognitive resources than those already available to the aforementioned avian species. Actually, among the great variety of sequences produced by different species of birds and by different species of (non-human) primates, the highest degrees of complexity (within the limits of the Chomskyan T3) are found within the former,[4] a fact that

[3] See note 6 below.

[4] On the attribution of types and degrees of computational complexity to patterns of animal behavior, see Pullum & Rogers (2006).

somehow contravenes any expectation based on grounds of evolutionary proximity.

The next crucial point in my argumentation is: How could we obtain a qualitative leap in the complexity of vocalizations in other respects similar to that of Campell's monkeys? Imagine that the system of composition introduces the constraint that a certain number of repetitions of a call unit (say, H_+) are to be followed by exactly the same number of repetitions of another unit (say, K_+). That is to say, that an imaginary organism were capable of discriminating, for example, sequences like $BBH_+H_+H_+K_+K_+K_+$ from sequences like $BBH_+H_+H_+K_+K_+K_+K_+$ or $BBH_+H_+H_+K_+$. This single fact will locate this system within the Chomskyan T2, because it would imply a memory device capable of maintaining for a while in active memory a set of units, until a certain point in the computation is finally reached: In our imaginary example, the system should maintain in working memory at least the first token of H_+ until the last token of K_+ is finally introduced, as required to compute that the same number of tokens of both units has been used.[5] The workings of such a system can be replicated by what is technically known as a "pushdown automata" (PDA). No vestiges of such a system has been found, however, in the different forms of "minimal syntax" (Újhelyi 1996) of primates thus far studied in the wild and there exists some experimental evidence that it could be beyond their cognitive capabilities (Fitch & Hauser 2004; see Perruchet & Rey 2005 for some critical comments).

Curiously enough, this type of computation has been attributed to certain experimentally induced behaviors of European starlings (*Sturnus vulgaris*). Apparently, after a period of training these birds are able to tell apart artificial sequences consisting in a certain number of repetitions of two different call units of the species (AB^n) from sequences consisting in a certain number of repetitions of one call unit followed by the same number of repetitions of another unit (A^nB^n) (Gentner *et al.* 2006). Note that if "n" refers to an arbitrarily large number of repetitions, then the processing of the second type of sequences requires a PDA. The bad news is that starlings fail to differentiate one type of sequence from the other when "n" equals or is higher than four, so the question remains indeterminate if they are using a PDA or a more elementary capacity of rote memory (Longa *et*

[5] Actually, it would require to maintain the three tokens of H_+ until the first one of K_+ is introduced, the first two tokens of H_+ until the second one of K_+ is introduced, and the first token of H_+ until the last one of K_+ is introduced. This is technically expressed by the name 'last-in-first-out' given to the memory regime of such a system.

al., In press). Anyway, it is significant that evidences of proximity to the computational complexity of language, even if dubious, are systematically found in unexpected species from the point of view of evolutionary descent.

A conclusion is in order. It is pretty clear that there exists a gap between the computational type and the cognitive resources of language, on the one hand, and those that we meet in other non-human forms of communicative behavior, on the other hand. No other animal behavior based on the transmission of signals thus far studied seems to reach a level of computational complexity higher than that of Chomskyan T3 languages. Language, in contrast, not only arrives at the level of complexity of T2 languages (structural embedding), but also shows long distance and crossing relations (agreement, co-reference, displacement, etc.) distinctive of Chomskyan T1 languages. This means that processing linguistic signals presupposes the workings of an "enhanced pushdown automaton" (PDA+) (Uriagereka 2008b), capable of maintaining in active memory a whole history (or context) of previous computational stages while the process is still running. This gap, represented in Fig. 8-4, does not completely deny that the computational system of language could be the result of a gradual process of evolution starting with the speciation of early hominids (Pinker & Bloom 1990), but it seriously put into question such a possibility.

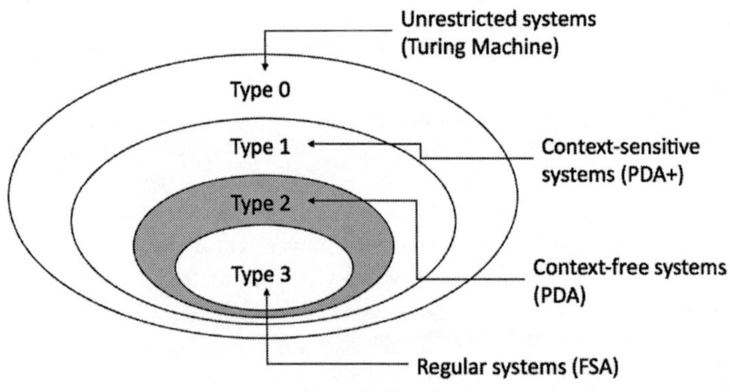

Fig. 8-4 Chomsky Hierarchy of formal complexity. Inclusion relations represent that languages of a higher order of complexity have access to the type of operations of languages of a lower level, but not the other way round. The shadow represents the zone of the Hierarchy to which apparently no animal system of communication belongs.

It is true that the vocalizations of monkeys like that of *Cercophitecidae* are meaningful in a way that birdsongs,[6] produced for attracting the attention and gaining the receptiveness of potential mating partners, are not (Marler 1998). This is evidently an aspect in which the similarity of the behavior of human and non-human primates is in agreement with their evolutionary closeness. However, that evolutionary expectations are fulfilled in this case actually increases the conflict that the gap in the computational side represents, because this gap goes in the opposite direction: It increases rather than decreases as we approach our closest relatives in the animal kingdom.

3. Does animal communication really exist?

The problem thus far settled reads like this. It is not exactly true that there is nothing that really resembles language. Something of the sort certainly exists. However: (1) there is a true gap in terms of computational complexity between language and any other form of animal communication, and (2) the gap does not decreases but rather increases, as a function of evolutionary proximity. Let us start moving towards a solution to this puzzle.

To start with, it is important to observe that the so-called "problem of continuity" is based on a problematic, actually ill founded, methodological premise: That of restricting the application of the comparative method in the evolutionary study of language to behaviors deemed "communicative" (Bickerton 1990, Hauser *et al.* 2002). And this actually rests on a further, and even more problematic, conceptual premise: That of admitting without argumentation that what we routinely call "animal communication" comprises a natural kind susceptible of motivating relevant generalizations from an evolutionary point of view. As this premise is untenable, the next logical move is to open the focus of comparisons and to extend it to any domain of animal experience that could prove informative for evolutionary

---

[6] The semantics of these vocalizations are, however, rather different from that of linguistic utterances, in that the former lacks the "compositionality" typical of the latter. In Campbell's monkeys, a call unit within a combination does not preserve the meaning that it has when used independently or within another combination, but it is modified or canceled. In the case of language, on the contrary, the meaning of the parts is preserved in the meaning of the combinations in which they participate. In any case, the fact that the vocalizations of Campbell's monkeys are simultaneously meaningful and computationally complex is an interesting piece of information from an evolutionary point of view.

concerns. But we need to start by showing that the premise in question is actually wrong.

Especially within the ethological tradition (Smith 1977 for a synthesis) but also within biologically oriented linguistics (starting with Hockett 1958), any form of behavior based on exteriorizing some type of signal falls under the rubric of "animal communication". Thus, calls for alarming, dissuading, distracting, seducing, etc., used by the most diverse species in interactions with individuals of the same or different species are customarily included into the inventory of animal communication systems (Hauser 1997, Slater 1999). Furthermore, given the advantages of these practices in terms of survival and reproduction, it becomes difficult to argue against the idea that they result from the action of Natural Selection (Smith 1977). However, what is clearly questionable, to begin with, is that all these skills are instances of one and the same form of biological behavior, call it "communication" or otherwise (Balari & Lorenzo 2009a).

To be clear, the problem is not the inclusion of language into a single catalogue with the rest of animal communication systems, as it currently done by the aforementioned ethological tradition (Eibl-Eibesfeldt 1984: ch. 6) and by different schools of functional linguistics (Dik 1978): The much more basic problem consists in dealing with "communication" as a biological function. That language is not a system "for" communication (Chomsky 1968) or that, even if it is, language most probably is not a descent of an ancestral system of communication (Bickerton 1990, Hauser *et al.* 2002), are ideas that modern generative grammar has firmly established in the last decades and are at the center of the "form-function" debate that language continues to incite (Newmeyer 1998). But the more radical idea that no animal communication system really is an animal communication system (Balari & Lorenzo 2009a) is perhaps a novelty in this debate. New and radical as this certainly is, there is a very strong conceptual reason for embracing it: chemical emissions, dances, songs, calls, utterances, etc., comprise such a heterogeneous collection of behaviors that they cannot be categorized as different instances of a single natural kind nor, crucially, as different outcomes of a single process of evolutionary diversification. Astounding as it might seem under this new light, these are however axiomatic stances of programs of investigation still centered around the idea that communication is a biological function extremely diversified under the influence of the most varied environmental pressures (Hauser 1997).

A caveat is in order: My criticism does not mean that natural skills such as those mentioned above cannot receive the attention of particularized investigations; nor that evolutionary links between some of them could eventually be unveiled. What the criticism exactly amounts to is: (i) that once these natural practices are characterized in practical terms (as a means for attracting, distracting, seducing, alarming, and so on), adding to the relevant characterization that, besides this, they are forms of fulfilling the "communicative function", from a biological point of view, means nothing at all; and (ii) that even if evolutionary links are discovered between some of these practices, as it is expected that they will be, it will still continue to be the case that they will not be linked *qua* communication systems.

Hopefully the argument of this section should have a liberating and refreshing effect on the study of language evolution: There is no reason for restricting our attention to any particular sphere of animal behavior or experience when applying the comparative method that the question calls for. Our attention must be directed to discovering whatever skills, no matter their utility or domain of application, that presuppose cognitive resources akin to those required by language. A nice insight in this respect is the observation due by Camps and Uriagereka (2006) that tying knots is an ability that actually requires a type of context sensitive grammar like those of Chomskyan T1 languages (Fig. 8-4): Note that when tying knots we apply new operations over certain specific points of the result of (non-immediate) previous operations, which must be stored in active memory until the complete figure is attained. Besides, Balari and Lorenzo (2008, 2009b) have added to this observation that some avian species tie knots when constructing their nests (Collias & Collias 1962), some of them very similar or even identical to certain human knots (Fig. 8-5). This is, unquestionably, an unexpected domain of animal experience to which attention is due when asking for the evolutionary origins of language. I will come back to knots, birds and language after some further criticism directed toward another ill-founded routine of current evolutionary linguistics.

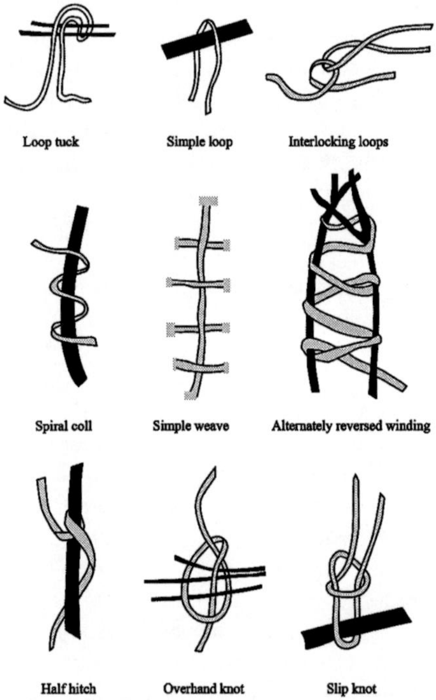

Fig. 8-5 A collection of knots made by different species of weaver birds (*Ploceidae*). Some of these presuppose a computational regime equivalent to that of Chomskyan T1 languages, to which human language also belongs. From Hansell (2000: 85).

4. Does behavior really evolve?

Once communication is rejected as an evolved and thus diversified biological function, evolutionary linguistics should not repeat the same mistake again, trying to identify another skill or capacity from which language could be declared a modified descendant. Thus my previous reasoning cannot be understood as establishing the position that language is an evolutionary variant of the constructive behaviors exhibited by other species or, for that matter, of any other form of behavior whatsoever. Actually, any assertion of the form "language is an evolutionary variant of behavior X" is ill formed, I contend, for the very reason that observable behaviors are not *per se* objects of evolutionary shaping (Klopfer 1973).

To properly understand this we need to clarify, following Love (2007), the ambiguity with which the concept of "function" is routinely used as referring to a distinct level of biological analysis: On the one hand, "function" is sometimes used to refer to the "activity" of a given organic system as a direct consequence of the parts it consists of and how these parts are organized; on the other hand, the concept of "function" is sometimes used to refer to the "use" given to the same structure as a consequence of how it is connected with other organic structures within the same organism and how this organism relates with some particular environmental conditions, including other organisms. In order to avoid ambiguities and clarify the level of analysis in which we move when using a functional vocabulary, from now on I will be exclusively using the concepts of "activity" and "use" instead of the equivocal concept of "function".

But what I am really interested in emphasizing here is that when we talk about the "activity" of an organ or an organic system, we are focusing on it from the point of view of its structure: In other (maybe misleading) words, from a non-functional point of view, because we are attending to the effects that follow from the way it is configured. The formal configuration and the activity of an organ, however conceptually discernible, are non-autonomous facets of a single (non-functional) level of biological analysis (Balari & Lorenzo 2009a, 2010). A properly speaking functional level of analysis is only reached when we further consider the usefulness of the organ in question to its holder within an organic, populational and environmental framework. However, such a properly functional level of analysis introduces certain conceptual difficulties that make it rather problematic from the point of view of evolutionary biology.

The first of such difficulties is that the description of organic structures in terms of their usefulness is usually based on categories that, on the one hand, come from different domains of human affairs and interests; and, on the other hand, are used as transcendental concepts whose usefulness is recognized as an *a priori* matter, so it is naturally accepted that they can become embodied in the most diverse types of organic structures. The former inevitably gives an anthropomorphic bias to research (Balari & Lorenzo, in press, Owen 1849, Searle 1995); the latter unconsciously opens the door to incorporeal categories in relation to which the specifics of the biological entities manifesting them becomes of secondary importance (Balari & Lorenzo 2009a). Functional analysis thus leads to a covert form of Platonism. The concept of "communication", as it is customarily used in evolutionary linguistics, is a clear illustration of that position, but many other common categories in biological analyses based on the observation of superficial behaviors, such as "alarm", "seduction", "deception", etc., also illustrate the same point: They are based on a human model and superimposed over the most varied kinds of organs and organisms (Balari & Lorenzo 2009a).

It is a truism that the usefulness of a given biological structure has an impact in the fitness of its holder and that it thus deserves attention from an evolutionary point of view: it is, no doubt, a part of the explanation of the fact that the organism in question *is there*, instead of having given way to other more efficient organisms. However, evolutionary explanations will be unmistakably ill founded if the following points are not at the same time taken into consideration:

1) Explaining that something *is there* is not the same thing as explaining *how it managed to be there* (Cummins 2002). This point is the subject of the next section.

2) Relative fitness is one thing, and a label with which we try to capture (Searle 1995) or to emphasize (Fodor & Piattelli-Palmarini 2010) a practice introducing different degrees of fitness is a very different and non-coextensive thing. Actually, these labels introduce a certain distortion of the facts, because they give place to numerous unjustified identities ("communication systems") and distinctions ("communication systems" *vs.* "seduction systems"), to which no explanatory power can thus be given.

3) Any such practice is, as explained above, the outcome of the convergence of different organic components in a particular type of environmental setting. This amounts to saying that practices or behaviors in which the "use" of organic structures and activities becomes manifest are not, properly speaking, objects of evolutionary modeling. Rather, they are indirect effects of complex, but in a certain way instantaneous, processes of integration of the real targets of evolution: organic structures and their associated activities. Those behaviors, whatever we decide to label them—a rather inconsequential question, as argued—have a direct repercussion on the proliferation and stabilization of the underlying structures, but are not in themselves naturally evolved phenomena.

Actually, the argument thus far put forward is in clear correspondence with the intuitions underlying the classic concept of "homology" (Owen 1843; Russell 1916, for a historical overview), with which relations of identity relevant from an evolutionary point of view have been traditionally captured. From Owen's (1943) definition on, the concept applies specifically to structures as well as to activities, once Love's (2007) qualification is attended to, deemed "identical" despite their formal and, crucially, functional diversification (Fig. 8-6). The contents of this section should thus be read as a plea for a strict application of the concept of homology to the evolutionary study of language: direct comparisons with observable practices or behavior, interesting as they can be from ecological or other concerns, are misleading when the aim is that of uncovering the evolutionary history of language.

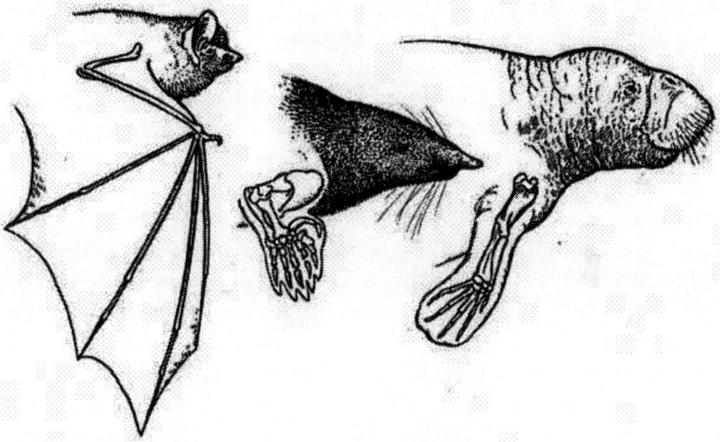

Fig. 8-6 The wings of bats, the paddles of moles and the fins of dugons are "the same organ under every variety of form and function" (Owen 1843: 379). This form of structural identity, known as "homology" from Owen's works, is the most distinctive signal of a shared evolutionary history.

5. Computing as an evolved organic activity

The previous section is an important first step in order to clarify the sense in which it is legitimate and *could* eventually proved true an evolutionary link between language and, *for example*, the constructive abilities of weaver birds. Note, however, that the underlying expressions introduce non-trivial qualifications regarding the ideas that I will put forward in this new section: Firstly, it is a link which has not been experimentally proven; secondly, it is not supposed to be a relation exclusively held between these two natural capacities. Most probably, it is a particular instance of an eventually huge family of "activity related" capacities. Were this point proven true, the next expectation would be that such a family comprises a very heterogeneous set of abilities from a functional (i.e., "use centered") point of view.

Needless to say, speaking and constructing nests are very disparate capacities from the point of view of their respective usefulness. However, the main conclusion of the previous section is that this is not an obstacle in order to establish a relevant evolutionary link between them attending to the fact that they require underlying computational capabilities of a similar level of complexity. It is then important to show that what I am calling "computation" could legitimately be the name of an organic "activity" associated to a certain organic structure, irrespective of the "uses" into which such an organ/activity translate in different organic and environmental contexts. My particular contention is thus that the "same" system of computation is present in the most diverse organisms, even if highly diversified form a formal point of view (for example, in terms of relative complexity) and connected with superficially very heterogeneous behaviors (singing, speaking, constructing nests, etc.). Some clarifications are thus still in order before accepting the idea that "computing" certainly refers to a natural kind of homologues of which evolutionary generalizations can be confidently asserted (Balari & Lorenzo, in progress).

"Computing", in the sense relevant to my argumentation, means sequencing and relating items according to the different potential of each particular system (Fig. 8-4): If the system only obeys lineal constraints, then we have a T3 system; if the constraints also apply to groups or sets of items, then we have a T2 system; if the constraints affect items at long and potentially crossing positions, then we have a T1 system. A forth type of system exists, T0, which can ideally deal with any kind of constraint, but, as I explain below, it is not relevant from a naturalistic perspective. What can then be said to be common to all computing systems is a "sequencer" or "pattern generator" (Lieberman 2006), while differences between such systems have to do with the potential of the "operative" or "working memory" capacity accessible to each type: It does not extend further than to the prior item of the sequence in the case of T3; it can contain a set of prior items in the case of T2; and it can contain a specification of the previous history (or "context") of the process in the case of T1. The power to establish relations between the components of the sequence as it is generated is correspondingly restricted. T0, as already observed, has no memory restrictions, but this is biologically irrelevant. Attending to these considerations, I conclude that natural systems of computation share a common structural design: "Pattern generator" + "working memory space" (Lieberman 2006), the latter being the *locus* of their range of variation (Balari & Lorenzo 2008, 2009b). Note that this characterization amounts to saying that the systems of computation underlying the songs of finches or nightingales, the alarm calls of monkeys, the nest constructing activities of weaver birds, or human syntax are homologous despite their diversification in terms of memory capacity. What is intriguing from an evolutionary point of view is the apparent departure of the latter two abilities in a parallel direction (T1), notwithstanding the phylogenic distance of their holders, a question to which I come back in the next section.

According to the point of view defended in the previous section, if we talk of "computing" as a distinct organic activity, this ultimately means that there must exist a particular organic system of which computing is the proper activity. I will refer to it as the "computational system" and, following Lieberman's (2006) proposal specifically directed to the case of language, I suggest that each separate aspect of the computational activity correlates with the following anatomical structures (see Benítez-Burraco, this volume):

1) The "sequencer" or "pattern generator" is located in the basal ganglia (Tettamanti *et al.* 2005), a group of structures defined by Lieberman as a repetitive sequencing machine underlying many different cognitive tasks, from motor to conceptual (Fig. 8-7);

2) The "working memory space" is located in Broca's area (Embick *et al.* 2000, Grodzinsky 2000, Moro *et al.* 2001, Musso *et al.* 2003), most plausibly forming networks with other frontal and parieto-occipital cortical areas (Aboitiz & García 1997, Aboitiz *et al.* 2006).

Caution is in order, but these are nevertheless very promising correlates for which certain empirical support can already be offered. An avian structure (area X) homologous to the human basal ganglia has been experimentally proved to underlie birdsong, and certain homologous genes have also been pointed out which are active during the development and activity of those structures (Ferland *et al.* 2003, Haesler *et al.* 2004, Haesler *et al.* 2007, Lai *et al.* 2003, Liégois *et al.* 2003, Miller *et al.* 2008, Rochefort *et al.* 2007, Takahashi *et al.* 2003, Teramitsu *et al.* 2004, White & Teramitsu 2006). Mutant versions of these genes lead to difficulties in the maturation and exercising of adult songs in the case of birds and to delays in language acquisition and difficulties relating to phonetics as well as to morphology and syntax in the case of humans (Gopnik 1997, Lai *et al.* 2001, Piattelli-Palmarini & Uriagereka 2009, Benítez-Burraco, this volume). These facts invite to extend the relevant experimental paradigms to other abilities, namely complex call vocalization or nest construction.

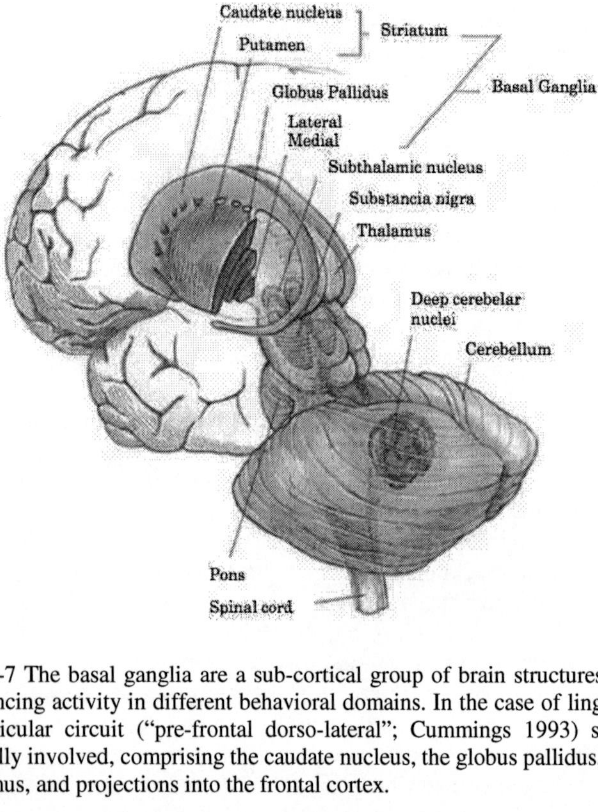

Fig. 8-7 The basal ganglia are a sub-cortical group of brain structures underlying sequencing activity in different behavioral domains. In the case of linguistic tasks, a particular circuit ("pre-frontal dorso-lateral"; Cummings 1993) seems to be crucially involved, comprising the caudate nucleus, the globus pallidus, parts of the thalamus, and projections into the frontal cortex.

The suggestion that I am putting forward in this section is also supported by the fact that the basal ganglia have been described as a highly conservative, almost unmodified, group of structures along the amniotes (Reiner *et al.* 1984), while the cortical part of their brains has been subject to the most wide-ranging evolutionary modifications in both close and distantly related species (Striedter 2005). This is obviously consistent with the identification of the cortex as the *locus* of main evolutionary novelties in terms of computational capabilities (Balari & Lorenzo 2008, Balari & Lorenzo 2009b).

As a point of clarification, it is important to remember that birds have a poorly developed anatomical homologue of neo-cortex, and that activities supported by the cortex in other species of vertebrates are mainly provided by the "dorsal ventricular ridge" in the case of birds (Striedter 2005).

Anyway, once the conceptual distinction has been established between "structure" and "activity" (Love 2007), it seems reasonable to admit that different structures can support the same (i.e., homologous) activity, like acting as a working memory device, providing that certain relevant properties are upheld, like maintaining the required connectivity and, crucially, a link with a sequencing machine. Therefore, this anatomical departure does not break the homology. Rather, it represents one particular parameter of the predictable "variety of forms" (Owen 1843) of computational systems (Balari & Lorenzo 2009a).

6. How do structures evolve?

As a matter of fact, nothing hitherto said explains the problem of continuity described in the second section. So far, what has been established is the existence of a set of homologous (i.e., identical despite formal and functional differences) systems of computation. It has also been observed that the degree of formal similitude, on grounds of computational complexity, does not correspond with the degree of evolutionary kinship of the holders of those systems, in that, for example, nest construction by some birds is computationally closer to language than, among other practices, the complex alarm calls of some monkeys. According to a particular (historical) point of view on the homology question,[7] this single piece of evidence should be enough to dispense with the idea that the underlying systems to this collection of behaviors are true homologues. In this section, however, I will put forward the idea that all this confusion can be cleared up by adopting: (1) the perspective introduced by the "biological concept of homology"; and (2) some general assumptions of current "developmental evolutionary biology" (Evo-devo).

The biological concept of homology (Wagner 1989a, 1989b) basically ascertains that two or more organic structures can be safely deemed true homologues if there exist common developmental constraints acting on their individuality and identity. It is not required, of course, that the factors obeying those constraints are the same in each particular situation, since variation at this level is the obvious source of the expected range of variation within families of homologues. The factors in question are actually an assortment thereof, ranging from genes to environmental inputs, whose strong links and concerted action guarantees deep

[7] See Lauder (1994), Rieppel (1994) and Stevens (1984), among other sources; Balari & Lorenzo (in progress) for a critique.

similarities of the developmental paths irrespective of the evolutionary distance.[8] In any event, given the present state of knowledge concerning such complex systems, the genetic level of analysis continues to be the most reliable source of data in order to corroborate homologies in the sense of the biological concept.

The central contention of current Evo-devo biology (Carroll 2005, Hall 1999, Minelli 2007) is that perturbations on the systems governing individual development are the main source of evolutionary novelties and of species diversification. A more specific Evo-devo claim is that, given the complexity of developmental systems in terms of number and type of factors, levels of organization and paths of interactions, relatively small perturbations on particular points of the system can bring about complex chain reactions with wide-ranging effects on the organic structures attained. Actually, a well-established fact due to Evo-devo inspired research programs is that developmental systems are highly conservative (the basic corporal pattern of flies and mice are based on quasi identical systems of development), and that small modifications against this uniform background are responsible for dramatic superficial differences (the body of flies and mice are remarkably different).

Note that these two approaches jointly point to a possible answer to the problem of continuity, because correlations between formal affinities and evolutionary closeness are not necessarily expected from the point of view of Evo-devo. If evolutionary diversification obtains from a strong background of developmental homogeneity, then distantly related species can converge non-problematically in specific design features. We can thus take as possible that certain qualitative divergences, like having *vs.* not having access to a certain degree of computational complexity, result from relatively minor quantitative perturbations on specific factors governing the development of computational systems and, accordingly, that different computational homologues could have undergone similar perturbations and gone through parallel evolutionary paths, no matter the genealogy of their holders.[9]

I contended in the previous section that the *locus* of variation of qualitatively different systems of computation (T3, T2, T1) is to be related to the cortical structure (or its equivalent in the avian case) where the

---

[8] See Gilbert & Epel (2008), Jablonka & Lamb (2005), Oyama (2000a) and Robert (2004), among others, to get a picture of the complexity of the factors implied and the interactions held within systems of development.

[9] This case would thus fall under the more general phenomenon of "syngeny" (Butler & Saidel 2000) or "deep homology" (Shubin *et al.* 2009).

working memory device is most probably located.[10] In the case of humans, the exceptional enlargement of the cortical structure can be safely put in relation to certain evolutionary modifications in the chronology and the intensity of its development relative to other closely related species. The three main phenomena that have been pointed out as the mechanistic causes of this overdevelopment of the human brain are: (1) Increments in the production of precursors cells, (2) the extension of the stage of asymmetric divisions, and (3) the amplification of cytogenesis (the stage starting with the production of precursors and ending with a maximum of cell divisions in which one of the cells does not divide further, but migrates to the emergent cortical structure). This latter mechanism, by the way, triggers an exponential growth of the attained brain structure, which in the case of primates is especially noteworthy in the cortical structure relative to other brain structures, such as the spinal cord and, meaningfully enough, the basal ganglia (Fig. 8-8). Besides, cortical overdevelopment is further strengthened by the second mechanism, triggering, in the case of humans, a neural population eight to sixteen times larger than in other primate species (Kaskan & Finlay 2001, Rakic & Kornack 2001).

These observations lead to a picture that coheres very well with the ideas put forward up to this point. In the case of humans, a conspiracy of developmental modifications has triggered an exponential and somehow abnormal growth of the cortex and,[11] correspondingly, an enhancement of circuits serving as working memory devices, paving the way to the access to qualitatively new computational subtleties. Such a picture is especially consistent with a scenario of concerted point mutations targeting certain regulatory genes acting on the proliferation, division, migration, growth and death of neuronal cells by orchestrating the activity of myriad other genes and protein products. Mutations affecting regulatory mechanisms, even if quantitatively minute, are the most obvious candidates when outcomes are qualitatively phenotypic novelties (Balari & Lorenzo 2008, Balari & Lorenzo 2009b).[12] This does not purport to say, however, that other non-genetic developmental factors cannot be given an important role in this evolutionary process. Actually, processes of phenotypic plasticity (West-Eberhard 2003) have most probably been involved in the fixation of

[10] As it would require an unlimited space of operative memory, T0 is a naturally non-viable computational type (Balari & Lorenzo 2009b).

[11] Technically, this is a case of "delayed offset peramorphosis" or "hypermorphosis". See Balari (2006), Balari & Lorenzo (2009b), and references therein.

[12] See Benítez-Burraco (2009: ch.5, this volume) and Balari *et al.* (in press), for a list of some specific candidate genes.

such an exceptionally enlarged structure, whose complexity escapes any intent of explaining its organization exclusively by means of genetic activity. Anyway, the search for point mutations such as those commented on above seems a good first step in order to begin to understand the mechanistic bases of this undoubtedly complex evolutionary process.

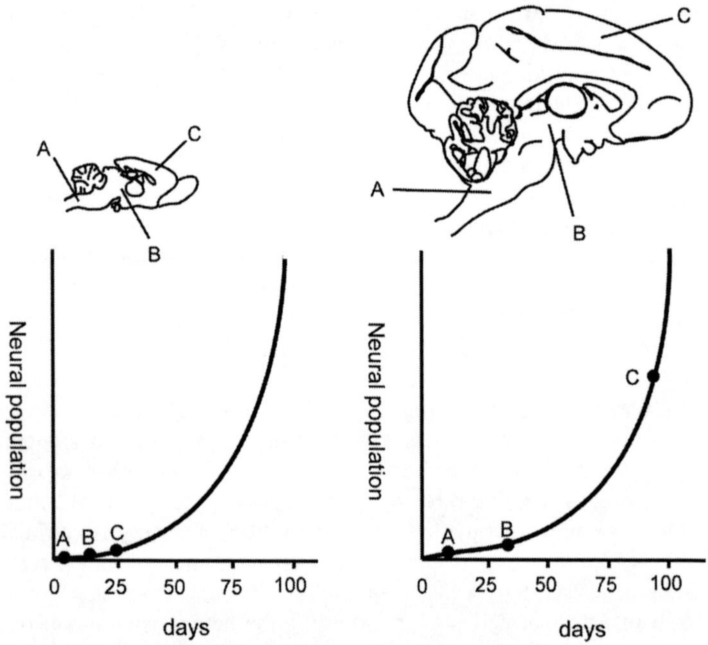

Fig. 8-8 The enlargement of cytogenesis triggers an exponential cortical growth (C), an effect especially more noteworthy in the case of primates (right) than in other species like rodents (left). Other cerebral structures such as the spinal cord (A) and the basal ganglia (B) respond in a more lineal fashion to this enlargement. As a consequence of this, the cortex is an abnormally outsized structure, especially in the case of humans. From Kaskan & Finlay (2001).

7. Computing is not speaking (singing, screaming or nesting)

Computing is obviously not speaking. As we have seen, a computational system can be used to sing like a nightingale, to scream like a Cambpell's monkey or to construct nests like a weaver bird. This means that a complete evolutionary explanation of language will not be fulfilled until a model is provided to explain how this core computational apparatus integrates into a more encompassing organic context, so a "faculty of language" properly speaking really obtains.

Actually, this aspect of the evolutionary explanation of language is that which more directly relates to the superficial differences of the various abilities I have been contrasting throughout this chapter. Avian skills like singing and nest building, on the one hand, do not show "referenciality" (or "semanticity") as one of their properties;[13] primate capabilities like screaming and speaking, on the other hand, do show it. This means that the respective computational systems are in contact with different cognitive components in each case, thus there emerge correspondingly different complex faculties, suitable to rather different practical uses (Balari & Lorenzo 2008, 2009, Hauser *et al.* 2002). Another parameter with an obvious impact on the differentiation of the resulting faculties is the degree of complexity independently attained in the sensorimotor and conceptual domains accessed by the computational system. In this respect, the contrast between human and non-human primates reveals that differences in both cognitive sides (sensorimotor and conceptual) are probably as remarkable as those observed in the respective computational capabilities.[14]

In this regard, the only observation I will make in this chapter is that the same Evo-devo model put forward in the previous section to explain the evolutionary emergence of language's T1 computational system could also serve to explain its evolutionary connection with the other cognitive areas that make up the faculty of language proper. The base of this contention is that the same developmental means underlying cortical growth could also translate into certain spontaneous cortical reorganization processes, the evolutionary significance of which has been independently defended: (1) the formation of independent areas or nuclei (Ebesson 1980), favoring the specialization and complexification of activities; and (2) the mutual invasion of those areas (Deacon 1990a, 1990b), favoring

[13] This does not mean that birds generally lack the capacity of semantic reference. Actually, the alarm systems of some avian species apparently show it (Griesser 2008). In any event, all known cases involve the emission of disconnected units, so the referential capacity is disconnected form the computational apparatus. In fact, this seems also to be case of most primate alarm systems.

[14] There in fact seem to exist more remarkable differences in the conceptual (Bickerton 1990, 1995) and motor (Ploog 2002) domains that in the perceptual domain (Hauser & Fitch 2003).

the configuration of networks of shared activity (Balari & Lorenzo 2008). As aptly argued by Striedter (2005), these are universal principles of vertebrate brain development and evolution, which most probably obtain via phenotypic plasticity means, eventually leading to point mutations favoring the stabilization of the attained pattern of cortical organization (West-Eberhard 2003).

8. Tying up loose ends

It is extremely difficult to establish the point in human evolution at which the primeval forms of complex language (i.e., already associated to a T1 computational system) could possibly have appeared. Nowadays, debates are mostly centered on the question whether it is an evolutionary innovation of Anatomical Modern Humans (AMH) or a shared character with Neanderthals, and thus inherited from Archaic sapiens, the last common ancestor of both species. Even if language happens to be a recent evolutionary novelty within both scenarios, it would still be about 100000 years old according to the former hypothesis and 300000 according to the latter.

The debate has recently been fuelled by the discovery that Neanderthals shared with AMH two point mutations of the *FOXP2* gene, responsible for a transcription factor involved in linguistic development and performance, which had been previously considered specifically human (Krause *et al.* 2007). The datum, however, is not conclusive. As explained in Benítez-Burraco *et al.* (2008) and Benítez-Burraco (this volume), *FOXP2* is a regulatory gene whose activity relates to hundreds of other genes, which means that the said mutations could have very different consequences in different molecular contexts. The fact that in the case of humans *FOXP2* has an impact in the regulation of language development is thus not enough evidence to infer that the same regulatory activity is something to be expected in the case of any other closely related species. A much more detailed knowledge concerning the system of target genes of *FOXP2* in the case of humans (Spiteri *et al.* 2007) and a similar amount of knowledge concerning the putative neanderthal system would be required to solve the matter by this approach to the question. Furthermore, there are still reasons to be cautious about the technical aspects of dating procedures used in these studies (Coop *et al.* 2008, Enard *et al.* 2002) and even doubts concerning whether these specific mutations are actually relevant to the case of language (Ptak *et al.* 2009).

Contrary to the uncertainties of this pathway of exploration, the archeological record shows a clear and strong asymmetry between

Neanderthals and AMH in almost every behavioral domain which can be reasonably connected to language, namely, ritual practices, symbolic objects, notation techniques, and so on. On the other hand, Neanderthal culture seems to have been extremely static, contrary to the dynamism and creativity of that of AMH, which can easily be related to the creativity of language itself and its capacity to rapidly transmit innovations. Curiously enough, the only attested "explosion" in Neanderthal culture is almost coincident with the arrival of AMH in Europe, so the most reasonable interpretation is that it was the result of an intent to emulate the practices of the latter species, the pressure of which Neanderthals were however unable to resist.[15]

The debate concerning Neanderthal cognitive capacities is certainly full of open questions. Anyway, some of the ideas explored in this chapter can be directly applied as new tools for discussing the asymmetries between Neanderthals and AMH, as well as their relation with the possession or absence of complex language in the former, by testing the computational complexity of the cognitive processes underlying the behaviors reflected in the archeological record. A very important suggestion in this direction has been previously put forward in Camps and Uriagereka (2006), where the existence of a context-sensitive grammar in the mind of Neanderthals is seriously put into question.

Camps and Uriagereka observe that one of the most remarkable asymmetries between the Neanderthal and the modern archeological records is the absence in the former of items that presuppose knot tying (small projectiles, decorated necklaces, and so on). As already mentioned, knot tying implies operations that apply to parts of previously executed operation, which have to be maintained in a rather powerful working memory space. It is thus an activity that, as far as it shows low automatization and signs of creativity, can be said to presuppose context sensitivity (Piattelli-Palmarini 2005). Camps and Uriagereka conclude that, at least in this domain, Neanderthals seems far from having attained a necessary cognitive requisite of complex language.

More recently, however, some archeological remains supposedly associated to Neanderthals have been identified that allegedly could motivate a reconsideration of the question (Zilhão *et al.* 2010) (Fig. 8-9). The evidence is comprised of some marine shells from Aviones (Murcia, Spain), which present the following features:

---

[15] See Balari *et al.* (2008) for a synthesis of all these questions, as well as for a comprehensive bibliography.

(1) They show remnants of pigments of different colors, indicative of an ornamental use;

(2) They show perforations, indicative of their being used as beads; and

(3) They have been dated at 50000 years, thus before the arrival of AMH in the Iberian Peninsula.

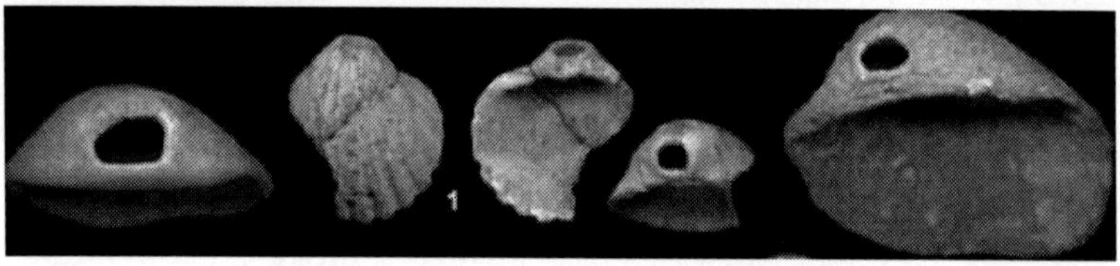

Fig. 8-9 A collection of pigmented and perforated marine shells from Aviones (Murcia, Spain): An improbable first neandertal necklace. From Zilhão *et al.* (2010).

This is obviously a finding worthy of serious consideration: it could be indicative of complex manual abilities, such as those required to thread beads and make knots; besides, the fact that the beads are decorated could be indicative of some associated symbolic values. Putting all the pieces together, we could find ourselves before a Neanderthal necklace based on a sort of complex "syntax" applied to "meaningful" units. It would be the closest that the Neanderthal mind has ever been shown to be of demonstrating language-like cognitive abilities.

Unfortunately, there are strong reasons to doubt this interpretation. The most important one is that marine shells more than 70000 years old found in Blombos (South Africa), associated to AMH, were deliberately perforated (d'Errico *et al.* 2005);[16] the shells from Aviones, on the other hand, were perforated due to natural causes (Zilhão *et al.* 2010). It is argued, in any case, that the latter could have been deliberately selected on the basis of the size of perforations, thus still allowing the contention that they were used as beads. However, from the computational perspective adopted in this chapter, it seems very implausible that individuals capable of threading and knotting had relied on accidental findings the collection of these appreciated items, given that the techniques required to puncture them, as well as the associated visuo-manual and computational cognitive

[16] More ancient perforated shells have been found (Bouzouggar *et al.* 2007, Vanhaeren *et al.* 2006), but the nature of the perforations is disputed.

requisites, are obviously much more elemental. This does not imply that a symbolic character cannot be attributed to these items,[17] but it does weaken the possibility that they were the units of knotted sequences, presupposing high-level computational capabilities.

We obviously do not know for sure whether the mind of Neanderthals reached the degree of computational complexity proper to modern language; maybe it did, but its computational system was disconnected from the capacity of referring symbolically (Balari *et al.*, in progress). What seems clear given the amount of information so far achieved, is that there existed an evident gap between their minds and the unquestionably linguistic minds of AMH.

9. Conclusion

Language simultaneously is and is not an exceptional biological feature. It is very probable that its evolutionary emergence was coincident with the speciation of AMH, in which case we would be the only species ever capable of composing and sharing mental representations with the expressive power of language. It seems clear that language is at the bottom of the uniqueness of human social relations and the rapidity of human cultural progress. It is also a crucial ingredient of human self-conscience and, with it, of our capacity for feeling "unique". However, the building blocks of language can be found in almost every corner of nature: in the techniques of seduction of nightingales, in the noisy alarm calls of monkeys, in the tangled plots of the nests of weaver birds, and so on… and, obviously, also in their brains and, ultimately, in their genes.

The key to evolution is to be found in the way individuals develop: When this development is somehow perturbed, organisms reorganize themselves and new forms of resisting environmental pressures come out. Language cannot be an exception to this principle and unquestionably it connects with ancestral developmental systems that, altered under exceptional circumstances, have given place to new and unique behavioral repertories, and this seems difficult to relate to that of other species. This is what makes language and humans so exceptional: Not more exceptional, however, than any other species and any other form of natural behavior.

[17] Both pigmentation and the fact that the shells belong to non-edible species seem convincing enough in this respect.

CHAPTER NINE

ANIMAL MINDS AND THE ROOTS OF HUMAN LANGUAGE

BRIDGET SAMUELS

1. Introduction

Given that language is unique to humans, to a first approximation, speech must be as well. This "speech is special" hypothesis (see Liberman 1996 for an overview) found some support from failures to teach apes how to speak and from the apparent absence of perceptual magnet effects in primates. But animal cognition studies since the 1970's have rapidly eroded the evidence for the uniqueness of both speech perception and production, and even more recently with the rapid growth of the biolinguistic movement, it has been hypothesized that what is special to language is not speech but rather syntax. However, there remains an unresolved question: is phonology, situated as it is at the interface of speech production/perception and syntax, itself special?

In keeping with the idea that language is designed more optimally for efficient computation than effective communication, recent Minimalist writings argue that phonology is an "ancillary" module, and that phonological systems are "doing the best they can to satisfy the problem they face: To map to the [Sensory-Motor system] interface syntactic objects generated by computations that are 'well-designed' to satisfy [Conceptual-Intentional system] conditions" but unsuited to externalization (Chomsky 2008: 136). Phonology is on this view an afterthought, a link to the outside world applied to an already fully-functional internal language system. If it is true that syntax has made do with a solution for externalization cobbled together from pre-existing parts of the sensory-motor systems, phonology might be less domain-specific—less "special" in the relevant sense—than has previously been thought, making use of evolutionarily older abilities that already found applications in other cognitive domains at the time externalized language emerged.

This view accords with the evolutionary scenario developed by Hauser, Chomsky, & Fitch (2002a) and Fitch, Chomsky, & Hauser (2005), who suggest that language may have emerged suddenly as a result of minimal genetic changes with far-reaching consequences (cf. Pinker & Jackendoff 2005 and Jackendoff & Pinker 2005, who see language as manifesting complex design). Particularly relevant is the distinction that Hauser *et al.* (2002) make between the "Faculty of Language – Broad Sense" (FLB), including all the systems that are recruited for language but need not be unique to language, or to humans, and the "Faculty of Language – Narrow Sense" (FLN), which is the subset of FLB that is unique to our species and to language. At present, the leading hypothesis among proponents of this view is that FLN is very small, perhaps consisting only of some type of recursion (i.e., Merge) and/or lexicalization[1] plus the mappings from narrow syntax to the interfaces. Pinker & Jackendoff (2005: 212) claim that phonology constitutes a problematic counterexample to this hypothesis because "major characteristics of phonology are specific to language (or to language & music), [and] uniquely human." In this chapter, I investigate the extent to which Pinker & Jackendoff's criticism is viable, first by examining what abilities animals have which are relevant to phonology, and then by sketching out a phonological theory which is consistent with the view that FLN is quite limited (developed more fully in Samuels 2011). Such an endeavor is part of the larger Minimalist and biolinguistic project which, in the words of Fujita (2007: 83), is

> a research program for building an evolutionarily adequate theory of UG through factorization of the language faculty to its bare minimum and a principled explanation of what is truly unique to humans and human language.

---

[1] Hauser *et al.* focused on the idea that recursion might be the crucial component in FLN. However, it has proven difficult to pinpoint what is meant by recursion in the relevant sense, such that it may be unique to humans and to language. Another hypothesis to which I am sympathetic has been proposed by authors such as Spelke (2003), Ott (2009), and Boeckx (2011). On their view, it is not recursion but rather lexicalization — the ability to embed any concept in a 'lexical envelope' which allows it to be recursively Merged — that arose uniquely in our species. For the purposes of the present inquiry, we may simply note that both of these hypotheses exclude phonology from FLN.

2. What Does Phonology Require?

Looking at the perception and production of speech, there is little to suggest that these abilities are unique to humans. For example, Hauser *et al.* (2002: 1573) list a number of approaches to investigating the Sensorimotor system's properties—shown below in (1), and these are all taken to fall outside FLN.

(1) a. *vocal imitation and invention*
Tutoring studies of songbirds, analyses of vocal dialects in whales, spontaneous imitation of artificially created sounds in dolphins
b. *neurophysiology of action-perception systems*
Studies assessing whether mirror neurons, which provide a core substrate for the action-perception system, may subserve gestural and (possibly) vocal imitation
c. *discriminating the sound patterns of language*
Operant conditioning studies of the prototype magnet effect in macaques and starlings
d. *constraints imposed by vocal tract anatomy*
Studies of vocal tract length and formant dispersion in birds and primates
e. *biomechanics of sound production*
Studies of primate vocal production, including the role of mandibular oscillations
f. *modalities of language production and perception*
Cross-modal perception and sign language in humans versus unimodal communication in animals

However, as I have already noted, the difference between speech and phonology cannot be underestimated; investigations of (1), while surely worthwhile, leave the major questions about phonology as a computational system unanswered.

Two papers by Yip (2006a, 2006b) outline a set of research aims that are more directly relevant to the evaluation of phonology with respect to the FLN/FLB distinction. She suggests that, if we are to understand whether "animal phonology" is possible, we should investigate whether other species are capable of the following:[2]

[2] Yip mentions two additional items which also appear on Hauser *et al.*'s list: Categorical perception/perceptual magnet effects and accurate production of sounds (mimicry).

(2) a. Grouping by natural classes
b. Grouping sounds into syllables, feet, words, phrases
c. Calculating statistical distributions from transitional probabilities
d. Learning arbitrary patterns of distribution
e. Learning/producing rule-governed alternations
f. Computing identity (total, partial, adjacent, non-adjacent)

This list can be divided roughly into three parts (with some overlap): (2a–b) are concerned with how representations are organized, (2c–d) are concerned with how we arrive at generalizations about the representations, and (2e–f) are concerned with the operations that are used to manipulate the representations. I would add three more areas to investigate in non-linguistic domains and in other species:

(2) g. Exhibiting preferences for contrast/rhythmicity
h. Performing numerical calculations (parallel individuation and ratio comparison)
i. Using computational operations: search, copy, concatenate, delete

In the sections to follow, I will present evidence that a wide range of animal species are capable of the tasks in (2a–i), although it is likely that there is no single species (except ours) in which all these abilities cluster in exactly this configuration—in other words, that human phonology relies on a unique *combination* of abilities, although the individual abilities themselves are found in many other species (as Charles Hockett already suspected). There is already a substantial amount of literature demonstrating this, and it seems reasonable to conclude on this basis that no part of phonology, as conceived in my ongoing work (see Samuels 2011), is part of FLN. In section 3, I focus on the abilities which underlie (2a,b,h)—that is, how phonological material is grouped. Next, in section 4, I turn to (2c-g), or the ability to identify and produce patterns. Finally, in section 5, I discuss (2e,i), the abilities which have to do with symbolic computation.

Before turning to these tasks, though, I would like to preempt a potential concern: how do we know that the animal abilities for which I provide evidence are truly comparable to the representations and operations found in human phonology, and what if these abilities are only analogous, not homologous? It is worth emphasizing that discovering analogs to human cognitive capabilities can have important implications for how we understand the evolution of language. In connection with this,

I would like to highlight the following statement from Hauser *et al.* (2002: 1572):

> Despite the crucial role of homology in comparative biology, homologous traits are not the only relevant source of evolutionary data. The convergent evolution of similar characters in two independent clades, termed 'analogies' or 'homoplasies,' can be equally revealing [(Gould 1976)]. The remarkably similar (but non-homologous) structures of human and octopus eyes reveal the stringent constraints placed by the laws of optics and the contingencies of development on an organ capable of focusing a sharp image onto a sheet of receptors. [...] Furthermore, the discovery that remarkably conservative genetic cascades underlie the development of such analogous structures provides important insights into the ways in which developmental mechanisms can channel evolution [(Gehring 1998)]. Thus, although potentially misleading for taxonomists, analogies provide critical data about adaptation under physical and developmental constraints. Casting the comparative net more broadly, therefore, will most likely reveal larger regularities in evolution, helping to address the role of such constraints in the evolution of language.

In other words, analogs serve to highlight "third-factor" principles, or general properties of biological/physical design (Chomsky 2005, 2007) which might be at play, and help us to identify the set of constraints which are relevant to the evolutionary history of the processes under investigation. For example, both human infants and young songbirds undergo a babbling phase in the course of the development of their vocalizations. Even though we do not want to claim that the mechanisms responsible for babbling in the two clades are homologous, nevertheless the resemblances between human and songbird vocal learning have been recognized since Darwin, and it is now understood that

> [T]heir core components share a deeply conserved neural and developmental foundation: Most aspects of neurophysiology and development—including regulatory and structural genes, as well as neuron types and neurotransmitters—are shared among vertebrates. That such close parallels have evolved suggests the existence of important constraints on how vertebrate brains can acquire large vocabularies of complex, learned sounds. Such constraints may essentially force natural selection to come up with the same solution repeatedly when confronted with similar problems. (Hauser *et al.* 2002: 1572)

Investigating the homologies and analogies between the mechanisms which underlie human and animal cognition therefore allows us to probe the constraints on cognitive evolution (see Bolhuis *et al.* 2010 for an excellent start with respect to babbling). The present study, then, provides a place for us to begin this investigation in the domain of human phonological computation. I also want to emphasize that the components of phonology in (1)–(2) are intended to be as theory-neutral as possible, though in section 6 I give a brief overview of Samuels (2011), a theory which I argue is especially well-suited to Hauser *et al.*'s hypotheses regarding the evolution of language, and also congenial to the Minimalist conception of the architecture of grammar. Furthermore, the basic argument I present against Pinker & Jackendoff (2005)—namely, that phonology does not constitute a major problem for Hauser *et al.* or for the Minimalist Program—can certainly hold even if one does not adopt my particular view of phonology.

3. Grouping

Since the hypothesis put forward by Hauser *et al.* (2002) takes recursion to be the central property of FLN (along with the mappings from narrow syntax to the conceptual-intentional and Sensorimotor interfaces), much attention has been paid to groupings, particularly recursive ones, in language. While phonology is widely considered to be free of recursion,[3] nevertheless grouping (of features, of segments, and of larger strings) is an integral part of phonology, and there is evidence that infants perform grouping or "chunking" in non-linguistic domains as well; see Feigenson & Halberda (2004). Additionally, segmenting the speech stream into words or morphemes (or syllables) also depends on what is essentially the converse of grouping, namely edge detection. We will discuss edge detection and the extraction of other patterns in section 4.

Many studies beginning with Kuhl & Miller (1975) show that mammals (who largely share our auditory system) are sensitive to the many of the same acoustic parameters as define phonemic categories in human language (see further discussion in Samuels 2011: sections 3.2 and 5.2). Experiments of this type provide the most direct comparanda to the groupings called natural classes (groups of segments that share phonetic

---

[3] Some authors have argued for recursion in the higher levels of the prosodic hierarchy (e.g., at the Prosodic Word level or above). See Truckenbrodt (1995) for a representative proposal concerning recursion at the Phonological Phrase level. Even if this is correct (though see Samuels 2011: chapter 4), the recursive groupings in question are mapped from syntactic structure, and are therefore not created by the phonological system alone.

features and sometimes undergo phonological alternations together). The concept of "natural classes" has also been studied in non-auditory domains to a certain degree, though not in those terms, if we think of natural classes as multiple ways of grouping the same objects into sets according to their different properties (i.e., features). Alex the parrot had this skill: He could sort objects by color, shape, or material (Pepperberg 1999).

Such results are valuable because they shed light on the origins of biases in perception which give rise to phonological patterns (see Blevins 2004 and Samuels 2011: chapter 2 for more details). Studies of categorical perception in domains other than language and in species other than ours show that the construction, modification, and perception of speech categories appears to make use of domain-general mechanisms such as the warping of cortical maps and cross-modal perceptual grounding. Because the human auditory system is largely shared with other mammals (and other, more distantly related clades), many other animals can also make the category distinctions (such as VOT contrasts, used to distinguish voiced from voiceless consonants) utilized in our speech, as has been known since Kuhl and Miller's (1975) pioneering work on chinchillas. Brown and Sinnott (2006), reviewing a large number of animal and human discrimination studies, found that humans and non-humans perceive similar categorical boundaries for seventeen of twenty-seven tested phonemic contrasts. More recently, with single-neuron recordings from the auditory cortex of ferrets (who have hearing very similar to ours) listening to human speech, we are just beginning to understand how individual cells react to formant transitions, frication noise, and other basic properties of speech (Mesgarani *et al.* 2008).

Another type of phonological grouping involves tones, which are relevant to all languages in one way or another: some have lexical contrasts based on pitch/tone, but all use intonation (prosody) to express differences in meaning. Rhesus monkeys are sensitive to pitch classes—they, like us, treat a melody which is transposed by one or two octaves to be more similar to the original than one which is transposed by a different interval (Wright *et al.* 2000). They can also distinguish rising pitch contours from falling ones, which is an ability required to perceive pitch accent, lexical tone, and intonational patterns in human speech (Brosch *et al.* 2004). However, animals are generally more sensitive to absolute pitch than they are to relative pitch; the opposite is true for humans (see Patel 2008).

Not only can animals process auditory input in ways that are utilized by human speech, their own vocalizations (particularly songs, as opposed to calls) can contain rich internal structure. The organization of bird song is particularly clear, though it is not obvious exactly whether/how analogies between song structure and human language should be made. The songs of many passerine songbirds consist of a sequence of one to three notes (or "songemes" as Coen 2006 calls them) arranged into a "syllable". Yip (2006a) discusses zebra finch song, building on work by Doupe & Kuhl (1999) and others. The syllables in zebra finch songs, which can be up to one second in length, are organized into motifs which Yip considers to be equivalent to prosodic words but others equate with phrases. A single song contains multiple motifs, combining to form a structure like the one in (3), where *M* stands for motif, σ stands for syllable, and *n* stands for note (modified from Yip 2006a):

(3)

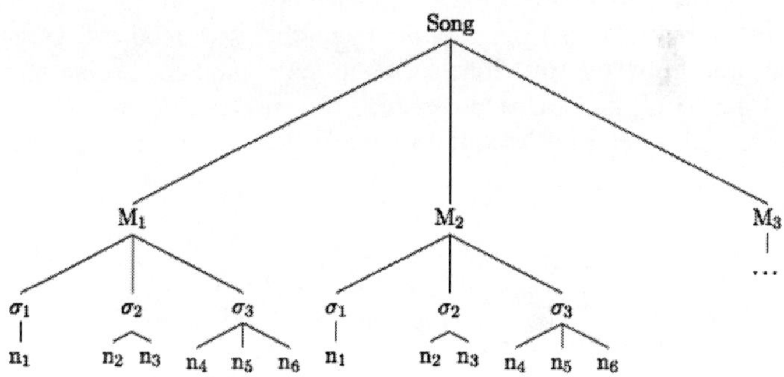

There are a few important differences between this birdsong structure and those found in human phonology, some of which are not apparent from the diagram. First, as Yip points out, there is no evidence for binary branching in this structure, which suggests that the combinatory mechanism used by birds cannot be equated with binary Merge, but it could be more along the lines of adjunction or concatenation, which creates a flat structure; see section 6 and Samuels & Boeckx (2009). Second, the definition of a "syllable" in birdsong is a series of notes/songemes bordered by silence (Williams & Staples 1992, Coen 2006). This is very unlike syllables, or indeed any other phonological categories, in human language. Third, the examples from numerous species in Slater (2000) show that the motif is typically a domain of repetition (as I have represented it above); the shape of a song is $((a^x)(b^y)(c^z))^w$ with a string of syllables *a, b, c* repeated in order. This is quite reminiscent of reduplication. Payne (2000) shows that virtually the same can be said of humpback whale songs, which take the shape (a … $n)^w$, where the number of repeated components, n, can be up to around ten.

Both birdsong and whalesong structures have a limited depth of embedding, as shown in (4)—and I argue in Samuels (2011) along with Neeleman & van de Koot (2006) that this is also true for human phonology. The "flat" nature of song does not appear to be accidental; across multiple domains, only shallow hierarchies are found in animals' cognitive representations, consistent with the view that they are unable to produce infinitely recursive structures of the type encountered in human language syntax (see Lorenzo, this volume). For example, Cheney & Seyfarth (2007) argue that baboon social knowledge is also flat or "linearly hierarchical," as they describe it. Baboons within a single tribe (of up to about eighty individuals) obey a strict, transitive dominance hierarchy. The hierarchy is divided by matrilines; individuals from a single matriline occupy adjacent spots in the hierarchy, with mothers, daughters, and sisters from the matriline next to one another. So an abstract representation of their linear dominance hierarchy would look something like this, with each x representing an individual and parentheses defining matrilines:

(4) (xxx)(xx)(xxxx)(xxx)(xxxxxxx)(xxx)(x)(xxxx)

The difference between the baboon social hierarchy and birdsong, which I translate into this sort of notation below, is merely the repetition which creates a motif (think of baboon individuals as corresponding to songemes and matrilines as corresponding to syllables):

(5)

$$\underbrace{(n_1)(n_2n_3)(n_4n_5n_6)}_{\text{motif}_1}\ \underbrace{(n_1)(n_2n_3)(n_4n_5n_6)}_{\text{motif}_2}$$

There is evidence to suggest that, as in phonology (but strikingly unlike narrow syntax), the amount of hierarchy capable of being represented by animals is quite limited.[4] In the wild, apes and monkeys

[4] I view the situation as an exact parallel to the difference between humans and animals in the domain of numerical cognition; perhaps the two dichotomies are indeed manifestations of the same cognitive difference, namely that only humans

very seldom spontaneously perform actions which are hierarchically structured with sub-goals and sub-routines, and this is true even when attempts are made to train them to do so. Byrne (2007) notes one notable exception, namely the food processing techniques of gorillas. Byrne provides a flow chart detailing a routine, complete with several decision points and optional steps, which mountain gorillas use to harvest and eat nettle leaves. This routine comprises a minimum of five steps, and Byrne reports that the routines used to process other foods are of similar complexity. Byrne further notes that

> all genera of great apes acquire feeding skills that are flexible and have *syntax-like* organisation, with hierarchical structure. [...] Perhaps, then, the precursors of linguistic syntax should be sought in primate *manual* abilities rather than in their vocal skills (Byrne 2007: 12; emphasis his).

I concur that manual routines provide an interesting source of *comparanda* for the syntax of human language, broadly construed (i.e., including the syntax of phonology). Fujita (2007, 2009) has suggested along these lines the possibility that Merge evolved from an "action grammar" of the type which would underlie apes' foraging routines.[5] This begs a question, however: why haven't the computations underlying these foraging routines been deployed in other motor domains, including communicative facial and body gestures, as well as vocal gestures? It may be that the capacity for representing such routines is locked into a particular context, unable to be extended to other domains. Perhaps only in the more immediate human lineage has the ability to produce hierarchical actions been "unlocked" for more general use, including in language.

Still, even manual routines appear limited in their complexity. For example, Johnson-Pynn *et al.* (1999) used bonobos, capuchin monkeys, and chimpanzees in a study similar to one done on human children by

---

have a recursive engine (Merge), as suggested by Hauser, Chomsky, and Fitch (2002). While many animals (and young human children) seem to be able to represent small numerals, only suitably mature (and, perhaps, suitably linguistic) humans go on to learn the inductive principle, which allows them to count infinitely high. See discussion later in this section and in section 5 for more discussion and references on numeracy in animals.

[5] This possibility has also been suggested by Greenfield, Nelson, & Saltzmann (1972) and Greenfield (1991, 1998). Interestingly, Greenfield (1991) considers the possibility that this type of action grammar could correspond to the combination of phonemes into words, although in other works she suggests that it is more akin to the combination of words. See Fujita (2009: 138).

Greenfield *et al.* (1972) (see also discussion of these two studies by Conway & Christiansen 2001). These experiments investigated how the subjects manipulated a set of three nesting cups (call them A, B, C in increasing order of size). The subjects' actions were categorized as belonging to the "pairing", "pot", or "subassembly" strategies, which exhibit varying degrees of embedding:[6]

(6) a. *Pairing strategy*: place cup B into cup C. Ignore cup A.
b. *Pot strategy*: first, place cup B into cup C. Then place cup A into cup B.
c. *Subassembly strategy*: first, place cup A into cup B. Then place cup B into cup C.

The pairing strategy is the simplest, requiring only a single step. This was the predominant strategy for human children up to twelve months of age, and for all the other primates—but the capuchins required watching the human model play with the cups before they produced even this kind of combination. The pot strategy requires two steps, but it is simpler than the subassembly strategy in that the latter, but not the former, requires treating the combination of cups A + B as a unit in the second step. (We might consider the construction of the A + B unit as being parallel to how complex specifiers and adjuncts are composed "in a separate derivational workspace" in the syntax; see Fujita 2007.) Human children use the pot strategy as early as eleven months (the youngest age tested) and begin to incorporate the subassembly strategy at about twenty months. In stark contrast, the non-human primates continued to prefer the pairing strategy, and when they stacked all three cups, they still relied on the pot strategy even though the experimenter demonstrated only the subassembly strategy for them. Though we should be careful not to discount the possibility that different experimental methodologies or the laboratory context is responsible for the non-humans' performance, rather than genuine cognitive limitations, the results are consistent with the hypothesis that humans have the ability to represent deeper hierarchies than other

---

[6] The situation is actually substantially more complicated than this, because the subjects need not put the cups in the nesting order. To give a couple examples, putting cup A into cup C counts as the pairing strategy; putting cup A into cup C and then placing cup B on top counts as the pot strategy. I refer the reader to the original studies for explanations of each possible scenario. The differences between the strategies as I have described them in the main text suffice for present purposes.

primates. This is predicted if only humans are endowed with the recursive engine that allows for infinite syntactic embedding.

Many other types of experimental studies have also been used to investigate how animals group objects and manipulate those groupings arithmetically. It is well known that a wide variety of animals, including rhesus monkeys, have the ability to perform comparisons of analog magnitude with small numbers (<4). They can discriminate between, for instance, groups of two and three objects and pick the group with more objects in it, which requires grouping the objects into distinct sets, then comparing the cardinality of those sets. Further data comes from the capacity for making transitive inferences found in capuchin monkeys (Addessi *et al.* 2008), baboons (Cheney & Seyfarth 2007), and non-mammals such as pinyon jays (Cheney & Seyfarth 2007) as well. For example, Schusterman & Kastak (1993) taught a California sea lion named Rio to associate arbitrary visual stimuli (cards with silhouettes of various objects printed on them). On the basis of being taught to select card B when presented with card A, and also to select card C when presented with card B, Rio transitively learned the A-C association. Rio also made symmetric associations: when presented with B, she would select A, and so forth. We might consider these groups Rio learned to be akin to learning arbitrary pairings such as which phonemes participate in a given alternation (A and C bear the same relation to B), or in which contexts a particular process occurs (choose A in the context of B; choose B in the context of C).

As regards the ability to group objects, then, I conclude that animals—especially birds and primates—are capable of the basic grouping abilities which phonology requires. They perceive sounds categorically like we do (even some of the sounds used in human language); their vocalizations show linearly hierarchical groupings like ours; they can assign objects arbitrarily to sets like we do; they can categorize objects into natural classes or overlapping sets according to different attributes like we do, and we are beginning to understand these processes at the neural level. Animals' main limitations seem to be in the area of higher-degree embedding, but phonological hierarchy seems to be limited in the same way.

4. Patterns

Pattern detection is closely related to grouping: a pattern is essentially a relation between multiple groups, or different objects within the same group. Thus, the ability to assign objects to a set or an equivalence class is a prerequisite for finding any patterns in which those objects participate; the abilities discussed in the previous section are very much relevant to this one as well.

Several experimental studies on animal cognition more generally bear on the issue of abstract pattern learning. One such study, undertaken by Hauser & Glynn (2009), tested whether wild rhesus monkeys could extract simple algebraic rules like same-different-different (ABB) or same-same-different (AAB) comprised of their species-specific vocalizations. They performed an experiment very similar to one run on infants by Marcus *et al.* (1999). The habituation material was either in the form of ABB or AAB with novel test material that either matched the habituation stimuli or mismatched. Specifically, after habituating the infants/rhesus to one of these conditions, they tested them on two novel test items: one from the same class to which they had been habituated, and a second from the other class. If subjects extracted the pattern during habituation, then they should respond more to the mismatched test trial than to the novel, but matching test trial. Both infants and rhesus evidenced learning of these simple patterns; they were more likely to dishabituate to the item with the mismatched pattern.

This type of pattern-extraction ability could serve phonology in several ways, such as the learning of phonological rules or phonotactic generalizations. Heinz (2007) showed that phonotactics (restrictions on the co-occurrence of segments, such as at the beginnings or ends of words) can be captured without any exceptions if three segments at a time are taken into account, so it seems on the basis of the rhesus monkey's success in the Hauser & Glynn experiment, together with other work by Murphy *et al.* (2008) on rats, that learning phonotactics would not be out of their range of capabilities. Furthermore, phonotactics (and all attested phonological rules) can be modeled with finite-state grammars, as has been known since Johnson (1970). In theory, then, if primates or other animals can learn finite-state patterns, we would expect that they could learn any attested phonotactic restriction or phonological rule. The findings of Fitch & Hauser (2004) are also relevant here: at least under one interpretation of the data obtained by Fitch and Hauser, tamarins succeed at learning finite-state grammars but fail to learn more complicated phrase-structure grammars The conclusion that animals are limited to finite-state computation is also supported by van Heijningen *et al.*'s (2009) nuanced testing of finches on similar patterns, which casts strong suspicion on Gentner *et al.*'s (2006) claim that starlings can go beyond finite-state patterns to learn a recursive context-free rule; the interpretation of these data are still under debate (ten Cate *et al.* 2010, Gentner *et al.* 2010).

One of the most important obstacles facing a language learner/user falls into the category of pattern-extraction. This difficult task is parsing the continuous speech stream into discrete units (be they phrases, words, syllables, or segments). This speaks directly to (2b–c). Obviously, segmenting speech requires some mechanism for detecting the edges of these units. Since the 1950s, it has been recognized that one way to detect the edges of words is to track transitional probabilities, usually between syllables. If Pr(AB) is the probability of syllable B following syllable A, and P(A) is the frequency of A, then the transitional probability between A and B can be represented as:

(7) $TP\ (A \rightarrow B) = Pr(AB)\ /\ Pr(A)$

The transitional probabilities within words are typically greater than those across word boundaries, so the task of finding word boundaries reduces to finding the local minima in the transitional probabilities. Numerous experimental studies suggest that infants do in fact utilize this strategy (among others) to help them parse the speech stream, and that statistical learning is not unique to the linguistic domain but is also utilized in other areas of cognition (see references in Gambell & Yang 2005). With respect to the availability of this strategy in non-humans, Hauser *et al.* (2001) found that tamarins are able to segment a continuous stream of speech into three-syllable CVCVCV "words" based solely on the transitional probabilities between the syllables. Rats are also sensitive to local minima in transitional probabilities, after a period of operant conditioning (Toro *et al.* 2005).

While transitional probabilities between syllables are strictly local calculations (i.e., they involve adjacent units), some phonological (and syntactic) dependencies are non-adjacent. This is the case with vowel harmony, for instance, and is also relevant to languages with "templatic" morphology, such as Arabic, in which a triconsonantal root is meshed with a different group of vowels depending on the part of speech which the root instantiates in a particular context. Comparing the results obtained by Newport & Aslin (2004) and Newport *et al.* (2004) provides a contrast between human and tamarin learning of such patterns. Newport *et al.* tested adult humans and cotton-top tamarins on learning artificial languages, all with three-syllable CVCVCV words, involving the three different kinds of non-adjacent depen-dencies which I list below.

(8) a. *Non-adjacent syllables*: the third syllable of each word was predictable on the basis of the first, but the second syllable varied.

b. *Non-adjacent consonants*: The second and third consonants of each word were predictable on the basis of the first, but the vowels varied.
c. *Non-adjacent vowels*: The second and third vowels of each word were predictable on the basis of the first, but the consonants varied.

Both humans and tamarins succeeded at learning the languages tested in the non-adjacent vowel condition. Humans also succeeded at the non-adjacent consonant condition. These results are expected, at least for the humans, because both of these types of dependencies are attested in natural language (in the guises of vowel harmony and templatic morphology). Tamarins failed in the non-adjacent consonant condition, though it is clear from their performance on the non-adjacent vowel condition that they can construct non-adjacent dependencies; their difficulties seem to stem from difficulties in distinguishing consonant sounds. In other words, their failure may not be due to the pattern-detection mechanism, but rather due to the input which was available to that mechanism.

From a phonological perspective, perhaps the most intriguing result is that humans failed at this non-adjacent syllable condition, whereas tamarins succeeded. Newport *et al.* (2004: 111) ask:

> Why should non-adjacency—particularly syllable non-adjacency—be difficult for human listeners and relatively easy for tamarin monkeys? [...T]his is not likely to be because tamarins are in general more cognitively capable than adult humans. It must therefore be because human speech is processed in a different way by humans than by tamarins, and particularly in such a way that the computation of non-adjacent syllable regularities becomes more complex for human adults.

They go on to suggest that perhaps the syllable level is only indirectly accessible to humans because we primarily process speech in terms of segments (whereas tamarins process it in more holistic, longer chunks). This is a possible contributor to the observed effect, but other explanations are available. Perhaps, for example, tamarins fail to exhibit a minimality effect that humans exhibit in this circumstance. Let us interpret the tamarins' performance in the non-adjacent consonant condition as suggesting that they either ignore or simply do not perceive consonants. Then for them, the non-adjacent syllable task differs minimally from the non-adjacent vowel task in that the former involves learning a pattern which skips the middle vowel. So rather than paying attention to co-occurrences between adjacent vowels, they have to look at co-occurrences between vowels which are one away from each other. It seems likely, as Newport *et al.* also suggest, that the adjacent vs. one-away difference represents only a small increase in cognitive demand. But for humans, the non-adjacent syllable condition is crucially different—and this is true no matter whether we are actually paying attention to syllables, consonants, or vowels. These categories have no import for tamarins, but for humans, they are special. The dependency we seek in this condition is between two non-adjacent elements of the same category, which are separated by another instance of the same category. This is a classical minimality effect: if α, β, γ are of the same category and α > β > γ (> should be read for phonology as "precedes" and for syntax, "c-commands"), then no relationship between α and γ may be established. Perhaps tamarins succeed where humans fail because they do not represent the portions of the stimuli which they track as all belonging to the same abstract category "vowel" which is sufficient to trigger minimality effects for us.

A variety of other studies on primate cognition focus on the ability to learn sequences. Given that sequencing or precedence relationships are extremely important to language, particularly given Raimy's (2000) insistence on precedence as a primitive relation in morphophonology, these studies are intriguing from a linguist's perspective. As Endress *et al.* (2009) emphasize, temporal sequencing underlies everything from foraging routines to structured vocalizations (i.e., the order of segments, morphemes, and words in human speech as well as notes and motifs in animal songs) and requires a specific type of memory which encodes the positions of items within the sequence. Conway & Christiansen (2001) report on a number of studies which compare primates' performances on this kind of task. When presented with an "artificial fruit" requiring four arbitrary actions to open it and thereby reveal a treat, chimpanzees and human preschoolers perform similarly; both succeed at learning the sequence.

However, one apparent cognitive limitation of non-human primates relative to our species in the domain of pattern-learning is that they have extreme difficulty with non-monotonic sequences. This seems to be related to a difference in the way humans and other primates plan and perform sequential actions. One experiment undertaken by Ohshiba (1997) tested human adults, Japanese monkeys, and a chimpanzee on the ability to learn an arbitrary pattern: They were presented with a touch screen with four different-sized colored circles on it and had to touch each one in sequence to receive a reward; the circles disappeared when touched. All the species succeeded in learning a monotonic pattern: touch the circles in order from smallest to largest or largest to smallest. They also all succeeded, but were slower, at learning non-monotonic patterns.[7] But as I will discuss in section 5, measurements of reaction times suggest the humans and monkeys used different strategies in planning which circles to touch.

Rhythm, too, is a type of pattern. Rhythmicity, cyclicity, and contrast are pervasive properties of language, particularly in phonology. Everything that has been attributed to the Obligatory Contour Principle (Leben 1973) fits into this category. Walter (2007) argues that these effects should be described not with a constraint against repetition (see also Reiss 2008), but as emerging from two major physical limitations: the difficulty of repeating a particular gesture in rapid succession, and the difficulty of perceiving similar sounds (or other sensory stimuli) distinctly in rapid succession. These are both extremely general properties of articulatory and perceptual systems which we have no reason to expect would be unique to language or to humans.

To date, perhaps the most direct cross-species tests of the perception of human speech rhythm (prosody) comes from Ramus *et al.* (2000). In Ramus *et al.*'s experiment, human infants and cotton-top tamarins were tested on their ability to discriminate between Dutch and Japanese sentences under a number of conditions: one in which the sentences were played forward, one in which the sentences were played backward, and one in which the sentences were synthesized such that the phonemic inventory in each language was reduced to /s a l t n j/. The results of these experiments showed that both tamarins and human newborns were able to discriminate between these two unfamiliar and prosodically different languages in the forward-speech condition, but not in the backward-speech condition. A generous interpretation of these results would suggest "at least some aspects of human speech perception may have built upon preexisting sensitivities of the primate auditory system" (Ramus *et al.* 2000: 351). However, Werker & Voloumanos (2000) caution that we cannot conclude much about the processing mechanisms which serve these discrimination abilities; this is of particular concern given that the tamarins' ability to tell Dutch and Japanese apart was reduced in the

[7] In some situations, non-human primates fail entirely at learning non-monotonic patterns. For example, Brannon & Terrace (1998, 2000) found that while rhesus macaques taught the first four steps in a monotonic pattern could spontaneously generalize to later steps, they failed to learn a four-member non-monotonic pattern even with extensive training. It is not clear what to attribute the worse performance in the Brannon & Terrace studies to; there are too many differences between the paradigm they used and the one reported in the main text, including the species tested.

reduced phonemic inventory condition. This may indicate that tamarins rely more strongly on phonetic cues rather than prosodic ones. Given the apparent importance of prosody for syntactic acquisition in human children—specifically, babies seem to use prosodic information to help them set the head parameter—Kitahara (2003: 38) puts forth the idea that "cotton-top tamarins fail to discriminate languages on the basis of their prosody alone, because syntactic resources that require such prosodic-sensitive system [*sic*] might not have evolved for them." Though it is unclear how one might either support or disprove such a hypothesis, it is at the very least interesting to consider what prosody might mean for an animal which does not have the syntactic representations from which prosodic representations are built.

Another example of rhythmicity in speech is the wavelike sonority profile of our utterances, which is typically discussed in terms of syllable organization. Syllables range widely in shape across languages. In (9)-(10) I give examples from opposite ends of the spectrum: a series of three CV syllables in (9), and a syllable in (10) that has a branching onset as well as a coda, and additionally appendices on both ends. The relative heights of the segments in (9)–(10) represent an abstract scale of sonority (making no claim about the units of this scale).[8]

(9)

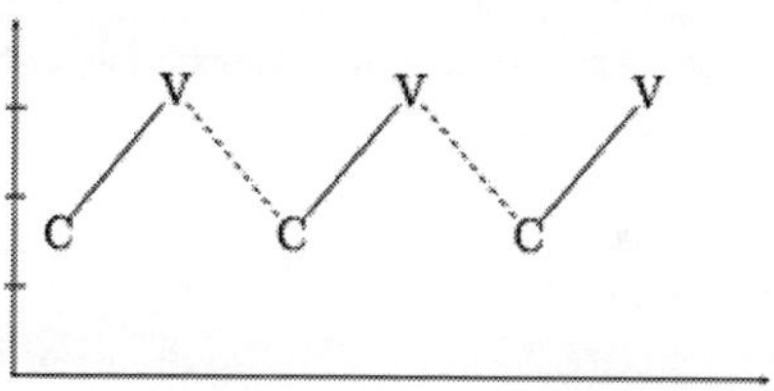

(10)

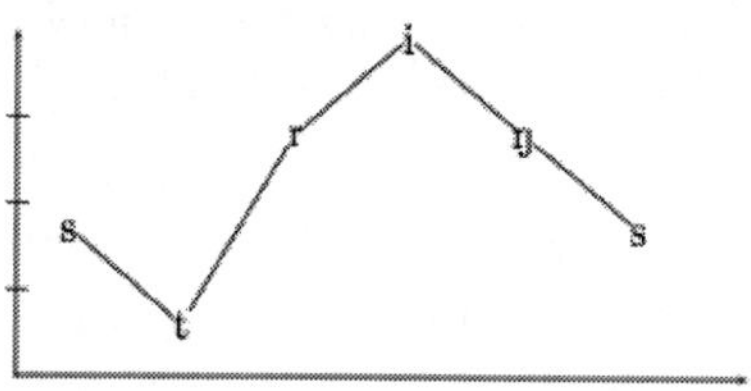

[8] I remain agnostic about the exact nature of sonority. However, see (among others) Ohala (1992) and Ohala & Kawasaki–Fukumori (1997) for arguments that it is a derived notion rather than a primitive one.

All syllables, from CV (9) to CCCVCC (10), combine to yield a sonority profile roughly as in (11):

(11)

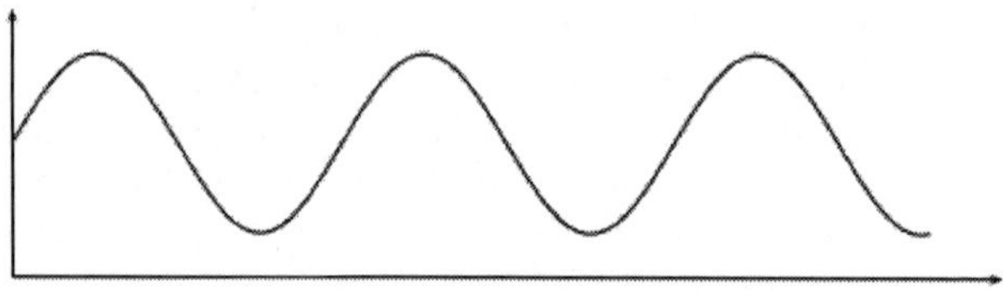

The peaks and troughs may not be so evenly dispersed, and they may not all be of the same amplitudes, but the general shape is the same no matter whether the sonority values being plotted come from syllables that are simple or complex. This is hardly a new observation; it is over a century old (e.g., Lepsius & Whitney 1865, de Saussure 1916). Ohala & Kawasaki–Fukumori (1997: 356) point out that it is virtually inevitable:

> Just by virtue of seeking detectable changes in the acoustic signal one would create as an epiphenomenon, i.e., automatically, a sequence showing local maxima and minima in vocal tract opening or loudness. In a similar way one could find "peaks" (local maxima) in a string of random numbers as long as each succeeding number in the sequence was different from the preceding one.

It has long been noted that the ability to break this wave up into periods (based partially on universal and partially on language-specific criteria) aids with the identification of word boundaries: they tend to fall at the local minima or maxima in the wave. And as we saw earlier in this section, we already know that both human infants and tamarins are sensitive to local minima (of transitional probabilities) in speech, which I believe suggests that this is a legitimate possibility.[9]

[9] In all of the studies on tamarins (and human infants) of which I am aware, the shape of syllables tested does not extend beyond CV. As a reviewer suggests, it would be most informative to see studies which test a variety of syllable shapes—but note that tamarins' difficulties with perceiving consonant sounds, as discussed earlier with regards to the Newport *et al.* (2004) experiments, would likely confound such investigations.

Animals from a wide variety of clades show preferences for rhythmicity in their vocalizations and other behaviors as well, though our own (non-musical) speech has no regular beat; while language does have a rhythm, it is not a primitive (see discussion in Patel 2008). Yip (2006b) mentions that female crickets exhibit a preference for males who produce rhythmic calls, and Taylor *et al.* (2008) discovered that female frogs prefer rhythmic vocalizations as well. Rhythmic behaviors, or the ability to keep rhythm, appear to be widespread in the animal kingdom. Gibbons produce very rhythmic "great calls", and while Yip (2006b: 443) dismisses this, saying that "the illusion of rhythm is probably more related to breathing patterns than cognitive organization", this should hardly disqualify the data. For example, the periodic modulation of sonority in our speech is closely connected to opening and closing cycle of the jaw (Redford 1999, Redford *et al.* 2001), and it is widely accepted that the gradual downtrend in pitch which human utterances exhibit has to do with our breathing patterns. So for humans, too, there is at least some purely physiological component; however, the fact that females of various species prefer rhythmic calls shows that at the very least, there is also a cognitive component to animals' perception of rhythmicity.

There are also some animals which synchronize the rhythms produced by multiple individuals. For example, frogs, insects, and bonobos all synchronize their calls; some fireflies synchronize their flashing, and crabs synchronize their claw-waving (see Merker 2000 and references therein). However, while elephants can be taught to drum with better rhythmic regularity than human adults, they do not synchronize their drumming in an ensemble (Patel & Iversen 2006).

Finally, we should note that it is extremely common for animals to exhibit "rule-governed" behavior in the wild, and in their communicative behavior in particular. Cheney & Seyfarth (2007) make the case that baboon vocalizations are rule-governed in that they are directional and dependent on social standing. That is, a baboon will make different vocalizations to a higher-ranked member of the group than she will to a lower-ranked member. By this same rubric, vervet monkey grunts and chimpanzee calls should also be considered rule-governed; a number of articles on species ranging from treefrogs to dolphins to chickadees in a special issue of the *Journal of Comparative Psychology* (August 2008, vol. 122.3) devoted to animal vocalizations further cement this point. And as we saw in the previous section, both bird and whale songs obey certain combinatorial rules—in other words, they have some kind of syntax (in the broad sense of the term). Here the distinction made by Anderson (2004) and suggested in earlier work by Peter Marler is useful: Plenty of animals have a "phonological" syntax to their vocalizations, but only humans have a "semantic" or "lexical" syntax which is compositional and recursive in terms of its meaning. Again, this reiterates the Minimalist view that what is special about human language is the mapping from syntax to the interfaces (and particularly the LF interface, as Chomsky emphasizes in recent writings; see, e.g., Chomsky 2004), not the externalization system.

5. Operations

The final set of abilities that we will discuss are those which pertain to the phonological operations for which I argue in Samuels (2009c, 2011): search, copy, and delete. It is particularly difficult to remain theory-neutral here because this is precisely the area in which phonological theories differ most: there are a number of quite different proposals, from Autosegmental Phonology to Optimality Theory to Government Phonology. I focus here on a small number of basic operations which any theory of phonology will have to admit (and which, if I am correct, exhaust the operations of phonological computation): a search mechanism that identifies an object to be manipulated based on some characteristic (feature) of its content, a means by which features may be copied from one object to another and/or deleted from an object, and a mechanism for combining or concatenating two objects. While these operations enjoy an elevated status in my work, *any* theory of phonology, or of language in general, will have to make use of these operations. For example, Hornstein (2001) argues that insertion of an element into a linguistic derivation is copying from the lexicon, and I would add that it is very difficult to see how this copying might be done without a prior search into the lexicon. I also discuss here a fourth operation, concatenation. By this I mean the ability to connect morphemes—a root and an affix, for example—in a manner that creates a linear structure, not the nested hierarchical structure of Merge.[10] This concatenation mechanism properly belongs to the syntax–phonology interface, but since it operates at a stage at which phonological material has already been added (see Idsardi and Raimy, In press), as we know since some affixes are sensitive to phonological properties such as the stress pattern of the stem, it is relevant to the present work.

[10] Whereas iterative applications of concatenate yield a flat structure, iterative applications of Merge yield a nested hierarchical structure: syntactic structures must be flattened, whereas linear order is a primitive in phonology (Raimy 2000). Samuels & Boeckx (2009) discuss this issue in greater detail.

Searching is ubiquitous in animal and human cognition. It is an integral part of foraging and hunting for food, to take but one example. The Ohshiba (1997) study of sequence-learning by monkeys, humans, and a chimpanzee is an excellent probe of searching abilities in primates because it shows that, while various species can perform the multiple sequential searches required to perform the experimental task (touching four symbols in an arbitrary order), they plan out the task in different ways. The humans were slow to touch the first circle but then touched the other three in rapid succession, as if they had planned the whole sequence before beginning their actions (a "collective search" strategy). The monkeys, meanwhile, exhibited a gradual decrease in their reaction times. It was as if they planned only one step before executing it, then planned the next, and so forth (a "serial search" strategy).

Perhaps most interestingly of all, the chimpanzee appeared to use the collective search strategy on monotonic patterns but the serial search strategy when the sequence was not monotonic. That chimpanzees employ collective searches is corroborated by the results of a similar experiment by Biro & Matsuzawa (1999). The chimp in this study, Ai, had extensive experience with numerals, and she was required to touch three numerals on a touch-screen in monotonic order. Again, her reaction times were consistently fast after the initial step. But when the locations of the two remaining numerals were changed after she touched the first one, her reactions slowed, as if she had initially planned all three steps but her preparation was foiled by the switch. It is not clear to me exactly what should be concluded from the disparity between humans, chimps, and monkeys, but notice that the search mechanism proposed by Mailhot & Reiss (2007) and extended by Samuels (2009c, 2011) operates in a manner consistent with the collective search strategy: scan the search space to find all targets of the operation to be performed, and then perform the operation to all targets in one fell swoop.

A close parallel to the copying operation in phonology, particularly the copying of a string of segments as in reduplication, would be the patterns found in bird and whale songs. As we saw in section 3, Slater (2000) shows that for many bird species, songs take the shape $((a^x)(b^y)(c^z))^w$: That is, a string of syllables *a, b, c,* each of them repeated, and then the whole string repeated. Whale songs are similarly structured (Payne 2000). With respect to the copying of a feature from one segment to another (as in assimilatory processes), the relevant ability might be transferring a representation from long-term memory to short-term memory: extracting a feature from a lexical representation and bringing it into the active phonological workspace. This seems like a pre-requisite for any task which involves the recall/use of memorized information, and perhaps can be seen as a virtual conceptual necessity arising from computational efficiency (a prime source of third-factor explanation; see Chomsky 2005, 2007).

In order to integrate a copied object with other material in the workspace, some sort of concatenative operation is necessary. Most notably, concatenation subserves the ability to connect morphemes—a root and an affix, for example. Concatenation serves not only this grouping function (including assigning objects to sets), but also the ability to chain together sequential actions. We have already seen that bird and whale songs have the kind of sequential organization which is indicative of concatenated chunks, and primates can perform multi-step actions with sub-goals. As alluded to earlier, concatenation may underlie the "number sense" common to humans and many other species as well (for overviews, see Dehaene 1997, Lakoff & Nuñez 2001, Devlin 2005, Shettleworth 2010). This is perhaps clearest in the case of parallel individuation /tracking, or the ability to represent in memory a small number of discrete objects (< 4; see Hauser *et al.* 2000 and references therein). The connection between parallel individuation and concatenation is suggested by the fact that the speed of recognizing the number of objects in a scene decreases with each additional object that is presented within the range of capability (Saltzman & Garner 1948). Such tasks require paying attention to each object in the array separately, yet they must be represented as a set in order to be manipulated arithmetically (Gelman & Gallistel 1978). Biro & Matsuzawa (1999) report that the chimpanzee Ai, when given rigorous training over a long period of time, was able to engage in basic counting, addition, and subtraction of natural numbers up to about ten. The ability to subtract has also been shown in pigeons (Brannon *et al.* 2001). These numerical tasks clearly involve the assignment of objects to sets, which is the fundamental basis of concatenation. Gallistel & Gelman (2005) and Shettleworth (2010) summarize the literature on counting and addition, which have been shown for species as diverse as parrots, rats, and chimpanzees.

Conversely, subtraction or removal of objects from a set could be seen as akin to the delete operation; the ability to subtract has also been shown in pigeons. This and a number of other studies showing that primates, rats, and birds can both count and add with a fair degree of precision are summarized in Gallistel & Gelman (2005).

6. Conclusions

The studies of animal cognition and behavior presented here provide evidence that the building blocks of phonology are present in a wide range of species: many animals can group objects, extract patterns from sensory input, perform sequential objects, perform searches, engage in copying behaviors, and manipulate sets through concatenation. Moreover, these capacities did not evolve for phonology, but rather, for other perceptual, cognitive and motor functions, many of which fall outside the realm of communication. This strongly suggests that our physiology and phonetic capabilities have shaped the structure of our speech (and sign), rather than vice versa (a point made previously by Piattelli-Palmarini 1989). There are many reasons to believe that this is the case for our perceptual systems, as we saw throughout this chapter. It is also increasingly known to be true for both the vocal and manual systems of articulation employed by spoken and signed language. For example, fossil evidence shows that the flexed basicranium indicative of a lowered, language-ready larynx and the flattened face responsible for the modern supralaryngeal vocal tract were both present hundreds of thousands of years before the emergence of language (Nishimura *et al.* 2006; see Bower 1989 for an overview). The emergence of bipedal walking and the shortening of the toes (with reduced grasping ability) appears to have led to the hands taking over increased area in the sensory-motor cortex, leading to increased dexterity that was subsequently exploited by gestural and sign systems (Bowers 2006).

In short, the sensory-motor systems were "language-ready" long before the key innovation that led to our unique semantic syntax. Phonology therefore provides little challenge to the idea that FLN is quite small. The abilities that underlie human phonological competence are found scattered across a wide range of animal species, though no single species besides ours may possess *all* of these abilities. In this sense, what may be unique to humans is our capacity to interface between these different components. With respect to the question of how phonology is "special," then, perhaps we might say that its parts are not special, but their sum, particularly in combination with human language syntax, is.

BIBLIOGRAPHY

Abe, K. and D. Watanabe 2011, "Songbirds possess the spontaneous ability to discriminate syntactic rules", *Nature Neuroscience* 14, 1067-1074

Aboitiz, F. and R. García 1997, "The evolutionary origins of the language areas in the human brain. A neuroanatomical perspective", *Brain Research Reviews* 25, 381-396

Aboitiz, F., R. García, E. Brunetti and C. Bosman 2006, "The origin of Broca's area and its connections from an ancestral working memory network", Y. Grodzinsky and K. Amunts (eds.) *Broca's region*, New York: Oxford University Press, 3-16

Abutalebi, J., S. F. Cappa and D. Perani 2001, "The bilingual brain as revealed by functional neuroimaging", *Bilingualism: Language and Cognition,* 4, 179-190

Abutalebi, J. and D. Green 2008, "Control mechanisms in bilingual language production: neural evidence from language switching studies", *Language and Cognitive Processes* 23, 557-582

Addessi, E., A. Mancini, L. Crescimbene, C. Padoa-Schiopp and E. Visalberghi 2008, "Preference transitivity and symbolic representation in capuchin monkeys (Cebusapella)" *PLoS ONE* 3, E2414

Aichert, I., and W. Ziegler 2004, "Syllable frequency and syllable structure in apraxia of speech", *Brain and Language* 88, 148-159

Alarcón, M., *et al.* 2008, "Linkage, association, and gene-expression analyses identify *CNTNAP2* as an autism-susceptibility gene", *Am. J. Hum. Genet.* 82, 150-159

Alberch, P. 1989, "The logic of monsters: Evidence for internal constraint in development and evolution", *Geobios* 12 (mémoire spécial no. 12), 21-57

—. 1991, "Del gen al fenotipo: Sistemas dinámicos y evolución morfológica", *Revista Española de Paleontología* (número extraordinario *El estudio de la forma orgánica y sus secuencias en Paleontología Sistemática, Paleontología y Paleontología Evolutiva*), 13-19

Alegría, J., J. Leybaert, B. Charlier and C. Hage 1992, "On the origin of phonological representations in the deaf: listening the lips and hands", J. Alegría, D. Holender, J. Morais and M. Radeau (eds.) *Analytic approaches to Human Cognition*, North-Holland, 107-132

Alter, S.G. 1999, *Darwinism and the Linguistic Image*. Baltimore and London: The John Hopkins University Press

Álvarez, C., M. Carreiras and M. Taft 2001, "Syllables and morphemes: contrasting frequency effects in Spanish", *Journal of Experimental Psychology: Learning, Memory and Cognition* 27, 158-169

American Psychiatric Association 1994, *Diagnostic and Statistical Manual of Mental Disorders (DSM-IV),* Washington: American Psychiatric Association

Amstrong, D. F., W. C. Stokoe and S.E. Wilcox 1995, *Gesture and the nature of language,* Cambridge: Cambridge University Press

Anderson, D. and J. Reilly 2002, "The MacArthur Communicative Development Inventory: Normative data for American Sign Language",*Journal of Deaf Studies and Deaf Education, 72*, 83-106

Anderson, S.R. 2004, *Dr. Dolittle's delusion: Animals and the uniqueness of human language*, New Haven: Yale University Press

Anderson, S.R. and D.W. Lightfoot 1999, "The human language faculty as an organ", *Annu Rev Physiol.* 62, 697-722

Anderson, S.R. and D.W. Lightfoot 2002, *The Language Organ. Linguistics as Cognitive Physiology*, Cambridge University Press: Cambridge

Andrewes, D. 2001, *Neuropsychology: From theory to practice*, Hove: Psychology Press

Andrews, S. 1989, "Frequency and neighborhood effects on lexical access: Activation or search?", *Journal of Experimental Psychology: Learning, Memory and Cognition* 15, 802-814

—. 1992, "Frequency and neighborhood effects on lexical access: Lexical similarity or orthographic redundancy?", *Journal of Experimental Psychology: Learning, Memory and Cognition* 18, 234-254

Ardila, A. 2010 "A proposed reinterpretation and reclassification of aphasic syndromes", *Aphasiology* 24, 363-394

Ariew, A. 1999, "Innateness is canalisation: In defense of a developmental account of innateness", V. Hardcastle (ed.) *Where Biology Meets Psychology. Philosophical Essays*. Cambridge (MA): MIT Press, 117-139

Arnold, K. and K. Zuberbühler 2006a, "Semantic combinations in primate calls", *Nature* 441, 303

Arnold, K. and K. Zuberbühler 2006b, "The alarm-calling system of adult male putty-nosed monkeys, Cercopithecus nictitans martini", *Animal Behaviour* 72, 643-653

Aronoff, M., I. Meir, C.A. Padden, and W. Sandler 2008, "The roots of linguistic organization in a new language", *Interaction Studies* 9:133-153

Arsenijevic, B. and W. Hinzen 2010, "Recursion as a human universal and as a primitive", *Biolinguistics* 4:165-173

Atlan, H. and M. Koppel 1990, "The cellular computer DNA: program or data", *Bulletin of Mathematical Biology* 52:335-348

Baars, B. J., M. T. Motley and D.G. MacKay 1975, "Output editing for lexical status in artificially elicited slips of the tongue", *Journal of Verbal Learning and Verbal Behavior* 14, 382- 391

Bagri, A., *et al.* 2002, "Slit proteins prevent midline crossing and determine the dorsoventral position of major axonal pathways in the mammalian forebrain", *Neuron* 33, 233-248

Baguñá, J. and García-Fernández, J. 2003, "Evo-Devo: The long and winding road", *Int. J. Dev. Biol.* 47, 705-713

Baker, M.C. 2001, *The Atoms of Language: The Mind's Hidden Rules of Grammar*, New York: Basic Books

—. 2003, "Language differences and language design", *Trends in Cognitive Sciencies* 7, 349-353

—. 2009, "Language universals: Abstract but not mythological", *Behavioral and Brain Sciences*, 32

Balaban, E. 2000, "Behavior genetics: Galen's prophecy or Malpighi's legacy?", R. Singh, C. Krimbas, D. Paul, and J. Beatty (eds.), *Thinking About Evolution: Historical, Philosophical, and Political Perspectives*, Cambridge: Cambridge University Press, 429-466

—. 2006, "Cognitive developmental biology: history, process and fortune's wheel", *Cognition* 101, 298-332

Balari, S. 2006. "Heterochronies in brain development and the origins of language. A coevolutionary scenario", J. Roselló and J. Martín (eds.), *The biolinguistic turn. Issues on language and biology*, Barcelona: PPU, 66-81

—. 2011. "La ciencia normal no hace revoluciones", *Ludus Vitalis* XIX (35), 177-180.

Balari, S., A. Benítez-Burraco, M. Camps, V.M. Longa, G. Lorenzo, and J. Uriagereka 2008, "¿Homo loquens neanderthalensis? En torno a las capacidades simbólicas y lingüísticas del Neandertal", *Munibe. Antropologia-Arkeologia* 59, 3-24

Balari, S., A. Benítez-Burraco, M. Camps, V.M. Longa, G. Lorenzo, and J. Uriagereka 2011, "The archaeological record speaks: bridging Anthropology and Linguistics" *IJEB* 2011: 382679

Balari, S., A. Benítez-Burraco, M. Camps, V.M. Longa, G. Lorenzo, and J. Uriagereka *In progress*, "Biolinguistics and Evolutionary Anthropology: Building Bridges", Ms., UAB, UH, GWU, USC, UniOvi and UMD

Balari, S., A. Benítez-Burraco, V.M. Longa and G. Lorenzo *In press*, "The fossils of language: What are they? Who has them? How did they evolve?", C. Boeckx and K. Grohmann (eds.), *The Cambridge handbook of Biolinguistics*. Cambridge: Cambridge University Press.

Balari, S. and G. Lorenzo 2008, "Pere Alberch's developmental morphospaces and the evolution of cognition", *Biological Theory* 3(4), 297-304

Balari, S. and G. Lorenzo 2009a "Comunicación. Donde la lingüística evolutiva se equivocó", Report de recerca CGT-09-10, Centre de Lingüística Teorica, UAB (English translation: "Communication. Where evolutionary linguistics went wrong", submitted to publication).

Balari, S. and G. Lorenzo 2009b, "Computational phenotypes: Where the theory of computation meets Evo-Devo", *Biolinguistics* 3(1), 2-60

Balari, S. and G. Lorenzo, *In press*, "¿Para qué sirve un ballestrinque? Reflexiones sobre el funcionamiento de artefactos y organismos en un mundo sin funciones", *Teorema. Revista Internacional de Filosofía.*

Balari, S. and G. Lorenzo *In progress*, "Computational homology", Ms., UAB and UniOvi

Balari, S. and G. Lorenzo *Submitted*, *Computational Phenotypes. Steps to an Evolutionary Developmental Bioliguistics*

Balota, D.A. 1990, "The role of meaning in word recognition", D.A. Balota, G.B. Flores d'Arcais and K. Rayner (eds.) *Comprehension processes in Reading*, Hillsdale NJ: Erlbaum, 9-32

Balota, D. A. and J. I. Chumbley 1984 "Are lexical decisions a good measure of lexical access? The role of lexical frequency in the neglected decision stage", *Journal of Experimental Psychology: Human Perception and Performance* 10, 340-357

Balota, D. A., M. Pilotti and M. J. Cortesse 2001, "Subjective frequency estimates for 2,938 monosyllabic words", *Memory and Cognition* 29, 639-647

Barrett, H.C. and R. Kurzban 2006, "Modularity in cognition: framing the debate", *Psychol. Rev.* 113, 628-647

Bartlett, C.W., J.F. Flax, M.W. Logue, V.J. Vieland, A.S. Bassett, P. Tallal and L.M. Brzustowicz 2002, "A major susceptibility *locus* for specific language impairment is located on 13q21", *Am. J. Hum. Genet.* 71, 45-55

Basso, A., A.R. Lecours, S. Moraschini, and M. Vanier 1985, "Anatomical correlations of the aphasias as defined through computerised tomography:Exceptions", *Brain and Language* 26, 201-229

Bateson, Patrick 2001, "Behavioral development and Darwinian evolution", S. Oyama, P.E. Griffiths and R.D. Gray (eds.), *Cycles of Contingencies. Developmental Systems and Evolution.* Cambridge MA: MIT Press, 149-166

Bateson, P. and M. Mamelli 2007, "The innate and the acquired: useful clusters or a residual distinction from folk Biology?" *Dev. Psychobiol.* 49, 818-831

Bathia, T.K. and W.C. Ritchie 2004, *The Handbook of Bilingualism*, Oxford: Backwell Publishing

Battison, R. 1978, *Lexical borrowing in American Sign Language,* Silver Spring MD: Linstok Press

Baumann, M. 1995, *The production of syllables in connected speech.* Ph.D. dissertation, Nijmegen University

Baus, C., A. Costa and M. Carreiras 2008, "Neighbourhood density and frequency effects in speech production: a case for interactivity", *Language and Cognitive Processes* 23, 866-888

Baus, C., E. Gutiérrez-Sigut, J. Quer and M. Carreiras 2008, "Lexical access in Catalan Signed Language (LSC) production", *Cognition* 108, 856-65

Behar, D.M., R. Villems, H. Soodyall, J. Blue-Smith, L. Pereira, E. Metspalu, R. Scozzari, H. Makkan, S. Tzur, D. Comas, *et al.* 2008, "The dawn of human matrilineal diversity", *The American Journal of Human Genetics* 82, 1130-1140

Belton, E., Salmond, C. H., Watkins, K. E., Vargha-Khadem, F., and D.G. Gadian 2003, "Bilateral brain abnormalities associated with dominantly inherited verbal and orofacial dyspraxia", *Hum. Brain Mapp.* 18, 194-200

Benítez-Burraco, A. 2008, "Más allá del TEL y de la dislexia: otros trastornos lingüísticos de base genética", *Neurología, Neurocirugía y Psiquiatría* 41, 21-36.

—. 2009, *Genes y lenguaje: aspectos ontogenéticos, filogenéticos y cognitivos,* Barcelona: Reverté

Benítez-Burraco, A. and V.M. Longa, 2010 "Evo-Devo —Of Course, But Which One? Some Comments on Chomsky's Analogies between the Biolinguistic Approach and Evo-Devo", *Biolinguistics* 4, 308-323

Benítez-Burraco, A., V.M. Longa, G. Lorenzo and J. Uriagereka 2008, "Also sprach Neanderthalis…or did she?", *Biolinguistics* 2, 225-232

Benson, D. F. 1979, *Aphasia, alexia and agraphia*, New York: Churchill Livingstone

Bergman, A. and Siegal, M. 2002, "Evolutionary capacitance as a general feature of complex gene networks" *Nature* 424, 549-552

Bernal, J. 2007, "Thyroid hormone receptors in brain development and function", *Nat. Clin. Pract. Endocrinol. Metab.* 3, 249-259

Berndt, R. 1987, "Symptom co-occurrence and dissociation in the interpretation of agrammatism", M. Coltheart, G. Sartori and R. Job (eds.), *The Cognitive Neuropsychology of Language*, Hove: Lawrence Erlbaum, 221-233

Berwick, R.C. 2011, "All you Need is Merge: Biology, Computation, and language from the Bottom-up", Di Sciullo and Boeckx (eds.), 2011, 461-491

Berwick, R.C. and N. Chomsky 2011, "The Biolinguistic program: the current state of its development", Di Sciullo and Boeckx (eds.), 2011, 19-41

Berwick, R.C., K. Okanoya, G. Beckers and J.J. Bolhuis 2011, "Songs to syntax: the linguistics of birdsong", *Trends in Cognitive Sciences* 15, 113-121

Bialystok, E. 1999, "Cognitive complexity and attentional control in the bilingual mind", *Child Development* 70, 636-644

—. 2001, *Bilingualism in development: Language, literacy, and cognition.* New York: Cambridge University Press

Bialystok, E., F.I.M. Craik, R. Klein and M. Viswanathan 2004, "Bilingualism, aging, and cognitive control: evidence from the Simon task", *Psychology and Aging* 19, 290-303

Bialystok, E., and M. Martin 2004, "Attention and inhibition in bilingual children: evidence from the dimensional change card sorting task", *Developmental Science* 7, 325-339.

Bickerton, Derek 1990, *Language and Species*, Chicago: The University of Chicago Press

—. 1995, *Language and human behavior*, Seattle: University of Washington Press

—. 2009, *Adam's tongue. How humans made language, how language made humans*, New York: Hill and Wang

Binder, J. and C.J. Price 2001, "Functional neuroimaging of language", R. Cabeza and A. Kingstone (eds.), *Handbook of functional neuroimaging of cognition*, Cambridge MA: MIT Press, 187-251

Birdsong, D. 1999, *Second language acquisition and the critical period hypothesis,* Mahwah NJ: Erlbaum

Birdsong, D. and M. Molis 2001, "On the evidence for maturational constraints in second-language acquisition", *Journal of Memory and Language* 44, 235-249

Biro, Dora and T. Matsuzawa 1999, "Numerical ordering in a chimpanzee (Pan troglodytes): planning, executing, and monitoring", *Journal of Comparative Psychology* 113, 178-185

Bishop, D.V.M. 2001, "Genetic and environmental risks for specific language impairment in children", *Philos. Trans. R. Soc. Lond., B, Biol. Sci.* 356, 369-380

—. 2002, "The role of genes in the etiology of specific language impairment", *J. Commun. Disord.* 35, 311-328

Bishop, D.V.M. and L. Leonard 2001, *Speech and language impairments in children: causes, characteristics, intervention and outcome*, Oxford: Psychology Press

Blevins, J. 2004, *Evolutionary Phonology: The Emergence of Sound Patterns*. Cambridge: Cambridge University Press

Bloom, P. 2000, *How Children Learn the Meanings of Words*. Cambridge MA: MIT Press

Bloomfield, T.C., T.Q. Gentner and D. Margoliash 2011, "What birds have to say about language", *Nature Neuroscience* 14, 947-948

Blumberg, M.S. 2005, *Basic Instinct. The Genesis of Behavior*. New York: Thunder's Mouth Press

Bock, J.K. 1991, "A sketchbook of production problems", *Journal of Psycholinguistic Research* 20, 141-160

Boeckx, C. 2006, *Linguistic Minimalism: Origins, Concepts, Methods, and Aims*, Oxford: Oxford University Press

—. 2008a, *Bare syntax*, Oxford: Oxford University Press

—. 2008b, "Did we really solve Plato's Problem (abstractly)?", Y. Otsu (ed.) *Proceedings of the 9th Annual Tokyo Conference on Psycholinguistics*, Tokyo: Hituzi Syobo

—. 2008c, "Islands", *Language and Linguistics Compass* 2, 151-167

—. 2009a, "Cartesian biolinguistics". SOLIFIC lecture, Sophia University, July 2009, Ms., ICREA-Universitat Autònoma de Barcelona

—. 2009b, *Language in Cognition: Uncovering mental structures and the rules behind them*, Malden: Wiley-Blackwell

—. 2009c, "Language Universals: Yesterday, Today, and Tomorrow" (with J.D. Fodor, L. Gleitman, and L. Rizzi), M. Piattelli-Palmarini, P. Salaburu and J. Uriagereka (eds.), *Of minds and language: A Basque encounter with N. Chomsky*, Oxford: Oxford University Press, 194-220

—. 2009d, "The locus of asymmetry in UG", *Catalan Journal of Linguistics* 8, 41-53

—. 2009e, "On the nature of Merge", M. Piattelli-Palmarini, P. Salaburu, and J. Uriagereka (eds.), *Of minds and language: A Basque encounter with N. Chomsky*, Oxford: Oxford University Press, 44-57

—. 2009f, "Some notes on the syntax-thought interface", *Proceedings of the Sophia University Linguistic Society* 24 , 92-103

—. 2009g, "When syntax meets the sensori-motor systems, phylogenetically and ontogenetically". Presented at the pre-ConSOLE Workshop on the syntax- morphology-phonology interface. Bellaterra, December 2009

—. 2010a, "Phases beyond explanatory adequacy". Ms., ICREA-Universitat Autònoma de Barcelona [Under review for A. Gallego (ed.) *Phase theory: developing the framework*, Berlin: Walter de Gruyter]

—. 2010b, "Syntactic order for free", Presented at the 10th European Conference on Complex Systems, Lisbon University Institute, Workshop on 'Modelling the non-separability of a very complex world'

—. 2010c, "Taming recursion", Ms., ICREA-Universitat Autònoma de Barcelona. Ms. based on Presentation at the UMass Conference on Recursion, 28 May 2009.

—. 2010d, "The Chomsky-Piaget encounter at Royaumont, 35 years after", CLT Seminar Presentation

—. 2010e, "What Principles and Parameters Got Wrong". Ms., ICREA and Universitat Autònoma de Barcelona. [Written version of Talk given at the Workshop on Variation in the Minimalist Framework, Barcelona, January 2010.].

—. 2010f, "Linguistic minimalism", B. Heine and H. Narrog (eds.) *Oxford Handbook of Linguistic Analysis*, Oxford: Oxford University Press, 485-505

—. 2010/*In progress, Elementary syntactic structures*. Ms., ICREA-UAB

—. 2011a, "Approaching Parameters from Below", Di-Sciullo and Boeckx (eds.), 205-221

—. 2011b, "The emergence of the language faculty, from a biolinguistic point of view", M. Tallerman and K. Gibson (eds.) *The Oxford Handbook of Language Evolution*, Oxford: Oxford University Press, 492-501

—. 2011c. "Some reflections on Darwin's Problem in the context of Cartesian Biolinguistics". Di Sciullo and Boeckx (eds.), 42-64

—. 2011d. "Biolinguistics: A divided or pluralist enterprise? A Reply to Jackendoff", Ms., ICREA/UAB.

—. (ed.) 2011, *Oxford handbook of linguistic minimalism*, Oxford: Oxford University Press

—. *In press*, *Syntactic Islands*, Cambridge: Cambridge University Press

—. *Submitted*, "What does minimalism have to say about syntactic locality?" Ms., ICREA-Universitat Autònoma de Barcelona. [Under review for *Proceedings volume of the Budapest conference on Minimalist Approaches to Syntactic Locality*].

Boeckx, C., and K.K. Grohmann 2007, "The Biolinguistics manifesto", *Biolinguistics* 1, 1-8

Boeckx, C. and V. Longa *In press*, "Lenneberg's views on language development and evolution and their relevance for modern biolinguistics", *Biolinguistics*

Boeckx, C. and M. Piattelli-Palmarini 2005, "Language as a natural object; linguistics as a natural science", *The Linguistic Review* 22, 467-471

Boeckx, C. and B. Samuels 2009, "What emerges from merge in phonology". Presented at the *6th Old World Conference on Phonology*, Edinburgh, UK.

Boeckx, C. and J. Uriagereka 2007, "Minimalism", G. Ramchand and C. Reiss (eds.) *The Oxford Handbook of Linguistic Interfaces*, Oxford: Oxford University Press, 541-573

Bolender, J. 2010, "Universal grammar as more than a programmatic label", *Lingua* 120, 2661-2663

Bolhuis, J.J., K. Okanoya and C. Scharff 2010, "Twitter evolution: Converging mechanisms in birdsong and human speech", *Nature Reviews: Neuroscience* 11, 747-759

Borer, H. 2005, *Structuring Sense* (2 vols.), Oxford: Oxford University Press

Borer, H. and K. Wexler 1987, "The maturation of syntax", T. Roeper and E. Williams (eds.) *Parameter Setting*, Dordrecht: Reidel, 123-172

Bornkessel, I. and M. Schlesewsky 2006, "The extendent argument dependency model: A neurocognitive approach to sentence comprehension across languages", *Psychological Review* 113 (4), 787-821

Boroditsky, L. 2009, "How does our language shape the way we think?", M. Brockman (ed.) *What's Next: Dispatches on the Future of Science*, Vintage Books: New York, 116-129

—. *et al.* 2003, "Sex, Syntax, and Semantics", D. Gentner and S. Goldin-Meadow (eds.) *Language in Mind: Advances in the Study of Language and Cognition*, Cambridge MA: MIT Press, 61-79

Borsley, R.D. 1991, *Syntactic Theory. A Unified Approach*, London: Eduard Arnold

Bosworth, R.G., and Emmorey, K. *In press*, "Effects of iconicity and semantic relatedness on lexical access in American Sign Language", *Journal of Experimental Psychology: Learning, Memory and Cognition*

Bouwman, J., A.S. Maia, P.G. Camoletto, G. Posthuma, E.W. Roubos, V.M. Oorschot, J. Klumperman and M. Verhage 2004, "Quantification of synapse formation and maintenance in vivo in the absence of synaptic release", *Neuroscience* 126, 115-126

Bouzouggar, A. N. Barton, M. Vanhaeren, F. D'errico, S. Collcut, T. Higham, E. Hodge, S. Parfitt, E. Rhodes, J.L. Schwenninger, C. Stringer, E. Turner, S. Ward, A. Moutmir, and A. Stambouli 2007, "82,000-year-old shell beads from North Africa and implications for the origins of modern human behavior", *Proceedings of the National Academy of Sciences USA* 104, 9964-9969

Bower, B. 1989 "Talk of ages: a tiny bone rekindles arguments over the roots of speech and language", *Science News* 24-26

Bowers, E.J. 2006, "A new model for the origin of bipedality", *Human Evolution* 21, 241-250

Brann, D.W., K. Dhandapani, C. Wakade, V.B. Mahesh and M.M. Khan 2007, "Neurotrophic and neuroprotective actions of estrogen: basic mechanisms and clinical implications", *Steroids* 72, 381-405

Brannon, E.M. and H.S. Terrace 1998, "Ordering of the numerosities 1 to 9 by monkeys", *Science* 282, 746-749

Brannon, E.M. and H.S. Terrace 2000, "Representation of the numerosities 1-9 by rhesus macaques (Macacamulatta)", *Journal of Experimental Psychology: Animal Behavior Processes* 26, 31-49

Brentari, D. 1998, *A prosodic model of sign language phonology.* Cambridge MA: MIT Press

Breuker, C.J., V. Debat and C.P. Klingenberg 2006, "Functional Evo-devo", *Trends Ecol. Evol.* 21, 488-492

Briscoe, T. 2002, "Introduction", T. Briscoe (ed.) *Linguistic Evolution Through Language Acquisition*, Cambridge: Cambridge University Press, 1-22

Bromberger, S. and M. Halle 1989, "Why Phonology is Different", *Linguistic Inquiry* 20, 51-70

Brosch, M., E. Selezneva, C. Bucks and H. Scheich 2004, "Macaque monkeys discriminate pitch relationships", *Cognition* 91, 259-272

Brown, A.S. 1991, "A review of the tip-of-the-tongue experience", *Psychological Bulletin* 109, 204-223

Brown, C.H. and J.M. Sinnott 2006, "Cross-species comparisons of vocal perception", S. Greenberg and W.A. Ainsworth (eds.) *Listening to speech: an auditory perspective,* Mahwah NJ: Lawrence Erlbaum Associates, 183-201

Brown, R., and D. McNeill 1966, "The 'tip-of-the-tongue' phenomenon", *Journal of Verbal Learning and Verbal Behavior* 5, 325-337

Bruce, H. A. and R.L. Margolis 2002, "*FOXP2*: novel exons, splice variants, and CAG repeat length stability", *Hum. Genet.* 111, 136-144

Brzustowicz, L. M. 1998, "Molecular genetic approaches to the study of language", *Hum. Biol.* 70, 325-345

Burbridge, T.J., Y. Wang, A.J. Volz, V.J. Peschansky, L. Lisann, A.M. Galaburda, J.J. Lo Turco and G.D. Rosen 2008, "Postnatal analysis of the effect of embryonic knockdown and overexpression of candidate dyslexia susceptibility gene homolog *Dcdc2* in the rat", *Neuroscience* 152, 723-733

Burke, D., D.G. MacKay, J.S. Worthley and E. Wade 1991, "On the tip of the tongue: What causes word finding failures in young and older adults?", *Journal of Memory and Language* 30, 542-579

Butler, A.B. and W.M. Saidel 2000, "Defining sameness: Historical, biological, and generative homology", *BioEssays* 22, 846-853

Byrne, R.W. 2007, "Clues to the origin of the human mind from primate observational field data", *The Japanese Journal of Animal Psychology* 57, 1-14

Cabeza, R. and L. Nyberg 1997, "Imaging Cognition: An Empirical Review of PET Studies with Normal Subjects", *Journal of Cognitive Neuroscience* 9 (1), 1-26

Cabeza, R. and L. Nyberg 2000, "Imaging Cognition II: An Empirical Review of 275 PET and fMRI Studies", *Journal of Cognitive Neuroscience* 12 (1), 1-47

Campbell, R., P. Martin and T. White 1992, "Forced choice recognition of sign in novice British Sign Language learners", *Applied Linguistics* 13, 185-201

Camps, M. and J. Uriagereka 2006, "The Gordian knot of linguistic fossils", J. Rosselló and J. Martín (eds.), *The biolinguistic turn. Issues on language and biology*, Barcelona, PPU, 34-65

Canguilhem, G. 1977, *Idéologie et rationalité dans l'histoire des sciences de la vie*, Paris: Vrin

Cannon, T. D. 2010, "Candidate gene studies in the GWAS era: the *MET* proto-oncogene, neurocognition, and schizophrenia", *Am. J. Psychiatry* 167, 369-372

Caramazza, A. 1997, "How many levels of processing are there in lexical access?", *Cognitive Neuropsychology* 14, 177-208

Caramazza, A. and M. Miozzo 1997, "The relation between syntactic and phonological knowledge in lexical access: Evidence from the 'tip-of-the-tongue' phenomenon", *Cognition* 64, 309-343

Caramazza, A. and E.B. Zurif, 1976, "Dissociation of algorithmic and heuristic processes in language comprehension", *Brain and Language* 3, 572-582

Cardon, L. R. and J.L. Bell 2001, "Association study designs for complex diseases", *Nat. Rev. Genet.* 2, 91-99

Carreiras, M. 2010, "Sign language processing", *Language and Linguistics Compass* 4 (7), 430-444

Carreiras, M., C. J. Álvarez and M. De Vega 1993 "Syllable-frequency and visual word recognition in Spanish", *Journal of Memory and Language* 32, 766-780

Carreiras, M., L. Ferrand, J. Grainger and M. Perea 2005, "Sequential effects of phonological priming in visual word recognition", *Psychological Science* 16, 585-589

Carreiras, M., E. Gutiérrez-Sigut, S. Baquero and D. Corina 2008, "Lexical processing in Spanish Sign Language (LSE)", *Journal of Memory and Language* 581, 100-122

Carreiras, M., and M. Perea 2004, "Naming pseudowords in Spanish: Effects of syllable frequency in production", *Brain and Language* 90, 393-400

Carreiras, M., M. Perea and J. Grainger 1997, "Effects of orthographic neighborhood in visual word recognition: Cross-task comparisons", *Journal of Experimental Psychology: Learning, Memory, and Cognition* 23, 857-871

Carroll, S.B. 2005, *Endless Forms Most Beautiful. The New Science of Evo Devo and the Making of the Animal Kingdom*. New York: W.W. Norton and Co.

Carruthers, P. 2006, *The Architecture of the Mind*, Oxford: Oxford University Press

Catani, M., D.K. Jones and D.H. Ffytche 2005, "Perisylvian language networks of the human brain", *Annales of Neurology* 57, 8-16

Catts, H.W., S.M. Adlof, T.P. Hogan and S.E. Weismer 2005, "Are specific language impairment and dyslexia distinct disorders?", *J. Speech Lang. Hear. Res.* 48, 1378-1396

Chen, L, H. Shu, Y. Liu, J. Zhao, P. Li 2007, "ERP signatures of subject-verb agreement in L2 learning", *Bilingualism, Language and Cognition* 10 (2), 161-174

Cheney, D.L., and R.M. Seyfarth 2007, *Baboon metaphysics: The evolution of a social mind.* Chicago: University of Chicago Press

Cheng, L. 1997, *On the typology of* wh-*questions*, New York and London: Garland

Cholin, J., W. J. M. Levelt and N. O. Schiller 2006, "Effects of syllable frequency in speech production", *Cognition* 99, 205-235

Chomsky, N. 1955, *The logical structure of linguistic theory*, Ms. Harvard University (Published in part in Chomsky 1975a)

—. 1956, "Three models for the description of language", *IRE Transactions on Information Theory* 2, 113-124

—. 1957, *Syntactic Structures*, The Hague: Mouton

—. 1959a, "A review of B. F. Skinner's *Verbal Behavior*", *Language* 35, 26-58

—. 1959b, "On certain formal properties of grammars", *Information and control* 2, 137-167

—. 1965, *Aspects of the theory of syntax*, Cambridge MA: MIT Press

—. 1967, "Recent contributions to the theory of innate ideas", *Synthese* 17, 2-11

—. 1968, *Language and Mind*, New York: Harcourt Brace Jovanovich

—. 1975a, *The Logical Structure of Linguistic Theory*, Chicago: University of Chicago Press

—. 1975b, *Reflections on Language*, New York: Pantheon

—. 1977, *Essays on Form and Interpretation*, New York: Elsevier-North Holland

—. 1980, *Rules and Representations*, Oxford: Basil Blackwell

—. 1981, *Lectures on Government and Binding*, Foris: Dordrecht

—. 1986, *Knowledge of Language. Its Nature, Origins and Use*, Praeger: New York

—. 1988, *Language and Problems of Knowledge: The Managua Lectures,* Cambridge MA: MIT Press

—. 1993, "A Minimalist Program for Linguistic Theory", K. Hale and S. J. Keyser (eds.) *The View from Building 20*, Cambridge MA: MIT Press, 1-52

—. 1995, *The Minimalist Program*, Cambridge MA: MIT Press

—. 2000a, "Minimalist inquiries: The framework", R. Martin, D. Michaels and J. Uriagereka (eds.), *Step by step. Essays in honor of Howard Lasnik*, Cambridge MA: MIT Press, 89-155

—. 2000b, *New horizons in the study of language and mind*, Cambridge: Cambridge University Press

—. 2000c, *The Architecture of Language*, New Delhi: Oxford University Press

—. 2001a, "Beyond Explanatory Adequacy", *MIT Occasional Working Papers in Linguistics* 20, 1-28 (reprinted in Chomsky 2004)

—. 2001b, "Derivation by phase", M. Kenstowicz (ed.) *Ken Hale: A Life in Language*, Cambridge MA: MIT Press, 1-52
—. 2002, *On Nature and Language*, Cambridge University Press: Cambridge
—. 2004, "Beyond Explanatory Adequacy", A. Belletti (ed.) *Structures and Beyond*, Oxford University Press: Oxford, 104-131
—. 2005, "Three factors in the language design", *Linguistic Inquiry* 36, 1-22
—. 2007, "Approaching UG from below", U. Sauerland and H.M. Gärtner (eds.) *Interfaces + Recursion = Language? Chomsky's minimalism and the view from Syntax-Semantics*, Berlin and New York: Mouton de Gruyter, 1-29
—. 2008, "On phases", C. Otero, R. Freidin and M.L. Zubizarreta (eds.) *Foundational issues in linguistic theory: Essays in honor of Jean-Roger Vergnaud*, Cambridge MA: MIT Press, 133-166
—. 2009, "Opening remarks", M. Piattelli-Palmarini, J. Uriagereka and P. Salaburu (eds.) *Of minds and language. A dialogue with N. Chomsky in the Basque Country*, Oxford: Oxford University Press, 13-43.
—. 2010a, "Poverty of Stimulus: Unfinished Business". Lecture presented in the Lecture Series "Sprache und Gehirn - Zur Sprachfähigkeit des Menschen" organized by Angela D. Friederici in the context of the Johannes Gutenberg endowed professorship summer 2010
—. 2010b, "Some simple evo devo theses: how true might they be for language?", R.K. Larson, V. Déprez and H. Yamakido (eds.) *The Evolution of Human Language. Biolinguistic Perspectives*, Cambridge: Cambridge University Press, 45-62
Chomsky, N. and H. Lasnik 1993, "The theory of principles and parameters", J. Jacobs, A. von Stechow, W. Sternefeld, and T. Vennemann (eds.) *Syntax. An International Handbook of Contemporary Research*, Berlin: Walter de Gruyter, 506-569
Clahsen, H. and C. Felser 2006, "How native-like is non-native language processing?", *Trends in Cognitive Sciences* 10 (12), 564-570
Clark, E.V. 1997, "Conceptual perspective and lexical choice in acquisition", *Cognition* 64, 1-37
Coen, M.H. 2006, *Multimodal dynamics: self-supervised learning in perceptual and motor systems*, Cambridge MA: MIT dissertation
Collias, N.E. and E.C. Collias 1962, "An experimental study of the mechanisms of nest building in a weaverbird", *The Auk* 79, 568-595
Collingridge, G.L. and J.T.R. Isaac 2003, "Functional roles of protein interactions with AMPA and kainate receptors", *Neurosci. Res.* 47, 3-15

Collins, A.M., and E.F. Loftus 1975, "A spreading-activation theory of semantic processing", *Psychological Review* 82, 407-428

Coltheart, M. 1999, "Modularity and cognition", *Trends Cogn. Sci.* 3, 115-120

Coltheart, M., B. Curtis, P. Atkins and M. Haller 1993, "Dual-route and parallel distributed-processing approaches", *Psychological Review* 100, 589-608

Coltheart, M., E. Davelaar, T. Jonansson and D. Besner 1977, "Access to the internal lexicon", S. Dornic (ed.) *Attention and performance VI,* London: Academic Press

Coltheart, M., K. Rastle, C. Perry, R. Langdon and J.C. Ziegler 2001, "DRC: A Dual Route Cascaded model of visual word recognition and reading aloud", *Psychological Review* 108, 204-256

Comrie, B. 2003, "Reconstruction, Typolgy and Reality", R. Hickey (ed.) *Motives for Language Change*, Cambridge University Press: Cambridge, 243-257

Conrad, M., M. Carreiras, S. Tamm and A.M. Jacobs 2009, "Syllables and bigrams: Orthographic redundancy and syllabic units affect visual word recognition at different processing levels", *Journal of Experimental Psychology: Human Perception and Performance* 35, 461-479

Conrad, M. and A.M. Jacobs 2004, "Replicating syllable-frequency effects in Spanish in German: One more challenge to computational models of visual word recognition", *Language and Cognitive Processes* 19 (3), 369-390

Conrad, M., P. Stenneken and A.M. Jacobs 2006, "Associated or dissociated effects of syllable frequency in lexical decision and naming", *Psychonomic Bulletin and Review* 13 (2), 339-345

Conrad, M., S. Tamm, M. Carreiras and A. Jacobs 2010, "Simulating syllable frequency effects within an interactive activation framework", *European Journal of Cognitive Psychology*, 22 (5), 861-89 (DOI:10.1080/09541440903356777)

Conway, C.M. and M.H. Christiansen 2001 "Sequential learning in non-human primates", *Trends in Cognitive Sciences* 5, 539-546

Coop, G., K. Bullaughey, F. Luca and M. Przeworski 2008, "The timing of selection at the human *FOXP2* gene", *Molecular Biology and Evolution* 25, 1257-1259

Cook, E. H., V. Lindgren, B.L. Leventhal, R. Courchesne, A. Lincoln, C. Shulman, C. Lord and E. Courchesne 1997, "Autism or atypical autism in maternally but not paternally derived proximal 15q duplication", *Am. J. Hum. Genet.* 60, 928-934

Corballis, M.C. 2007, "The uniqueness of human recursive thinking". *American Scientist* 95, 240-248

Corina, D. and K. Emmorey 1993, "Lexical priming in American Sign Language", Paper presented at the Psychonomics Society.

Corina, D. and U. Hildebrandt 2002, "Psycholinguistic investigations of phonological structure in ASL", R. P. Meier, K. Cormier and D. Quinto-Pozos (eds.), *Modality and structure in signed and spoken languages*, Cambridge: Cambridge University Press, 88-111

Corina, D.P. and H. Knapp 2006, "Lexical retrieval in American Sign Language production", L. Goldstein, D. Whalen and C.T. Best (eds.) *Papers in laboratory phonology 8: Varieties of phonological competence*, Berlin: Mouton de Gruyter, 213-240

Corina, D. and W. Sandler 1993, "On the nature of phonological structure in sign language", *Phonology* 10, 165-207

Corning, P.A. 2002, "The re-emergence of 'emergence': A venerable concept in search of a Theory", *Complexity* 7(6), 18-30

Costa, A. 2005, "Lexical access in bilingual production", J. F. Kroll and A. M. B. de Groot (eds.) *Handbook of bilingualism: Psycholinguistic approaches*, New York: Oxford University Press, 308-25

Costa, A., B. Albareda and M. Santesteban 2008 "Assessing the presence of lexical competition across languages: Evidence from the stroop task", *Bilingualism: Language and Cognition* 11, 121-131

Costa, A. and A. Caramazza 1999, "Is lexical selection in bilingual speech production language-specific? Further evidence from Spanish-English and English-Spanish bilinguals", *Bilingualism: Language and Cognition* 2(3), 231-244

Costa, A., A. Caramazza and N. Sebastián-Gallés 2000, "The cognate facilitation effect: implications for models of lexical access", *Journal of Experimental Psychology: Learning Memory and Cognition* 26, 1283-96

Costa, A., M. Hernández and N. Sebastián-Gallés 2008, "Bilingualism aids conflict resolution: Evidence from the ANT task", *Cognition* 106, 59-86

Costa, A., M. Miozzo and A. Caramazza 1999, "Lexical selection in bilinguals: do words in the bilingual's two lexicons compete for selection?", *Journal of Memory and Language* 41, 365-97

Costa, A. and M. Santesteban 2004, "Lexical access in bilingual speech production: Evidence from language switching in highly proficient bilinguals and L2 learners", *Journal of Memory and Language* 50, 491-511

Costa, A., M. Santesteban and A. Cano 2005, "On the facilitatory effects of cognate words in bilingual speech production", *Brain and Language* 94, 94-103

Costa, A., M. Santesteban and I. Ivanova 2006, "How do highly proficient bilinguals control their lexicalization process? Inhibitory and Language-Specific Selection mechanisms are both functional", *Journal of Experimental Psychology: Learning, Memory and Cognition* 32, 1057-1074

Costa, A. and N. Sebastián-Gallés 1998, "Abstract phonological structure in language production: Evidence from Spanish", *Journal of Experimental Psychology: Learning, Memory and Cognition* 24, 886-903

Cowie, F. 1997, "The logical problem of language acquisition", *Synthese* 111, 17-51

—. 1999, *What's Within. Nativism Reconsidered*, Oxford: Oxford University Press

Coyne, J. 2009, *Why evolution is true*, New York: Viking

Craik, F. I. M. and E. Bialystok 2006, "Cognition through the lifespan", *Trends in Cognitive Sciences*, 10(3)

Craik, F.I.M., E. Bialystok and M. Freedman 2010, "Delaying the onset of Alzheimer's disease: Bilingualism as a form of cognitive reserve", *Neurology* 75, 1726-1729

Crain, S. 1991, "Language acquisition in the absence of experience", *Behav. Brain Sci.* 14, 597-650

Crain, S. and M. Nakayama 1987, "Structure dependence in grammar formation". *Language* 63, 522-542

Crain, S. and P. Pietroski 2002, "Why language acquisition is a snap". *The Linguistic Review* 19, 163-183

Crinion, J. *et al.* 2006, "Language Control in the Bilingual Brain", *Science* 312, 1537

Cuetos, F. 1998, *Evaluación y rehabilitación de las afasias*. Madrid: Editorial Médica Panamericana

Cuetos, F. 2003, *Anomia: la dificultad para recuperar las palabras*. Madrid: TEA Ediciones

Cuetos, F., M. González-Nosti, L. Martínez, N. Mantiñán, A. Olmedo and A. Dioses 2010, "¿Síndromes o síntomas en la evaluación de los pacientes afásicos?" *Psicothema* 22

Culicover, P.W. and R. Jackendoff 2005, *Simpler syntax*, Oxford: Oxford University Press

Cummings J.L. 1993, "Frontal-subcortical circuits and human behavior", *Archives of Neurology* 50, 873-880

Cummins, R. 2002 "Neo-Teleology", A. Ariew, R. Cummins and M. Perlman (eds.) *Functions. New Essays in the Philosophy of Psychology and Biology*, Oxford: Oxford University Press, 157-172

Cutler, A., J.M. McQueen, D. Norris and A. Somejuan 2001, "The roll of the silly ball", E. Dupoux (ed.) *Language, brain and cognitive development: Essays in honor of Jacques Mehler,* Cambridge MA: MIT Press, 181-194

Cutting, J.C., and V.S. Ferreira 1999, "Semantic and Phonological information flow in the production lexicon", *Journal of Experimental Psychology: Learning, Memory, and Cognition* 25, 318-344

Cuxac, C. 2000, "Compositionalité sublexicale morphémique /iconique en Langue des Signes Française", *Recherches Linguistiques de Vincennes* 29: 55-72

Dallas, A. and E. Kaan 2008, "Second Language Processing of Filler-Gap Dependencies by Late Learners", *Language and Linguistics Compass* 2(3), 372-388

Damasio, A.R. 1992, "Aphasia", *New England Journal of Medicine* 326, 531-539

Damasio, H. and A.R. Damasio 1980, "The anatomical basis of conduction aphasia", *Brain* 103, 337-350

Darwin, C. 1859, *On the origin of species by means of natural selection,* London: John Murray

—. 1871, *The Descent of Man*, London: D. Appleton and Co.

Davidson, E.H. 1986, *Gene Activity in Early Development.* Orlando: Academic Press

Dawkins, R. 2009, *The Greatest Show On Earth: The evidence for evolution,* New York: Free Press

De Renzi, A. and L. A. Vignolo 1962, "Token Test: A sensitive test to detect receptive disturbances in aphasics", *Brain: A Journal of Neurology* 85: 665-678

de Waal, F., and P.F. Ferrari 2010, "Towards a bottom-up perspective on animal and human cognition", *Trends in Cognitive Sciences* 14, 201-207

Deacon, T.W. 1990a, "Fallacies of progression in theories of brain-size evolution", *International Journal of Primatology* 11, 193-236

—. 1990b, "Problems of ontogeny and phylogeny in brain size evolution", *International Journal of Primatology* 11, 237-282.

—. 1997, *The Symbolic Species: the Co-Evolution of Language and the Brain*, New York: W.W. Norton

—. 2003, "The hierarchic logic of emergence: Untangling the interdependence of evolution and self-organization", B. Weber and D.

Depew (eds.) *Evolution and learning: The Baldwin effect reconsidered*, Cambridge MA: MIT Press, 273-308

—. 2006, "Emergence: The Hole at the Wheel's Hub". P. Clayton and P. Davies (eds.) *The re-emergence of emergence: The emergentist hypothesis from science to religion*, Oxford: Oxford University Press, 111-150

Dean, C., F.G. Scholl, J. Choih, S. DeMaria, J. Berger, E. Isacoff and P. Scheiffele 2003, "Neurexin mediates the assembly of presynaptic terminals", *Nat. Neurosci.* 6, 708-716

Dediu, D. and D.R. Ladd 2007, "Linguistic *tone* is related to the population frequency of the adaptive haplogroups of two brain size genes, *ASPM* and *Microcephalin*", *PNAS* 104, 10944-10949 (Quoted from preprint version)

Dehaene, S. 1997, *The number sense: how the mind creates mathematics*. Oxford: Oxford University Press

—. 2005, "Evolution of human cortical circuits for reading and arithmetic: The 'neuronal recycling' hypothesis", S. Dehaene, J.-R. Duhamel, M. Hauser and J. Rizzolatti (eds.) *From monkey brain to human brain*, Cambridge MA: MIT Press, 133-157

Dehaene, S., E. Dupoux, J. Mehler, L. Cohen, E. Paulesu, D. Perani *et al.* 1997, "Anatomical variability in the cortical representation of first and second language", *Neuroreport* 817, 3809-3815

del Viso, S., J.M. Igoa and J.E. García-Albea 1987, *Corpus de errores espontáneos del español*. Unpublished manuscript. Oviedo: Universidad de Oviedo

Dell, G.S., 1986 "A spreading-activation theory of retrieval in sentence production", *Psychological Review* 93, 283-321

Dell, G.S., and P.A. Reich 1981, "Stages in sentence production: An analysis of speech error data", *Journal of Verbal Learning and Verbal Behavior* 20, 611- 629

Dennett, D. 1995, *Darwin's dangerous idea: evolution and the meanings of life,* New York: Simon and Schuster

d'Errico, F., C. Henshilwood, M. Vanhaeren and K. van Niekerk 2005 "*Nassarius kraissianus* shell beads from Blombos cave: evidence for symbolic behaviour in the Middle Stone Age", *Journal of Human Evolution* 48, 3-24

Descartes, R. 1648, "Comments on a certain broadsheet", J. Cottingham, R. Stoothoff and D. Murdoch (eds.) 1985, *The Philosophical Writings of Descartes, Vol. I.* Cambridge: Cambridge University Press, 294-311

Desmet T. and D. Wouter 2007, "Bilingual Language Processing", *Language and Linguistics Compass* 1(3): 168-194

Desmond, J. E. and J.A. Fiez 1998, "Neuroimaging studies of the cerebellum: language, learning and memory", *Trends Cogn. Sci.* 2: 355-362

Deutscher, G. 2010, *Through the Language Glass: How Words Colour your World*, Portsmouth: Heinemann

Devlin, K. 2005, *The math instinct*, New York: Thunder's Mouth Press

Di Sciullo, A.M. and C. Boeckx 2011, "Introduction: Contours of the Biolinguistic research agenda", Di Sciullo and Boeckx (eds), 1-16

Di Sciullo, A.M. and C. Boeckx (eds) 2011, *The biolinguistic enterprise. New perspectives on the evolution and nature of the human language faculty,* Oxford: Oxford University Press

Diamond, J.M. 1992, *The third chimpanzee: The evolution and future of the human animal*, New York: HarperCollins

Díaz, B., C. Baus, C. Escera, A. Costa and N. Sebastián-Gallés 2008, "Brain potentials to native phoneme discrimination reveal the origin of individual differences in learning the sounds of a second language", *Proceedings of the National Academy of Sciences* 105 (42), 16083-16088

Díaz, B., N. Sebastián-Gallés, K. Erdocia, J. Mueller, I. Laka 2011, "On the cross-linguistic validity of electrophysiological correlates of morphosyntactic processing: A study of case and agreement violations in Basque", *Journal of Neurolinguistics* 24, 357-373

Dijkstra,T., M. Timmermans and H. Schriefers 2000, "On being blinded by your other language: effects of task demands on interlingual homograph recognition", *Journal of Memory and Language* 42, 445-64

Dik, S.C. 1978, *Functional Grammar*, Amsterdam: North-Holland

Doupe, A.J. and P.K. Kuhl 1999, "Birdsong and human speech: common themes and mechanisms", *Annual Review of Neuroscience* 22, 567-631

Dricot, L., B. Sorger, C. Schiltz, R. Goebel and B. Rossion 2008, "Evidence for individual face discrimination in non-face selective areas of the visual cortex in acquired prosopagnosia", *Behav. Neurol.* 19, 75-79

Dronkers, N.F. 1996, "A new brain region for coordinating speech articulation", *Nature* 384, 159-161

—. 2000, "The pursuit of brain-language relationships", *Brain and Language* 71, 59-61

Dronkers, N.F., O. Plaisant, M.T. Iba-Zizen and E.A. Cabanis 2007, "Paul Broca's historic cases : high resolution MR imaging of the brains of Leborgne and Lelong", *Brain* 130, 1432-1441

Dronkers, N.F., B.B. Redfern and R. Knight 2000, "The neural architecture of language disorders", M.S. Gazzaniga (ed.) *The New Cognitive Neuroscience,* Cambridge MA: MIT Press, 949-958

Duffau, H. 2008, "The anatomo-functional connectivity of language revisited. New insights provided by electrostimulation and tractography", *Neuropsychologia* 46, 927-934.

Duffield, N. 2010, "Roll up for the Mystery Tour!", *Lingua* 120: 2673-2675

Duyck, W. 2005, "Translation and associative priming with cross-lingual pseudohomophones: evidence for nonselective phonological activation in bilinguals", *Journal of Experimental Psychology: Learning Memory and Cognition* 31, 1340-59

Duyck, W., K. Diependaele, D. Drieghe and M. Brysbaert 2004, "The size of the cross-lingual masked phonological priming effect does not depend on second language proficiency", *Experimental Psychology* 51, 1-9

Dye, M.W.G. and S. Shih 2006, "Phonological priming in British Sign Language", L.M. Goldstein, D.H. Whalen and C.t. Best (eds.) *Papers in Laboratory of Phonology 8*, Berlin: Mouton de Gruyter

Ebbeson, S.O.E. 1980, "The parcellation theory and its relation to interespecific varuabiluty in brain organization, evolutionary and ontogenetic development", *Cell and Tissue Research* 213, 179-212

Eibl-Eibesfeldt, I. 1984. *Die Biologie des Menschlinhen Verhaltnes: Grundiss der Humanethologie* (Spanish translation: *Biología del comportamiento humano. Manual de etología humana*. Madrid: Alianza, 1993)

Ellis, A.W. and A.W. Young 1988, *Human Cognitive Neuropsychology*, London: Lawrence Erlbaum

Embick, D., A. Marantz, Y. Miyashita, W. O'Neil and K.L. Sakai 2000, "A syntactic specialization for Broca's area", *Proceedings of the National Academy of Sciences USA* 97, 6150-6154

Emmorey, K. 1991, "Repetition priming with aspect and agreement morphology in American Sign Language", *Journal of Psycholinguistic Research* 20, 365-388

Emmorey, K. and D. Corina 1990, "Lexical recognition in sign language: effects of phonetic structure and morphology", *Perceptual and Motor Skills* 71, 1227-1252

Emmorey, K., T. Grabowski, S. McCullough, H. Damasio, L. Ponto, R. Hichwa and U. Bellugi 2004, "Motor-iconicity of sign language does not alter the neural systems underlying tool and action naming", *Brain and Language* 891, 27-37

Enard, W., *et al.* 2002, "Molecular evolution of FOXP2, a gene involved in speech and language", *Nature* 418, 869-872

—. *et al.* 2009, "A humanized version of *Foxp2* affects cortico-basal ganglia circuits in mice", *Cell* 137, 961-71

Endress, A.D., D. Cahill, S. Block, J. Watumull and M.D. Hauser 2009, "Evidence of an evolutionary precursor to human language affixation in a nonhuman primate", *Biology Letters* 5(6), 749-751

Endress, A.D., S. Carden, E. Versace and M.D. Hauser 2010, "The apes' edge: Positional learning in chimpanzees and humans", *Animal Cognition* 13(3), 483-495

Endress, A.D., G. Dehaene-Lambertz and J. Mehler 2007, "Perceptual constraints and the learnability of simple grammars", *Cognition* 105(3), 577-614

Endress, A.D. and M.D. Hauser 2010, "Word segmentation with universal prosodic cues", *Cognitive Psychology* 612, 177-199

Endress, A.D. and M.D. Hauser 2011, "The influence of type and token frequency on the acquisition of affixation patterns: Implications for language processing", *Journal of Experimental Psychology: Learning, Memory and Cognition* 371, 77-95

Endress, A.D. and J. Mehler 2009, "Primitive Computations in Speech Processing", *Quarterly Journal of Experimental Psychology* 62, 2187-2209

Endress, A.D. and J. Mehler2010, "Perceptual Constraints in Phonotactic Learning", *Journal of Experimental Psychology: HP&P* 36 (1), 235-250

Endress, A.D., M. Nespor and J. Mehler 2009, "Perceptual and Memory Constraints on Language Acquisition", *Trends in Cognitive Sciences* 13 (8), 348-353

Evans, N. and S.C. Levinson 2009a, "The Myth of Language Universals: Language diversity and its importance for cognitive science", *Behavioral and Brain Sciences* 32, 429-448

Evans, N. and S.C. Levinson 2009b, "With diversity in mind: Freeing the language sciences from Universal Grammar", *Behavioral and Brain Sciences* 32: 472-483

Evans, N. and S.C. Levinson 2010, "Time for a sea-change in linguistics: Response to comments on 'The Myth of Language Universals'", *Lingua* 120, 2733-2758

Everett, D.L. 2005: "Cultural Constraints on Grammar and Cognition in Pirahã", *Current Anthropology* 46: 621-646

—. 2007, "Cultural Constraints on Grammar in Pirahã. A Replay to Nevins, Pesetsky, and Rodrigues" (http://ling.auf.net/lingbuzz)

Evinck, S. 1997, *Production de la parole en français: Investigation des unités impliquées dans l'encodage phonologique des mots*, Unpublished Ph.D. dissertation, Bruxelles University

Falconer, D.S. and T.F.C. Mackay 1996, *Introduction to Quantitative Genetics*, London: Longmann

Feigenson, L. and J. Halberda 2004, "Infants chunk object arrays into sets of individuals", *Cognition* 91, 173-190

Felleman, D.J. and D.C. Van Essen 1991, "Distributed hierarchical processing in the primate cerebral cortex", *Cereb. Cortex* 1, 1-47

Ferland, R.J., T.J. Cherry, P.O. Preware, E.E. Morrisey and C.A. Walsh 2003 "Characterization of *Foxp2* and *Foxp1* mRNA and protein in the developing and mature brain", *Journal of Comparative Neurology* 460, 266-279

Ferrand, L. and J. Grainger 1994, "Effects of orthography are independent of phonology in masked form priming", *Quarterly Journal of Experimental Psychology* 47A, 365-382

Ferrand, L., J. Segui and J. Grainger 1996, "Masked priming of word and picture naming: The role of syllable units", *Journal of Memory and Language* 35, 708-723

Ferrand, L., J. Segui and G.W. Humphreys 1997, "The syllable's role in word naming", *Memory and Cognition* 25, 458-470

Feyerabend, P.K. 1970, "Against method: Outline of an anarchistic theory of knowledge", *Minnesota Studies in the Philosophy of Science* IV, 17-130

—. 1975, *Against Method*, London: Verso

Filipek, P.A., *et al.* 2003, "Mitochondrial dysfunction in autistic patients with 15q inverted duplication", *Ann. Neurol.* 53, 801-804

Filipovich, L. 2011, "Speaking and remembering in one or two languages: bilingual vs. monolingual lexicalization and memory for motion events", *International Journal of Bilingualism* 15 (2)

Fisher S.E. 2006, "Tangled webs: tracing the connections between genes and cognition", *Cognition* 101, 270-97

Fisher, S.E., C.S. Lai and A.P. Monaco 2003, "Deciphering the genetic basis of speech and language disorders", *Annu. Rev. Neurosci.* 26, 57-80

Fisher, S.E. and C. Scharff 2009 "*FOXP2* as a molecular window into speech and language", *Trends Genet.* 25, 166-177

Fisher S.E., F. Vargha-Khadem, K.E. Watkins, A.P. Monaco and M.E. Pembrey 1998, "Localisation of a gene implicated in a severe speech and language disorder", *Nature Genet.* 18 (2), 168-70

Fitch, W.T. 2009a, "Biology of Music: Another One Bites the Dust", *Current Biology* 19, R403-R404

—. 2009b, "Prolegomena to a Future Science of Biolinguistics", *Biolinguistics* 3, 283-320

—. 2010, *The evolution of language*, Cambridge: Cambridge University Press

—. 2011, "Unity and diversity in human language", *Philosophical Transactions of the Royal Society. Biological Sciences* 366, 376-388

Fitch, W.T. and M.D. Hauser 2004, "Computational constraints on syntactic processing in a nonhuman primate" *Science* 303, 377-380

Fitch, W.T., M.D. Hauser and N. Chomsky 2005, "The evolution of the language faculty: clarifications and implications", *Cognition* 97, 179-210

Fitch, W., L. Huber and T. Bugnyar 2010, "Social Cognition and the Evolution of Language: Constructing Cognitive Phylogenies", *Neuron* 65, 795-814

Flege, J.E. 1999, "Age of learning and second-language speech", D. Birdsong (ed.) *Second language acquisition and the critical period hypothesis,* Mahwah NJ: Erlbaum, 101-132

Flege, J.E., G.H. Yeni-Komshian and S. Liu 1999, "Age constraints on second-language acquisition", *Journal of Memory and Language* 41, 78-104

Fodor, J.A. 1975, *The Language of Thought*. New York: Crowell

—. 1983, *The Modularity of Mind,* Cambridge MA: MIT Press

—. 1998, *In critical condition: Polemical essays on cognitive science and the philosophy of mind,* Cambridge MA: MIT Press

Fodor, J.A. and M. Piattelli-Palmarini 2010, *What Darwin Got Wrong*, New York: Farrar, Straus and Giroux

Fodor, J.A. and Z. Pylyshyn 1988, "Connectionism and cognitive architecture: A critical analysis", *Cognition* 28, 3-71

Forster, K.I. 1979, "Levels of processing and the structure of the language procesor", W. E. Cooper and E. Walker (eds.) *Sentence processing: Psychological studies presented to Merrill Garrett*, Hillsdale NJ: Erlbaum, 27-85

Forster, K.I. and S.M. Chambers 1973, "Lexical access and naming time", *Journal of Verbal Learning and Verbal Behavior* 12, 627-635

Forster, K.I. and D. Shen 1996, "No enemies in the neighborhood: absence of inhibitory neighborhood effects in lexical decision and semantic categorization", *Journal of Experimental Psychology: Learning, Memory and Cognition* 22, 696-713

Francks, C., I.L. MacPhie and A.P. Monaco 2002, "The genetic basis of dyslexia", *Lancet Neurol.* 1, 483-490

Franklin, T.B. and I.M. Mansuy 2010, "The prevalence of epigenetic mechanisms in the regulation of cognitive functions and behaviour", *Curr. Opin. Neurobiol.* 20, 1-9

Frauenfelder, U.H. and L.K. Tyler 1987, "The process of spoken word recognition: an introduction", *Cognition* 251 (2), 1-20

Frauenfelder, U.H. G. Peeters 1990, "Lexical segmentation in TRACE: an exercise in simulation", G.T.M. Atlmann (ed.) *Cognitive models of processing: Psycholinguistic and computational perspectives,* Cambridge MA: MIT Press, 50-86

Freidin, R. 1991, "Linguistic theory and language acquisition: A note on structure dependence", *Behavioral and Brain Sciences* 14, 618-619

French, C.A., M. Groszer, C. Preece, A.M. Coupe, K. Rajewsky and S.E. Fisher 2007, "Generation of mice with a conditional *Foxp2* null allele", *Genesis* 45, 440-446

Friederici, A.D. 2002, "Towards a neural basis of auditory sentence processing", *Trends in Cognitive Sciences*, 6 (2), 78-84

Fujita, E., Y. Tanabe, A. Shiota, M. Ueda, K. Suwa, M.Y. Momoi and T. Momoi 2008, "Ultrasonic vocalization impairment of *Foxp2* (R552H) knockin mice related to speech-language disorder and abnormality of Purkinje cells", *Proc. Natl. Acad. Sci. U. S. A.* 105, 3117-3122

Fujita, K. 2007, "Facing the logical problem of language evolution. Review of Jenkins 2004, *Variation and universals in biolinguistics*", *English Linguistics* 24, 78-108

—. 2009, "A Prospect for Evolutionary Adequacy: Merge and the Evolution and Development of Human Language", *Biolinguistics* 3, 128-153

Fukui, N. and Y. Takano 1998, "Symmetry in syntax: Merge and Demerge", *Journal of East Asian Linguistics* 7, 27-86

Galaburda, A.M., J. LoTurco, F. Ramus, R.H. Fitch and G.D. Rosen 2006, "From genes to behavior in developmental dyslexia", *Nat. Neurosci.* 9, 1213-1217

Gallistel, C.R. 2009, "The foundational abstractions", M. Piattelli-Palmarini, P. Salaburu, and J. Uriagereka (eds.) *Of minds and language: A Basque encounter with N. Chomsky*, Oxford: Oxford University Press, 58-73

Gallistel, C.R. and R. Gelman 2005, "Mathematical cognition", K. Holyoak and R. Morrison (eds.) *The Cambridge handbook of thinking and reasoning*, Cambridge: Cambridge University Press, 559-588

Gallistel, C.R. and A.P. King 2009, *Memory and the Computational Brain. Why Cognitive Science Will Transform Neuroscience*, New York: Blackwell/Wiley

Gambell, T. and C.D. Yang 2005, "Word segmentation: quick but not dirty" Ms., Yale University

Garbin, G., *et al.* 2010, "Bridging language and attention: Brain basis of the impact of bilingualism on cognitive control", *NeuroImage* 53, 1272-1278

Garrett, M.F. 1980, "Levels of processing in sentence production", B. Butterworth (ed.) *Language production. Vol. 1: Speech and talk.* London: Academic Press

—. 2001, "Now you see it, now you don't: Frequency effects in language production", E. Dupoux (ed.) *Language, brain, and cognitive development: Essays in honor of Jacques Mehler,* Cambridge MA, MIT Press, 227-240

Gathercole, S.E. and A.D. Baddeley 1993, *Working Memory and Language*. Hillside: Lawrence Erlbaum Associates

Gelman, R. and C.R. Gallistel 1978, *The child's understanding of number*, Cambridge MA: Harvard University Press

Gentner, T.Q., K.M. Fenn, D. Margoliash and H.C. Nusbaum 2006, "Recursive syntactic pattern learning by songbirds", *Nature* 440, 1204-1207

Gentner, T.Q., K.M. Fenn, D. Margoliash and H.C. Nusbaum 2010, "Simple stimuli, simple strategies", *PNAS* 107, E65

Gervain, J. and J. Mehler 2010, "Speech Perception and Language Acquisition in the First Year of Life", *Annual Review of Psychology* 61, 191-218

Geschwind, N. 1965, "Disconnexion syndromes in animals and man", *Brain* 88, 237-294

—. 1967, "The varieties of naming errors", *Cortex* 3, 97-112

Geschwind, D.H. and G. Konopka 2009, "Neuroscience in the era of functional genomics and systems biology", *Nature* 461, 908-915

Gevins, A., H. Leong, M.E. Smith, J. Le and R. Du 1995, "Mapping cognitive brain function with modern high-resolution electroencephalography", *Trends in Neurosciences,* 18: 429-436

Gibson, C.J. and J.R. Gruen 2008, "The human lexinome: genes of language and reading", *J. Commun. Disord.* 41, 409-420

Gilbert, S.F. and D. Epel 2009, *Ecological developmental biology. Integrating epigenetics, medicine, and evolution*, Sunderland MA: Sinauer Associates Inc.

Gingeras, T. R. 2009, "Implications of chimaeric non-co-linear transcripts", *Nature* 461, 206-211

Givon, T. 2002, *Bio-Linguistics: The Santa Barbara Lectures*, Amsterdam: John Benjamins

Glaser, W.R., and F.J. Düngelhoff 1984, "The time course of picture-word interference", *Journal of Experimental Psychology: Human Perception and Performance* 10, 640-654

Glasser, M.F. and J.K. Rilling 2008, "DTI tractography of the human brain's language pathways", *Cerebral Cortex* 18, 2471-2482

Godfrey-Smith, P. 2007, "Information in biology", D. Hull and M. Ruse (eds.) *The Cambridge Companion to the Philosophy of Biology*, Cambridge: Cambridge University Press, 103-119

Goldstein, J. 1999, "Emergence as a Construct: History and Issues", *Emergence* 11, 49-72

Golestani N., F.X.Alario, S. Meriaux, D. Le Bihan, S. Dehaene and C. Pallier 2006, "Syntax production in bilinguals", *Neuropsychologia* 44, 1029-1040

Gollan, T., T.J. Slattery, D. Goldenberg, E. van Assche, W. Duyck and K. Rayner *In press*, "Frequency drives lexical access in reading but not in speaking: the frequency-lag hypothesis", *Journal of Experimental Psychology: General.*

Goodman, C. S. and B.C. Coughlin 2000, "Introduction. The evolution of Evo-devo biology", *Proc. Natl. Acad. Sci. U.S.A.* 97, 4424-4425

Goodwin, B. 1994, *How the Leopard Changed its Spots. The Evolution of Complexity*. New York: Charles Scribner's Sons

Gopnik, M. 1990, "Feature-blind grammar and dysphasia", *Nature* 344: 715

—. (ed.) 1997, *The inheritance and innateness of grammars*. Oxford: Oxford University Press

Gottesman, I.I. and T.D. Gould 2003, "The endophenotype concept in psychiatry: etymology and strategic intentions", *Am. J. Psychiatry* 160, 636-645

Gottlieb, G. 1997, *Synthesizing Nature-Nurture*. Mahwah NJ: Lawrence Erlbaum

—. 2001, "A developmental psychobiological systems view: Early formulation and current status", S. Oyama, P.E. Griffiths and R.D. Gray (eds.) *Cycles of Contingencies. Developmental Systems and Evolution*, Cambridge MA: MIT Press, 41-54

Gould, S.J. 1976, "In defense of the analog: A commentary to N. Hotton" R.B. Masterson, W. Hodo and H.J. Jerison (eds.) *Evolution, brain and behavior: persistent problems*, New York: Wiley, 175-179

—. 1996, "The Pattern of Life's History", J. Brockman (ed.) *The Third Culture. Beyond the Scientific Revolution*, New York: Simon & Schuster

—. 2002, *The Structure of Evolutionary Theory*, Cambridge (MA): Harvard University Press

Gould, S.J. and R.C. Lewontin 1979, "The spandrels of San Marco and the Panglossian paradigm: a critique of the adaptationist programme", *Proceedings of the Royal Society of London. Series B, Biological Sciences* 205, 581-598

Gould, T.D. and I.I. Gottesman 2006, "Psychiatric endophenotypes and the development of valid animal models", *Genes Brain Behav.* 5, 113-119

Gräff, J. and I.M. Mansuy 2008, "Epigenetic codes in cognition and behaviour", *Behav. Brain Res.* 192, 70-87

Grainger, J. 1990, "Word frequency and neighborhood frequency effects in lexical decision and naming", *Journal of Memory and Language* 29, 228-244

Grainger, J. and A.M. Jacobs 1996, "Orthographic processing in visual word recognition: A multiple read-out model", *Psychological Review* 103, 518-565

Grainger, J., and J. Seguf 1990, "Neighborhood frequency effects in visual word recognition: A comparison of lexical decision and masked identification latencies", *Perception and Psychophysics* 47, 191-198

Grainger, J., J,K. O'Regan, A.M. Jacobs and J. Seguí 1989, "On the role of competing word units in visual word recognition: The neighborhood frequency effect", *Perception and Psychophysics* 45, 189-195

Grant, S.G. 2003, "Systems biology in neuroscience: bridging genes to cognition", *Curr. Opin. Neurobiol.* 13, 577-582

Green, D. W. 1998, "Mental control of the bilingual lexico-semantic system", *Bilingualism: Language and Cognition* 1, 67-81

Greenberg, J.H. 1957, "Language and Evolutionary Theory", J. Greenberg, *Essays in Linguistics*, Chicago, The University of Chicago Press, 56-65

—. 1963, *Universals of Language*, Cambridge: MIT Press

Greenfield, P.M. 1991, "Language, tools and brain: The ontogeny and phylogeny of hierarchically organized sequential behavior", *Behavioral and Brain Sciences* 14, 531-595

—. 1998, "Language, tools, and brain revisited", *Behavioral and Brain Sciences* 21, 159-163

Greenfield, P.M., K. Nelson and E. Saltzmann 1972, "The development of rulebound strategies for manipulating seriated nesting cups: a parallel between action and grammar", *Cognitive Psychology* 3, 291-310

Griesser, M. 2008, "Referential calls signal predator behavior in a group-living bird species", *Current Biology* 18, 69-73

Griffiths, P.E. 2001, "Genetic information: A metaphor in search of a theory", *Philosophy of Science* 68, 394-412

—. 2007, "Evo-Devo meets the mind: towards a developmental evolutionary psychology", R. Brandon and R. Sansom (eds.) *Integrating Evolution and Development: Form Theory to Practice*, Cambridge MA, MIT Press, 195-225

Griffiths, P.E. and R.D. Gray 1994, "Developmental systems and evolutionary explanation", *Journal of Philosophy* 91, 277-304

Griffiths, P.E. and R.D. Gray 2001, "Darwinism and developmental systems", S. Oyama, P.E. Griffiths and R.D. Gray (eds) *Cycles of Contingencies. Developmental Systems and Evolution*, Cambridge MA: MIT Press, 195-218

Griffiths, P.E. and R.D. Gray 2004, "The developmental systems perspective: organism-environment systems as units of evolution", K. Preston and M. Pigliucci (eds.) *Phenotypic Integration: Studying the Ecology and Evolution of Complex Phenotypes*, Oxford: Oxford University Press, 409-431

Griffiths, P.E. and R.D. Knight 1998, "What is the developmentalist challenge?", *Philosophy of Science* 65, 253-258

Griffiths, P.E. and K. Stotz 2000, "How the mind grows: A developmental perspective on the biology of cognition", *Synthese* 122, 29-51

Grimbert, P. *et al.* 2003, "Truncation of C-mip (Tc-mip), a new proximal signaling protein, induces c-maf Th2 transcription factor and cytoskeleton reorganization", *J. Exp. Med.* 198, 797-807

Grodzinsky, Y. 2000, "The neurology of syntax: Language use without Broca's area", *Behavioral and Brain Sciences* 23, 1-71

Grosjean, F. 1981, "Sign and word recognition: a first comparison", *Sign Language Studies* 32, 195-219

—. 1982, *Life with Two Languages: An Introduction to Bilingualism*, Cambridge MA: Harvard University Press

Grosjean, F. 1989, "Neurolinguists, beware! The bilingual is not two monolinguals in one person", *Brain and Language* 36 (1), 3-15

Groszer, M. *et al.* 2008, "Impaired synaptic plasticity and motor learning in mice with a point mutation implicated in human speech deficits", *Curr. Biol.* 18, 354-362

Grote, K. and E. Linz 2003, "The influence of sign language iconicity on semantic conceptualization", W.G. Müller and O. Fischer (eds.) *From sign to signing: iconicity in language and literature*, Amsterdam: John Benjamins, 23-40

Guasti, M.T. 2002, *Language Acquisition. The Growth of Grammar*, Cambridge MA: MIT Press

Haesler S., C. Rochefort, B. Georgi, P. Licznerski, P. Osten and C. Scharff 2007, "Incomplete and inaccurate vocal imitation after knockdown of FoxP2 in songbird basal ganglia nucleus Area X", *PLoS Biol.* 5, e321

Haesler, S., K. Wada, A. Nshdejan, E.E. Morrisey, T. Lints, E.D. Jarvis and C. Scharff 2004, "Foxp2 expression in avian vocal learners and non-learners", *The Journal of Neuroscience* 24, 3164-3175

Hagen, K. 2008, "The Bilingual Brain: Human Evolution and Second Language Acquisition" *Evolutionary Psychology* 61, 43-63

Hahne, A. 2001, "What's Different in Second-Language Processing. Evidence from Event-Related Brain Potentials", *Journal of Psycholinguistic Research* 30 (3), 251-266

Hahne, A. and A.D. Friederici 2001, "Processing a second language late learners' comprehension mechanisms as revealed by event-related brain potentials", *Bilingualism: Language and Cognition* 4 (2), 123-141

Hagoort, P. 2003, "How the brain solves the binding problem for language: A neurocomputational model of syntactic processing", *Neuroimage* 20, 218-229

Hale, K., and S. J. Keyser 1993, "On argument structure and the lexical expression of grammatical relations", K. Hale and S. J. Keyser (eds.) *The view from Building 20*, Cambridge MA: MIT Press, 53-110

Hale, K., and S. J. Keyser 2002, *Prolegomenon to a theory of argument structure*, Cambridge MA: MIT Press

Hall, B.K. 1999, *Evolutionary Developmental Biology. Second edition*, Dordrecht: Kluwer Academic

Hall, B.K. and W.M. Olson 2003, "Introduction: Evolutionary developmental mechanisms", B.K. Hall and W.M. Olson (eds.) *Keywords and Concepts in Evolutionary Developmental Biology*, Cambridge: Harvard University Press, xiii-xvi

Halle, M. and A. Marantz 1993, "Distributed morphology and the pieces of inflection", K. Hale and S. J. Keyser (eds.) *The view from building 20*, Cambridge MA: MIT Press, 111-176

Halle, M. and A. Nevins 2004, "Intrinsic vs. Extrinsic Ordering, Rules, and Exceptions", Presented at the *CUNY Symposium on Phonological Theory*, February 20-21, 2004.

Hallgrímsson, B. and B.K. Hall (eds.) 2005, *Variation*, New York: Academic Press

Halsband, U., B.J. Krause, H. Sipila, M. Teraas and A. Laihinen 2002, "PET studies on the memory processing of word pairs in bilingual Finnish-English subjects", *Behavioural Brain Research* 132, 47-57

Hämäläinen, M., R. Hari, R. J. Ilmoniemi, J. Knuutila and O. V. Lounasmaa 1993, "Magnetoencephalography —theory, instrumentation, and applications to noninvasive studies of the working human brain", *Rev. Mod. Phys.* 65, 413-497

Hannula-Jouppi, K., N. Kaminen-Ahola, M. Taipale, R. Eklund, J. Nopola-Hemmi, H. Kaariainen and J. Kere 2005, "The axon guidance receptor gene *ROBO1* is a candidate gene for developmental dyslexia", *PLoS Genet.* 1: e50

Hansell, M.H. 2000, *Bird nests and construction behaviour*, Cambridge: Cambridge University Press

Harley, T.A. 1993, "Phonological activation of semantic competitors during lexical access in speech production", *Language and Cognitive Processes* 8, 291-309

Harley, T.A. and H.E. Brown 1998, "What causes a tip-of-the-tongue state? Evidence for lexical neighbourhood effects in speech production", *British Journal of Psychology* 89, 151-174

Hashimoto, T., M.J. de Hoon, S.M. Grimmond, C.O. Daub, Y. Hayashizaki and G.J. Faulkner 2009, "Probabilistic resolution of multi-mapping reads in massively parallel sequencing data using MuMRescueLite", *Bioinformatics* 25: 2613-2614

Haspelmath, M. 2008, "Parametric versus functional explanations of syntactic universals", T. Biberauer (ed.) *The Limits of syntactic variation,* Amsterdam: John Benjamins, 75-107

Hauser, M.D. 1997, *The evolution of communication*, Cambridge MA: The MIT Press

—. 2009a, "Origin of the mind", *Scientific American* September, 44-51

—. 2009b, "The possibility of impossible cultures" *Nature* 460, 190-196

Hauser, M.D. and T. Bever 2008, "A biolinguistic agenda", *Science* 322, 1057-1059

Hauser, M.D., S. Carey and L.B. Hauser 2000, "Spontaneous number representation in semi-free-ranging rhesus monkeys", *Proceedings of the Royal Society: Biological Sciences* 267, 829-833

Hauser, M.D., N. Chomsky and W.T. Fitch 2002, "The Faculty of Language: What is it, who has it, and how it evolved?", *Science* 298, 1569-1579

Hauser, M.D. and W.T. Fitch 2003, "What are the uniquely human components of the language faculty?", M.H. Christiansen and S. Kirby (eds.) *Language Evolution*, Oxford: Oxford University Press, 158-181

Hauser, M.D. and D.D. Glynn 2009, "Can free-ranging rhesus monkeys (Macaca mulatta) extract artificially created rules comprised of natural vocalizations?", *Journal of Comparative Psychology* 123, 161-167

Hauser, M.D., E.L. Newport and R.N. Aslin 2001, "Segmentation of the speech stream in a nonhuman primate: Statistical learning in cotton top tamarins", *Cognition* 78, B53-B64

Heine, B. and T. Kuteva 2007, *The Genesis of Grammar*, Oxford: Oxford University Press

Hermans, D., T. Bongaerts, K. De Bot and R. Schreuder 1998, "Producing words in a foreign language: Can speakers prevent interference from their first language?", *Bilingualism: Language and Cognition* 1, 213-29

Hernández, M., A. Costa, L.J. Fuentes, A.B. Vivas and N. Sebastián-Gallés 2010, "The impact of bilingualism on the executive control and orienting networks of attention", *Bilingualism: Language and Cognition* 13, 315-325

Hernandez, A., M. Dapretto, J. Mazziotta and S. Bookheimer 2001, "Language Switching and Language Representation in Spanish-English Bilinguals: An fMRI Study", *NeuroImage* 14, 510-520

Hickok, G. and D. Poeppel 2004, "Dorsal and ventral streams: a framework for understanding aspects of the functional anatomy of language", *Cognition* 92, 67-99

Hinzen, W. 2006, *Mind design and minimal syntax*, Oxford: Oxford University Press

—. 2007, *An Essay on Naming and Truth*, Oxford: Oxford University Press

—. 2009, The Successor Function + LEX = Human Language? K.K. Grohmann (ed.) *InterPhases: Phase-Theoretic Investigations of Linguistic Interfaces*, Oxford: Oxford University Press, 25-47

—. 2011a, "Emergence of a systemic semantics through minimal and underspecified codes", A.M. Di Sciullo and C. Boeckx (eds.) *The biolinguistic enterprise: New perspectives on the evolution and nature of the human language faculty*, Oxford: Oxford University Press, 417-439

—. 2011b, "Language and thought", C. Boeckx (ed.) *Oxford Handbook of Linguistic Minimalism*, Oxford: Oxford University Press, 499-522

Hirschfeld A.L. 2008, "The Bilingual Brain Revisited: A Comment on Hagen 2008", *Evolutionary Psychology* 61, 182-185

Hochmann, J.R., M. Azadpour and J. Mehler 2008, "Do humans really learn AnBn artificial grammars from exemplars?", *Cognitive Science* 32, 1021-1036

Hockett, C.F. 1958, *A course in modern linguistics*, New York: The Macmillan Company

Hofmann, H.A. 2003, "Functional genomics of neural and behavioral plasticity", *J. Neurobiol.* 54, 272-282

Hohenberger, A., D. Happ and H. Leuninger 2002, "Modality dependent aspects of signed language production: Evidence from slips of the hands and their repairs in German Sign Language", R.P. Meier, K. Cormier and D. Quinto-Pozos (eds.) *Modality and structure in signed and spoken language*, Cambridge: Cambridge University Press, 112-142

Holmberg, A. and I. Roberts 2010, "Macroparameters in the Minimalist Program", Presented at *Workshop on Linguistic Variation in the Minimalist Framework*, Universitat Autònoma de Barcelona (January 2010)

Hornstein, N. 2001, *Move! A minimalist theory of construal,* Oxford: Blackwell

—. 2009, *A Theory of Syntax. Minimal Operations and Universal Grammar*, Cambridge: Cambridge University Press

Hornstein, N. and A. Drummond *To appear*, "Review of Piattelli-Palmarini, Uriagereka, and Salaburu", *Biolinguistics*

Hornstein, N. and D. Lightfoot 1981, "Introduction", N. Hornstein and D. Lightfoot (eds) *Explanation in Linguistics. The Logical Problem of Language Acquisition*, London: Longman, 9-31

Hornstein, N., J. Nunes and K. Grohmann 2005, *Understanding Minimalism*, Cambridge: Cambridge University Press

Horwitz, B., J.M. Rumsey and B.C. Donohue 1998, "Functional connectivity of the angular gyrus in normal reading and dyslexia", *Proc. Natl. Acad. Sci. USA* 95, 8939-8944

Hull, R. and J. Vaid 2005, "Clearing the cobwebs from the study of the bilingual brain", J.F. Kroll and A.M.B.D. Groot (eds.) *Handbook of Bilingualism: Psycholinguistic Approaches*, Oxford: Oxford University Press,480-496

Hurford, J.R. 1992, "An Approach to the Phylogeny of the Language Faculty", J.A. Hawkins and M. Gell-Mann (eds.) *The Evolution of Human Languages*, Reading MA: Addison-Wesley, 273-303

—. 2002, "Expression/Induction Models of Language Evolution: Dimensions and Issues", T. Briscoe (ed.) *Linguistic Evolution Through*

Language Acquisition, Cambridge: Cambridge University Press, 301-344
—. 2007, *The Origins of Meaning*, Oxford: Oxford University Press
Hyams, N. 1986, *Language Acquisition and the Theory of Parameters*, Dordrecht: Reidel
—. 2002, "Clausal structure in child Greek: a reply to Varlokosta, Vainikka and Rohrbacher and a reanalysis", *The Linguistic Review* 19, 225-269
Idsardi, W.J. and E. Raimy *In press*, "Three types of linearization and the temporal aspects of speech", T. Biberauer and I. Roberts (eds.) *Principles of linearization.* Berlin: Mouton de Gruyter
Isles, A.R. and L.S. Wilkinson 2000, "Imprinted genes, cognition and behaviour", *Trends Cogn. Sci.* 4, 309-318
Ivanonva, I. and A. Costa 2008, "Does bilingualism hamper lexical access in speech production?", *Acta Psychologica* 127, 277-288
Jablonka, E. 2001, "The systems of inheritance", S. Oyama, P.E. Griffiths and R.D. Gray (eds.) *Cycles of Contingencies. Developmental Systems and Evolution*, Cambridge MA: MIT Press, 99-116
—. 2002, "Information: Its interpretation, its inheritance, and its sharing", *Philosophy of Science* 69, 578-605
Jablonka, E. and M.J. Lamb 2005, *Evolution in four dimensions. Genetic, epigenetic, behavioral, and symbolic variation in the history of life*, Cambridge MA: The MIT Press
Jackendoff, R. 1997, *The architecture of the language faculty*. Cambridge MA: MIT Press
—. 2002, *Foundations of language: brain, meaning, grammar, evolution*, Oxford: Oxford University Press
—. 2005, "Alternative minimalist visions of language", *Proceedings from the 41st Annual Meeting of the Chicago Linguistic Society* 2, 189-226
—. 2010, *Meaning and the Lexicon: The Parallel Architecture 1975-2010.* Oxford: Oxford University Press
—. 2011, "What is the human language faculty? Two views", *Language* 87 (3), 586-624
Jackendoff, R. and S. Pinker 2005, "The nature of the language faculty and its implications for evolution of language (Reply to Fitch, Hauser, and Chomsky)", *Cognition* 97, 211-225
Jacob, F. and J. Monod 1961, "Genetic regulatory mechanisms in the synthesis of proteins", *Journal of Molecular Biology* 3, 318-356
Jenkins, L. 1979, "The genetics of language", *Linguistics and Philosophy* 3, 105-119

—. 2000, *Biolinguistics: Exploring the Biology of Language,* Cambridge: Cambridge University Press

—. (ed.) 2004, *Variation and Universals in Biolinguistics*, Amsterdam: Elsevier

Jescheniak, J.D. and W.J.M. Levelt 1994, "Word frequency effects in speech production: Retrieval of syntactic information and phonological form", *Journal of Experimental Psychology: Learning, Memory, and Cognition* 20, 824-843

Jescheniak, J.D. and H. Schriefers 1998, "Serial versus cascaded processing in lexical access in language production: Further evidence from the coactivation of near-synonyms", *Journal of Experimental Psychology: Learning, Memory and Cognition* 24, 1256-1274

Joanisse, M.F. 2004, "Specific Language Impairments in Children", *Current Directions in Psychological Science* 13, 156-160

Joanisse, M.F., F.R. Manis, P. Keating and M.S. Seidenberg 2000, "Language deficits in dyslexic children: Speech perception, phonology, and morphology", *Journal of Experimental Child Psychology* 77, 30-60

Johnston, T. 1987, "The persistence of dichotomies in the study of behavioral development", *Developmental Review* 7, 149-182

Johnston, T. and L. Edwards 2002, "Genes, interactions, and the development of behavior", *Psychological Review* 109, 26-34

Johnson, C.D. 1970, *Formal aspects of phonological representation.* Berkeley CA: University of California dissertation

Johnson-Pynn, J., D.M. Fragaszy, E.M. Hirsh, K.E. Brakke and P.M. Greenfield 1999, "Strategies used to combine seriated cups by chimpanzees (Pan troglodytes), bonobos (Pan paniscus), and capuchins (Cebusapella)", *Journal of Comparative Psychology* 113, 137-148

Joos, M. (ed.) 1957, *Readings in Linguistics*, Washington: American Council of Learned Societies

Karmiloff-Smith, A. 1992, *Beyond Modularity: A Developmental Perspective on Cognitive Science*, Cambridge: MIT Press

—. 1998, "Development itself is the key to understanding developmental disorders", *Trends Cogn. Sci.* 2, 389-398

Kaskan, P.M. and B.L. Finlay 2001, "Encephalization and its developmental structure: How many ways can a brain get big?", D. Falk and K.R. Gibson (eds.) *Evolutionary Anatomy of the Primate Cerebral Cortex*, Cambridge: Cambridge University Press, 14-19

Kauffman, S.A. 1993, *The Origins of Order. Self-organization and selection in evolution*, London: Oxford University Press

—. 1995, *At Home in the Universe: The Search of the Laws of Self-Organization and Complexity*, New York: Oxford University Press
—. 2000, *Investigations*, New York: Oxford University Press
Kaufmann, W.E. and P.F. Worley 1999, "Neural activity and immediate early gene expression in the cerebral cortex", *Ment. Retard. Dev. Disabil. Res. Rev.* 5, 41-50
Kay, L. E. 2000, *Who Wrote the Book of Life: A History of the Genetic Code,* Stanford: Stanford University Press
Kayne, R.S. 1994, *The antisymmetry of syntax*, Cambridge MA, MIT Press
—. 2005, *Movement and silence*. Oxford: Oxford University Press
Keller, E.F. 2000, *The Century of the Gene*, Cambridge MA: Harvard University Press
Keller, R. 1990, *Sprachwandel: von der unsichtbaren Hund in der Sprache*, Tübingen: Francke
Kertesz, A., W. Harlock and R. Coates 1979, "Computer tomographic localization, lesion size and prognosis in aphasia and nonverbal impairment", *Brain and Language* 8, 34-50
Kim, K.H., N.R. Relkin, K.M. Lee and J. Hirsch 1997, "Distinct cortical areas associated with native and second languages" *Nature* 388, 171-174
Kimura, D. 1973, "The asymmetry of the human brain", *Scientific American* 228, 70-78
Kinsella, A.R. 2009, *Language evolution and syntactic theory*, Cambridge: Cambridge University Press
Kinsella, A.R. and G.F. Marcus 2009, "Evolution, Perfection, and Theories of Language", *Biolinguistics* 3, 186-212
Kirschner, M. and J. Gerhart 2005, *The Plausibility of Life: Resolving Darwin's Dilemma*, New Haven: Yale University Press
Kishino, T., M. Lalande and J. Wagstaff 1997, "UBE3A/E6-AP mutations cause Angelman syndrome", *Nature Genet.* 15, 70-73. *Erratum*: *Nature Genet.* 15, 411
Kitahara, H. 2003, "Acoustic perception in human newborns and cotton-top tamarins", *Reports of the Keio Institute of Cultural and Linguistic Studies* 35, 31-41
Kitano, H. 2004, "Biological robustness", *Nature Rev. Genet.* 5, 826-837
Klima, U., and U. Bellugi 1979, *The signs of language*, Cambridge MA: Harvard University Press
Klopfer, P.H. 1973, "Does behavior evolve?", *Annals of the New York Academy of Sciences* 223, 113-125

Koerner, K. (ed.) 1983, *Linguistics and Evolutionary Theory. Three Essays by August Schleicher, Ernst Haeckel, and Wilhelm Bleek*, Amsterdam/Philadelphia: John Benjamins

Kohn, S.E. and R.B. Friedman 1986, "Word-meaning deafness: A phonological-semantic dissociation", *Cognitive Neuropsychology* 3, 291-308

Koizumi, H. 2004, "The concept of 'developing the brain': a new natural science for learning and education", *Brain Dev.* 26, 434-441

Konopka, G. *et al.* 2009, "Human-specific transcriptional regulation of CNS development genes by FOXP2", *Nature* 462, 213-217

Kosik, K.S. 2003, "Beyond phrenology, at last", *Nat. Rev. Neurosci.* 4, 234-239

Koster, J. 2009, "Ceaseless, unpredictable creativity: Language as technology", *Biolinguistics* 3, 61-92

—. 2010, *Language and tools*, Ms., Universiteit Groningen

Kotz, S. 2009, "A critical review of ERP and fMRI evidence on L2 syntactic processing", *Brain and Language* 109, 68-74

Kovacs A.M. 2009, "Early bilingualism enhances mechanisms of false-belief reasoning", *Developmental Science* 12, 48-54

Kovelman, I., S.A. Baker and L.A. Petitto 2008, "Bilingual and monolingual brains compared using fMRI: Is there a neurological signature of bilingualism?", *Journal of Cognitive Neuroscience* 201, 1-17

Krause, J. *et al.* 2007, "The derived Foxp2 variant of modern humans was shared with Neandertals", *Current Biology* 17, 1908-1912

Kroll, J. F., S. C. Bobb, and Z. Wodniecka 2006, "Language selectivity is the exception, not the rule: arguments against a fixed locus of language selection in bilingual speech", *Bilingualism: Language and Cognition* 9, 119-35

Kroll, J.F. and A.M.B.D. Groot (eds.) 2005, *Handbook of Bilingualism: Psycholinguistic Approaches*, Oxford: Oxford University Press

Kroll, J.F. and E. Stewart 1994, "Category interference in translation and picture naming-evidence for asymmetric connections between bilingual memory representations", *Journal of Memory and Language* 33, 149-74

Kucera, H. and W.N. Francis 1967, *Computational analysis of present-day American English,* Providence RI: Brown University Press

Kuhl, P.K. and J.D. Miller 1975, "Speech perception by the chinchilla", *Science* 190, 69-72

Kuhn, T.S. 1970a, "Logic of discovery or psychology of research?" I. Lakatos and A. Musgrave (eds.) *Criticism and the Growth of Knowledge*, Cambridge: Cambridge University Press, 1-23

Kuhn, T.S. 1970b, *The Structure of Scientific Revolutions*. Second edition. Chicago: The University of Chicago Press

Kwasnicka-Crawford, D.A., A.R. Carson, W. Roberts, A.M. Summers, K. Rehnstrom, I. Jarvela and S.W. Scherer 2005, "Characterization of a novel cation transporter ATPase gene (*ATP13A4*) interrupted by 3q25-q29 inversion in an individual with language delay", *Genomics* 86, 182-194

Laganaro, M. 2005, "Syllable frequency effect in speech production: evidence from aphasia", *Journal of Neurolinguistics* 18, 221-235

Laganaro, M. and F.X. Alario 2006, "On the locus of the syllable frequency effect in speech production", *Journal of Memory and Language* 55, 178-196

La Heij, W. 1988, "Components of Stroop-like interference in picture naming", *Memory and Cognition* 16, 400-410

—. 2005, "Selection processes in monolingual and bilingual lexical access", J.F. Kroll and A.M.B. de Groot (eds.) *Handbook of bilingualism: Psycholinguistic approaches*, New York: Oxford University Press, 289-307

Lakoff, G. and R. Nuñez 2001, *Where mathematics comes from: how the embodied mind brings mathematics into being*, New York: Basic Books

Lai, C.S., S.E. Fisher, J.A. Hurst, F. Vargha-Khandem and A.P. Monaco 2001, "A forkhead-domain gene is mutated in a severe speech and language disorder", *Nature* 413, 519-523

Lai, C.S., D. Gerrelli, A.P. Monaco, S.E. Fisher and A.J. Copp 2003, "FOXP2 expression during brain development coincides with adult sites of pathology in a severe speech and language disorder", *Brain* 126, 2455-2462

Lander, E. and L. Kruglyak 1995, "Genetic dissection of complex traits: guidelines for interpreting and reporting linkage results", *Nat. Genet.* 11, 241-47

Lander, E.S. and R.A. Weinberg 2000, "Genomics: Journey to the center of Biology", Science 287, 1777-1782

Langer, S.K. 1960, "The origins of speech and its communicative function", *The Quarterly Journal of Speech* XLVI (2), 121-134

Larson, R.K., V. Déprez and H. Yamakido (eds.) 2010, *The Evolution of Human Language. Biolinguistic Perspectives*, Cambridge: Cambridge University Press

Lasnik, H. 2000, *Syntactic structures revisited,* Cambridge MA: MIT Press

—. 2005, "Grammar, levels, and biology", J. McGilvray (ed.) *The Cambridge Companion to Chomsky*, Cambridge: Cambridge University Press, 60-83

Lasnik, H. and J. Uriagereka 2002, "On the poverty of the challenge", *The Linguistic Review* 19, 147-150

Lass, R. 1997, *Historical Linguistics and Language Change*, Cambridge: Cambridge University Press

Lauder, G.V. 1994, "Homology, form, and function", B.K. Hall (ed.) *Homology: The hierarchical basis of comparative biology*, San Diego: Academic Press, 151-196

Leben, W. 1973*, Suprasegmental phonology*, Ph.D. dissertation, MIT

Legate, J.A. and C.D. Yang 2002, "Empirical re-assessment of stimulus poverty arguments", *The Linguistic Review* 19, 151-162

Lehrman, D.S. 1953, "A critique of Konrad Lorenz's theory of instinctive behaviour", *Quarterly Review of Biology* 28, 337-363

Lenin, V.I. 1920, "Left wing' communism: An infantile disease", In *Collected Works*, vol. 31. 1964 edition. Moscow: Progress Publishers, 17-118

Lenneberg, E.H. 1967, *Biological Foundations of Language*, New York: John Wiley and Sons

Leonard, L.B. 2002, *Children with Specific Language Impairment.* Boston: MIT Press

Lepsius, R. and W.D. Whitney 1865, "On the relation of vowels and consonants", *Journal of the American Oriental Society* 8, 357-373

Levelt, W.J.M. 1989, *Speaking: from intention to articulation*, Cambridge MA, MIT Press.

Levelt, W.J.M., A. Roelofs and A.S. Meyer 1999, "A theory of lexical access in speech production", *Behavioral and Brain Sciences* 22*,* 1-75

Levenson, J.M. and J.D. Sweatt 2006, "Epigenetic mechanisms: a common theme in vertebrate and invertebrate memory formation", *Cell. Mol. Life Sci.* 63: 1009-1016

Levinson, S.C. 2003, *Space in language and cognition: Explorations in cognitive diversity*, Cambridge: Cambridge University Press

Lewontin, R.C. 2000a, *The Triple Helix: Gene, Organism, and Environment*, Cambridge MA: Harvard University Press

—. 2000b, "Foreword", S. Oyama 2000, *The ontogeny of information. Developmental systems and evolution. Second edition.* Durham (NC): Duke University Press, vii-xv

Li G., J. Wang, S.J. Rossiter, G. Jones and S. Zhang 2007, "Accelerated FoxP2 evolution in echolocating bats", *PLoS ONE* 2 (9), e900

Li, S., J. Weidenfeld and E.E. Morrisey 2004, "Transcriptional and DNA binding activity of the Foxp1/2/4 family is modulated by heterotypic and homotypic protein interactions", *Mol. Cell. Biol.* 24, 809-822

Liberman, M. 2006. Starlings, http://itre.cis.upenn.edu/~myl/languagelog/archives/003076.html.

Liberman, A.M. 1996, *Speech: a special code,* Cambridge MA: MIT Press

Lickliter, R. 2008, "The growth of developmental thought: Implications for a new evolutionary psychology", *New Ideas in Psychology* 26, 353-369

Lidz, J., S. Waxman and J. Freedman 2003, "What infants know about syntax but couldn't have learned: experimental evidence for syntactic structure at 18 months", *Cognition* 89, B65-B73

Lieberman, P. 2000, *Human Language and Our Reptilian Brain. The Subcortical Bases of Speech, Syntax and Thought*, Cambridge MA: Harvard University Press

—. 2002, "On the nature and evolution of the neural bases of human language", *Am. J. Phys. Anthropol.* 45, 36-62

—. 2006, *Toward an Evolutionary Biology of Language*. Cambridge MA: Harvard University Press.

Liégeois, F., *et al.* 2003, "Language fMRI abnormalities associated with *FOXP2* gene mutation", *Nat. Neurosci.* 6: 1230-1237

Lightfoot, D. 1982, *The Language Lottery: Toward a Biology of Grammars*, Cambridge: MIT Press

Livingstone, M. S., G.D. Rosen, F.W. Drislane and A.M. Galaburda 1991, "Physiological and anatomical evidence for a magnocellular defect in developmental dyslexia", *Proc. Natl. Acad. Sci. U. S. A.* 88, 7943-7947

Longa, V.M. 2008, "Una visión crítica sobre la noción de 'programa genético' desde la biología y la lingüística: consecuencias para la conceptualización de la ontogenia del lenguaje", *Verba* 35, 347-385

—. 2009, "Sobre el efecto Baldwin y la noción de herencia", *Signos Filosóficos* XI, 43-72

Longa, V.M. and G. Lorenzo 2008, "What about a (really) minimalist theory of language acquisition?", *Linguistics* 46, 541-570

Longa, V.M. and G. Lorenzo, *In press*, "¿Reduce, completa o elimina? Sobre el estatus del Programa Minimalista en la Gramática Generativa", *Revista de la Sociedad Española de Lingüística*

Longa, V.M and C. Reynares 2010, "La jerarquía en el lenguaje: necesidad conceptual y auto-organización", *Entornos* 23, 105-116

Longa, V.M., G. Lorenzo and J. Uriagereka 2011, "Minimizing language evolution. The Minimalist Program and the evolutionary shaping of language", C. Boeckx (ed.) *The Oxford Handbook of Linguistic Minimalism*, Oxford: Oxford University Press, 595-616

Longobardi, G. 2003, "Methods in Parametric Linguistics and Cognitive History", *Linguistic Variation Yearbook* 3: 101-138

—. 2005, "A minimalist program for parametric linguistics?", H. Broekhuis, N. Corver, R. Huybregts, U. Kleinhenz and J. Koster (eds.) *Organizing grammar*, Berlin: Mouton de Gruyter, 407-414

Longobardi, G. and I. Roberts 2010, "Universals, diversity and change in the science of language: Reaction to 'The Myth of Language Universals and Cognitive Science'", *Lingua* 120: 2699-2703

Lorenzo, G. 2010, "¿Se pueden contar las lenguas? reflexiones sobre la diversidad lingüística desde una perspectiva biológica". Ms., Universidad de Oviedo.

Lorenzo, G. and V.M. Longa 2003, "Minimizing the genes for grammar. The minimalist program as a biological framework for the study of language", *Lingua* 113: 643-657

Lorenzo, G. and V.M. Longa 2009, "Beyond generative geneticism. Rethinking language acquisition from a developmentalist point of view", *Lingua* 119: 1300-1315

Loureda Lamas, O. 2010, "Marcadores del discurso, pragmática experimental y traductología: horizontes para una nueva línea de investigación (I)", *Pragmalingüística* 18: 74-107

Love, A.C. 2007, "Functional homology and homology of function: biological concepts and philosophical consequences", *Biology and Philosophy* 22, 691-708

Lovegrove, W.J., A. Bowling, D. Badcock and M. Blackwood 1980, "Specific reading disability: differences in contrast sensitivity as a function of spatial frequency", *Science* 210: 439-440

Luce, P. A., S.D. Goldinger, E.T. Auer and M.S. Vitevitch 2000, "Phonetic priming, neighborhood activation, and PARSYN", *Perception and Psychophysics* 62, 615-625

Luce, P.A. and D.B. Pisoni 1998, "Recognizing spoken words: the neighborhood activation model", *Ear Hearing* 191, 1-36

Lucy, J. 1992, *Grammatical categories and thought: A case study of the linguistic relativity hypothesis*, Cambridge: Cambridge University Press

Lupker, S.J. 1979, "The semantic nature of response competition in the picture-word interference task", *Memory and Cognition* 7, 485-495

—. 1988, "Picture naming: an investigation of the nature of categorical priming", *Journal of Experimental Psychology: Learning Memory, and Cognition* 14, 444-455

Luria, A.R. 1976, *Fundamentals of neurolinguistics*, New York, Basic Books

MacDermot K.D. *et al.* 2005, "Identification of FOXP2 truncation as a novel cause of developmental speech and language deficits", *Am. J. Hum. Genet.* 76 (6), 1074-80

Maclaurin, J. 2002, "The resurrection of innateness", *The Monist* 85, 105-130

MacLusky, N.J., T. Hajszan, J. Prange-Kiel and C. Leranth2006, "Androgen modulation of hippocampal synaptic plasticity", *Neuroscience* 138, 957-965

MacWhinney, B. 2005 "A unified model of language acquisition", J.F. Kroll and A.M.B. De Groot (eds.) *Handbook of bilingualism: Psycholinguistic approaches*, Oxford: Oxford University Press, 49-67

Magenis, R.E., *et al.* 1990, "Comparison of the 15q deletions in Prader-Willi and Angelman syndromes: specific regions, extent of deletions, parental origin, and clinical consequences", *Am. J. Med. Genet.* 35, 333-349

Maienschein, J. 2005, "Epigenesis and preformationism", *Stanford Enclyclopedia of Philosophy*, http://plato.stanford.edu/archives/fall2008/entries/epigenesis/#7

Mailhot, F. and C. Reiss 2007, "Computing long-distance dependencies in vowel harmony", *Biolinguistics* 1, 28-48

Mameli, M. 2004, "Nongenetic selection and nongenetic inheritance", *British Journal for the Philosophy of Science* 55, 35-71

—. 2005, "The inheritance of features", *Biology and Philosophy* 20, 365-399

Mandel, M. 1977, "Iconic devices in American Sign Language", L. Friedman (ed.) *On the other hand: New perspectives on American Sign Language,* New York: Academic Press, 57-107

Manolio, T.A., F. Collins, N. Cox *et al.* 2009, "Finding the missing heritability of complex diseases", *Nature* 461, 747-53

Marantz, A. 2005 "Generative linguistics within the cognitive neuroscience of language", *The Linguistic Review* 22, 429-445

Marcus, G.F. 1993, "Negative evidence in language acquisition", *Cognition* 46, 53-85

—. 1998, "Rethinking eliminative connectionism", *Cognitive Psychology* 37, 243-282

—. 2004, *The birth of the mind: How a tiny number of genes creates the complexities of human thought*, New York: Basic Books

—. 2006. "Cognitive architecture and descent with modification" *Cognition* 101, 443-465

—. 2008, *Kluge*, Boston: Houghton Mifflin

Marcus, G. F. and S.E. Fisher 2003, "*FOXP2* in focus: what can genes tell us about speech and language?", *Trends Cogn. Sci.* 7, 257-262

Marcus, G.F., *et al.* 1999, "Rule learning by seven-month-old infants", *Science* 283, 77-80

Marian, V., M. Spivey and J. Hirsch 2003, "Shared and separate systems in bilingual language processing: converging evidence from eyetracking and brain imaging", *Brain and Language* 86, 70-82

Marler, P. 1998, "Animal communication and human language", N.G. Jablonski and L.C. Aiello (eds.) *The origin and diversification of language*, San Francisco: Memoirs of the Academy of Sciencies, 1-19

Marler, P. and H. Slabberkoorn (eds) 2004, *Nature's music. The science of birdsong*, San Diego: Elsevier Academic Press

Marr, D. 1982, *Vision*, San Fransisco: Freeman

Marshall, J.C. 1986, "The description and interpretation of aphasic language disorder", *Neuropsychologia* 24, 5-24"

Marslen-Wilson, W. and L.K. Tyler 1980, "The temporal structure of spoken language understanding", *Cognition* 8, 1-71

Marslen-Wilson, W., P. Zwitserlood 1989, "Accessing Spoken Words: The Importance of Word Onsets", *Journal of Experimental Psychology: Human Perception and Performance* 15, 576-585

Mathey, S., and D. Zagar 2000, "The neighborhood distribution effect in visual word recognition: Words with single and twin neighbors", *Journal of Experimental Psychology: Human Perception and Performance* 26,184-205

Matthews, P.H. 2001, *A Short History of Structural Linguistics*, Cambridge: Cambridge University Press

Mattick, J.S. and I.V. Makunin 2006, "Non-coding RNA", *Hum. Mol. Genet.* 15, R17-29

Mattick, J.S., R.J. Taft and G.J. Faulkner 2009, "A global view of genomic information: moving beyond the gene and the master regulator", *Trends in Genetics* 26: 21-28

Mayberry, R.I., E. Lock and H. Kazmi 2002, "Linguistic ability and early language exposure", *Nature*, 417, 38

Mayberry, R.I., and P. Witcher 2005, "Age of Acquisition Effects on Lexical Access in ASL: Evidence for the Psychological Reality of

Phonological Processing in Sign Language", *Paper presented at the 30th Boston University Conference on Language Development.*

Maynard-Smith, J. and E. Szathmáry 1999, *The Origins of Life: From the Birth of Life to the Origin of Language*. New York: Oxford University Press

Maynard-Smith, J. 2000, "The concept of information in Biology", *Philosophy of Science* 67, 177-194

Mayr, E. 1982, *The Growth of Biological Thought. Diversity, Evolution, and Inheritance*, Cambridge MA: Harvard University Press

Mazoyer, B., *et al.* 2001, "Cortical networks for working memory and executive functions sustain the conscious resting state in man", *Brain Res. Bull.* 54, 87-298

McArthur, G.M. and D.V.M. Bishop 2001, "Auditory perceptual processing in people with reading and oral language impairments: current issues and recommendations", *Dyslexia* 7: 150-170

McBrearty, S., and A.S. Brooks 2000, "The revolution that wasn't: a new interpretation of the origin of modern human behavior", *Journal of human evolution* 39, 453-563

McClelland, J.L., M.M. Botvinick, D.C. Noelle, D.C. Plaut, T.T. Rogers, M.S. Seidenberg and L.B. Smith 2010, "Letting Structure Emerge: Connectionist and Dynamical Systems Approaches to Cognition", *Trends in Cognitive Sciences* 14, 348- 356

McClelland, J.L. and J.L. Elman 1986, "The TRACE model of speech perception", *Cognitive Psychology* 181, 1-86

McClelland, J.L. and D.R. Rumelhart, D.E. 1981, "An interactive activation model of context effects in letter perception: Part 1. An account of basic finding", *Psychological Review* 55, 375-407

McGrath, L.M., S.D. Smith and B.F. Pennington 2006, "Breakthroughs in the search for dyslexia candidate genes", *Trends Mol. Med.* 12, 333-341

McLaughlin, J. *et al.* 2010, "Brain Potentials Reveal Discrete Stages of L2 Grammatical Learning", *Language Learning* 60 (Suppl. 2, December), 123-150

McMurray, B. and E, Wasserman 2009, "Variability in languages, variability in learning?", *Behavioral and Brain Sciences*, 32

Meader, C.L. and J.H. Muysken 1950, *Handbook of Biolinguistics*. Toledo: Herbert Weller

Mechelli A. *et al.* 2004, "Neurolinguistics: Structural plasticity in the bilingual brain", *Nature* 431, 757

Mehler, M.F. 2008, "Epigenetic principles and mechanisms underlying nervous system functions in health and disease", *Prog. Neurobiol.* 86, 305-341

Mehler, J. *et al.* 1981, "The syllable's role in speech segmentation", *Journal of Verbal Learning and Verbal Behaviour* 20, 298-305

—. *et al.* 1988, "A precursor of language acquisition in young infants", *Cognition* 29, 143-178

Mellars, P., K. Boyle, O. Bar-Yosef and C. Stringer (eds.) 2007, *Rethinking the human revolution*, Cambridge: McDonald Institute Monographs

Mendívil-Giró, J.L. 2006, "Languages and Species: Limits and Scope of a Venerable Comparison", J. Martín and J. Rosselló (eds.) *The Biolinguistic Turn. Issues on Language and Biology*, Barcelona: PPU, 82-118

—. 2009, *Origen, evolución y diversidad de las lenguas. Una aproximación biolingüística*, Frankfurt: Peter Lang

Meng, H. *et al.* 2005, "DCDC2 is associated with reading disability and modulates neural development in the brain", *Proc. Nat. Acad. Sci. USA* 102, 17053-17058

Merker, B. 2000, "Synchronous chorusing and human origins", N. Wallin, B. Merker and S. Brown (eds.) *The origin of music*, Cambridge MA: MIT Press, 315-328

Mesgarani, N., S.V. David, J.B. Fritz and S.A. Shamma 2008, "Phoneme representation and classification in primary auditory cortex", *Journal of the Acoustical Society of America* 1232, 899-909

Metter, E.J. and W.R.Hanson 1994, "Use of positron emission tomography to study aphasia", A. Kertescz (ed.) *Localization and neuroimaging in neuropsychology*, San Diego: Academic Press, 123-147

Meyer, D.E. and R.W. Schvaneveldt 1971, "Facilitation in recognizing pairs of words: Evidence of a dependence between retrieval operations", *Journal of Experimental Psychology* 90, 227-234

Miller, G.A. 1991, *The Science of words*, New York: Scientific American

Miller, J.E. *et al.* 2008, "Birdsong decreases protein levels of FoxP2, a molecule required for human speech", *Journal of Neurophysiology* 100, 2015-2025

Millotte, S. and A. Christophe 2009, "À la découverte des mots: le rôle de la prosodie dans l'acquisition du lexique et de la syntaxe", *Enfance* 61, 283-292

Minelli, A. 2003, *The Development of Animal Form: Ontogeny, Morphology, and Evolution*, Cambridge: Cambridge University Press

—. 2007, *Forme del divenire*, Torino: Einaudi (English translation: *Forms of becoming. The evolutionary biology of development*, Princeton and Oxford: Princeton University Press, 2009)

Miozzo, M. and A. Caramazza 1997, "The retrieval of lexical-syntactic features", *Journal of Experimental Psychology: Learning, Memory, and Cognition* 23, 1410-1423

Missiaen, L. *et al.* 2007, "Calcium in the Golgi apparatus", *Cell Calcium* 41, 405-416

Mithen, S.J. 1996, *The prehistory of the mind*, London: Thames and Hudson

—. 2005, *The singing neanderthals: The origins of music, language, mind and body*, London: Weidenfeld and Nicolson.

Monod, J. 1970, *Le hasard et la nécessité. Essai sur la philosophie naturelle de la biologie moderne*, Paris: Seuil (quoted from the English translation, *Chance and Necessity*. New York: Knopf, 1971)

Moore, D.S. 2001, *The Dependent Gene. The Fallacy of 'Nature vs. Nurture'*, New York: Henry Holt

Moreno, E., A. Rodriguez-Fornells and M. Laine2008, "Event-related potentials (ERPs) in the study of bilingual language processing", *Journal of Neurolinguistics* 21, 477-508

Moro, A. 2000, *Dynamic antisymmetry*, Cambridge MA: MIT Press

—. 2008, *The Boundaries of Babel*, Cambridge MA: MIT Press

—. *et al.* 2001, "Syntax and the brain: Disentagling grammar by selective anomalies", *NeuroImage* 13, 110-118

Morsella, E. and M. Miozzo 2002, "Evidence for a cascade model of lexical access in speech production", *Journal of Experimental Psychology: Learning, Memory, and Cognition* 28, 555-563

Morton, J. 1969, "The interaction of information in word recognition", *Psychological Review* 76, 165-178

Moss, L. 2003, *What Genes Can't Do*, Cambridge MA: MIT Press

Mueller, J., A. Hahne, Y. Fujii and A.D. Friederici 2005, "Native and nonnative speakers' processing of a miniature version of Japanese as revealed by ERPs", *Journal of Cognitive Neuroscience* 17 (8) 1229-1244

Mueller, J., M. Hirotani and A.D. Friederici 2007, "ERP evidence for different strategies in the processing of case markers in native speakers and non-native learners", *BMC Neuroscience* 8, 18

Mufwene, S. 2001, *The ecology of language evolution*. Cambridge: Cambridge University Press

Mukherji, N. 2010, *The Primacy of Grammar*, Cambridge MA: MIT Press

Müller, G.B. 2007 "Evo-devo: extending the evolutionary synthesis", *Nature Rev. Genet.* 8, 943-949

Müller, G.B. and S.A. Newman (eds) 2005, "Evolutionary innovation and morphological novelty", *J. Exp. Zool. B Mol. Dev. Evol.* 304, Special issue

Müller, G.B. and L. Olsson 2003, "Epigenesis and epigenetics", B.K. Hall and W.M. Olson (eds.) *Keywords and Concepts in Evolutionary Developmental Biology*, Cambridge MA: Harvard University Press, 114-123

Murphy, R.A., E. Mondragon and V.A. Murphy 2008, "Rule learning by rats", *Science* 319, 1849-1851

Murray, W. S. and K. Forster 2004, "Serial mechanisms in lexical access: The rank hypothesis", *Psychological Review* 111, 721-756

Musso, M. *et al.* 2003, "Broca's area and the language instinct", *Nature Neuroscience* 6, 774-781

Narita, H. 2010, "The Tension between Explanatory and Biological Adequacy. A Review of N. Fukui 2006, *Theoretical Comparative Syntax: Studies in Macroparameters*, Routledge, London and New York", *Lingua* 120:1313-1323

Narita, H. and K. Fujita 2010, "A Naturalist Reconstruction of Minimalist and Evolutionary Biolinguistics", *Biolinguistics* 4 (4), 356-376

Neeleman, A. and H. van de Koot 2006 "On syntactic and phonological representations", *Lingua* 116, 1524-1552

Neely, J. H. 1991, "Semantic priming effects in visual word recognition: A selective review of current findings and theories", D. Besner and G.W. Humphreys (eds.) *Basic processes in reading*, Hillsdale NJ: Lawrence Erlbaum Associates, 264-336

Nevins, A., D. Pesetsky and C. Rodrigues 2009, "Pirahã Exceptionality: A Reassessment", *Language* 85 (2), 355-404

Newbury, D. *et al.* 2009, "CMIP and ATP2C2 modulate phonological short-term memory in language impairment", *Am. J. Hum. Genet.* 85, 264-272

Newkirk, D., E.S. Klima, C. Canady Pedersen and U. Bellugi 1980, "Linguistic evidence from slips of the hand", V.A. Fromkin (ed.) *Errors in linguistic performance: Slips of the tongue, ear, pen and hand,* New York: Academic Press, 165-197

Newman, S.A. and W.D. Comper 1990, "'Generic' physical mechanisms of morphogenesis and pattern formation", *Development* 110: 1-18

Newman, S.A., G. Forgacs and G.B. Müller 2006, "Before programs: the physical origination of multicellular forms", *Int. J. Dev. Biol.* 50, 289-299

Newmeyer, F.J. 1997, "Genetic dysphasia and linguistic theory", *J. Neurolinguistics* 10, 47-73

—. 1998, *Language form and language function*, Cambridge MA: MIT Press

—. 2005, *Possible and Probable Languages. A Generative Perspective on Linguistic Typology*, Oxford: Oxford University Press

Newport, E.L. 1990, "Maturational constraints on language learning", *Cognitive Science* 14, 11-28

—. 1991, "Contrasting conceptions of the critical period for language", S.C.R. Gelman (ed.), *The epigenesis of mind,* Hillsdale NJ: Erlbaum, 111-130

Newport, E.L. and R.N. Aslin 2004, "Learning at a distance I. Statistical learning of non-adjacent dependencies", *Cognitive Psychology* 48, 127-162

Newport, E.L., M.D. Hauser, G. Spaepen and R.N. Aslin 2004, "Learning at a distance II. Statistical learning of non-adjacent dependencies in a non-human primate", *Cognitive Psychology* 49, 85-117

Newport, E.L. and R.P. Meier 1985, "The acquisition of American Sign Language", D. Slobin (ed.) *The cross-linguistic study of language acquisition,* Hillsdale NJ:Lawrence Erlbaum Associates, 881-938

Nicolson, R.I. and A.J. Fawcett 2006, "Do cerebellar deficits underlie phonological problems in dyslexia?" *Dev. Sci.* 9, 259-262

Nishimura, T., A. Mikami, J. Suzuki and T. Matsuzawa 2006, "Descent of the hyoid in chimpanzees: evolution of face flattening and speech", *Journal of Human Evolution* 51, 244-254

Nobre, A.C. and K. Plunkettt 1997, "The neural system of language: structure and development", *Curr. Opin. Neurobiol.* 7, 262-266

Nopola-Hemmi, J., *et al.* 2001, "A dominant gene for developmental dyslexia on chromosome 3", *J. Med. Genet.* 38, 658-664

Norbury CF. 1995 "Barking up the wrong tree? Lexical ambiguity resolution in children with language impairments and autistic spectrum disorders", *J. Exp. Child Psychol.* 90, 142-171

Norris, D. 1994, "A connectionist model of continuous speech recognition", *Cognition* 52, 189-234

Norris, D., J.M. McQueen and A. Cutler 2000, "Merging information in speech recognition: Feedback is never necessary", *Behavioral and Brain Sciences* 23, 299-380

O'Donnell, T., M.D. Hauser and W.T. Fitch 2005, "Using mathematical models of language experimentally", *Trends in Cognitive Sciences,* 9(6), 284-289

O'Grady, W. 2008, "The emergentist program", *Lingua* 118, 447-464

Ohala, J.J. 1992, "Alternatives to the sonority hierarchy for explaining segmental sequential constraints", *Papers from the CLS Parasession on the Syllable*, 319-338

Ohala, J.J. and H. Kawasaki-Fukumori 1997, "Alternatives to the sonority hierarchy for explaining segmental sequential constraints", S. Eliasson and E.H. Jahr (eds.) *Language and its ecology: Essays in memory of Einar Haugen*, Berlin: Mouton de Gruyter, 343-365

Ohshiba, N. 1997, "Memorization of serial items by Japanese monkeys, a chimpanzee, and humans", *Japanese Psychological Research* 39, 236-252

Ojima, S., H. Nakata and R. Kakigi 2005, "An ERP study of second language learning after childhood: Effects of proficiency", *Journal of Cognitive Neuroscience* 17 (8), 1212-122

Okanoya, K. 2002, "Sexual display as a syntactic vehicle: The evolution of syntax in bird song and human language through sexual selection", A. Wray (ed.) *The transition to language*, Oxford: Oxford University Press, 46-64

Oldfield, R.C. and A. Wingfield 1965, "Response latencies in naming objects", *Quarterly Journal of Experimental Psychology* 17, 273-281

Orfanidou, E., R. Adam, J.M. McQueen and G. Morgan 2009, "Making sense of nonsense in British Sign Language (BSL): The contribution of different phonological parameters to sign recognition", *Memory and Cognition* 37 (3), 302-315

Orlansky, M.D. and J.D. Bonvillian 1984, "The role of inconicity in early sign language acquisition", *The Journal of Speech and Hearing Disorders* 49(3), 287-292

Ormel, E., D. Hermans, H. Knooers and L. Verhoeven 2009, "The role of sign phonology and iconicity during sign processing: the case of deaf children", *Journal of Deaf Studies and Deaf Education* 14, 436-448

Otero, C. 1976, "The dictionary in a generative grammar", Presented at the 91st Annual Convention of the MLA, New York.

—. 1983, "Towards a model of paradigmatic grammar", *Quaderni di Semantica* 4, 134-144

—. 1996, "Head movement, cliticization, precompilation, and word insertion (comments on Uriagereka's paper)". R. Freidin (ed.) *Current issues in comparative grammar*, Dordrecht: Kluwer, 296-337

Ott, D. 2009, "The Evolution of I-Language: Lexicalization as the Key Evolutionary Novelty", *Biolinguistics* 3, 255-269

Ouattara, K., A. Lemasson and K. Zuberbühler 2009, "Campbell's monkeys concatenate vocalizations into context-specific call sequences", *Proceedings of the National Academy of Sciences USA*

(www.pnas.org/cgi/doi/10.1073/pnas.0908118106)

Owen, R. 1843, *Lectures on the Comparative Anatomy and Physiology of the Invertebrate Animals*. London: Longman, Brown, Green, and Longmans

—. 1849, *On the Nature of Limbs. A Discourse*, London: John Van Voorst

Oyama, S. 1976, "A sensitive period for the acquisition of a nonnative phonological system", *Journal of Psycholinguistic Research* 5, 261-283

—. 2000a, *The Ontogeny of Information. Developmental Systems and Evolution* 2nd edition, Durham NC: Duke University Press

—. 2000b, "Causal democracy and causal contributions in developmental systems theory", *Philosphy of Science* 67, S332-S347

—. 2000c, *Evolution's Eye: A System's View of the Biology-Culture Divide*, Durham NC: Duke University Press

Oyama S., P.E. Griffiths and R.D. Gray 2001, "Introduction: What is Developmental Systems Theory?", S. Oyama, P.E. Griffiths and R.D. Gray (eds.) *Cycles of Contingencies. Developmental Systems and Evolution*. Cambridge MA: MIT Press, pp. 1-11

Oyama, S., P.E. Griffiths and R.D. Gray (eds.) 2001, *Cycles of Contingencies. Developmental Systems and Evolution*. Cambridge MA: MIT Press

Paracchini, S. *et al.* 2006, "The chromosome 6p22 haplotype associated with dyslexia reduces the expression of *KIAA0319*, a novel gene involved in neural migration", *Hum. Mol. Genet.* 15, 1659-1666

Paradis, M. 2004, *A Neurolinguistic Theory of Bilingualism*, Amsterdam/Philadelphia: John Benjamins Publishing Co.

Parisi, D. 1987, "Grammatical disturbances of speech production", M. Coltheart, G. Sartori and R. Job (eds.) *The Cognitive Neuropsychology of Language*, Hove: Lawrence Erlbaum, 201-219

Parker, G.J., S. Luzzi, D.C. Alexander, C.A. Wheeler-Kingshott, O. Ciccarelli and M.A. Lambon Ralph 2005, "Lateralization of ventral and dorsal auditory-language pathways in the human brain", *Neuroimage* 24, 656-666

Patel, A. 2008, *Language and music*, Oxford: Oxford University Press

—. 2010, "Music, biological evolution, and the brain", M. Bailar (ed.) *Emerging disciplines*, Houston: Rice University Press, 91-144

Patel, A. and J.R. Iversen 2006, "A non-human animal can drum a steady beat on a musical instrument", *Proceedings of the 9th International Conference on Music Perception and Cognition*, 477.

Patel, A.D., J.R. Iversen, M.R. Bregman and I. Schulz 2009, "Experimental evidence for synchronization to a musical beat in a nonhuman animal", *Current biology* 19, 827-830

Paulesu, E. *et al.* 2001, "Dyslexia: cultural diversity and biological unity",*Science* 291, 2165-2167

Payne, K. 2000, "The progressively changing songs of humpback whales: a window on the creative process in a wild animal", N. Wallin, B. Merker and S. Brown (eds.) *The origins of music*, Cambridge MA: MIT Press, 135-150

Penn, D.C., K.J. Holyoak and D.J. Povinelli 2009, "Universal grammar and mental continuity: Two modern myths", *Behavioral and Brain Sciences*, 32

Pepperberg, I.M. 1999, *The Alex studies: cognitive and communicative abilities of grey parrots.* Cambridge MA: Harvard University Press

Perani, D. *et al.* 1998, "The bilingual brain: Proficiency and age of acquisition of the second language", *Brain* 121, 1841-1852

Perea, M. 1993, *Influencia de los factores de vecindad sobre el acceso léxico: Análisis experimentales y de simulación.* Ph.D dissertation, Universitat de València

Perea, M. and M. Carreiras, M. 1998, "Effects of syllable-frequency and syllable neighborhood frequency in visual word recognition", *Journal of Experimental Psychology: Human Perception and Performance* 24, 134-144

Pérez, E., J. Santiago, A. Palma and P. O'Seaghdha 2007, "Perceptual bias in speech error data collection: Insights from Spanish speech errors", *Journal of Psycholinguistic Research* 36, 207-235

Perlmutter, D. 1992, "Sonority and syllable structure in American Sign Language", *Linguistic Inquiry* 23, 407-442

Perniss, P., R. Thompson and G. Vigliocco *In press*, *Iconicity as a general property of language: Evidence from spoken and signed languages*

Perruchet, P. and A. Rey 2005, "Does the mastery of center-emebedded linguistic structures distinguish humans from non-humans", *Psychonomic Bulletin and Review* 12, 307-313

Peterson, R.R. and P. Savoy 1998, "Lexical selection and phonological encoding during language production: evidence for cascaded processing", *Journal of Experimental Psychology: Learning, Memory, and Cognition* 24, 539-557

Piattelli-Palmarini, M. (ed.) 1980, *Language and learning: the debate between Jean Piaget and N. Chomsky.* Cambridge MA: Harvard University Press

—. 1989, "Evolution, selection, and cognition: from 'learning' to parameter setting in biology and in the study of language", *Cognition* 31, 1-44

—. 2001, "Portrait of a 'classical' cognitive scientist: What I have learned from Jacques Mehler", E. Dupoux (ed.) *Language, brain and cognitive development: Essays in honor of Jacques Mehler*, Cambridge MA, MIT Press, 3-21

—. 2006, "The return of the laws of form". C.P. Manzu' (ed.) *Life on the edge (La vita in Bilico), vol. 2*, Rome: Centro Pio Manzu', 45-57

—. 2010, "What is language that it may have evolved, and what is evolution that it may apply to language", R.K. Larson, V. Déprez and H. Yamakido (eds.) *The evolution of human language: biolinguistic perspectives*, Cambridge: Cambridge University Press, 148-162

Piattelli-Palmarini, M. and J. Uriagereka 2005, "The evolution of the narrow faculty of language", *Lingue e Linguaggio* IV (1), 27-79

Piattelli-Palmarini, M. and J. Uriagereka 2011, "A geneticist's dream, a linguist's nightmare: the case of FOXP2", A.M. Di Sciullo and C. Boeckx (eds.) *The biolinguistic enterprise: New perspectives on the evolution and nature of the human language faculty*, Oxford: Oxford University Press, 100-125

Piattelli-Palmarini, M., J. Uriagereka and P. Salaburu (eds.) 2009, *Of Minds and Language: A Dialogue with N. Chomsky in the Basque Country*, Oxford: Oxford University Press

Pickering, M. J., and S. Garrod 2004, "Toward a mechanistic pychology of dialogue", *Behavioral Brain Sciences* 27, 169-190

Pietrandrea, P. 2002, "Iconicity and arbitrareness in Italian Sign Language", *Sign Language Studies* 2 (3), 296-321

Pietroski, P.M. 2005, *Events and semantic architecture*, Oxford: Oxford University Press

—. 2007, "Systematicity via monadicity". *Croatian Journal of Philosophy* 7, 343-374

—. 2008, "Minimalist Meaning, Internalist Interpretation", *Biolinguistics* 2, 317-341

—. 2011, "Minimal semantic instructions", C. Boeckx (ed.) *Oxford Handbook of Linguistic Minimalism*, Oxford: Oxford University Press, 472-498

—. *To appear*, *Semantics without truth values*, Oxford: Oxford University Press

Pigliucci, M. and G. Müller (eds.) 2010, *Evolution—The Extended Synthesis*, Cambridge MA: MIT Press

Pigliucci, M., C.D. Schlichting, C.S. Jones and K. Schwenk 1996, "Developmental reaction norms: the interactions among allometry, ontogeny, and plasticity", *Plant Species Biol.* 11, 69-85

Pinker, S. 1994, *The Language Instinct. How the Mind Creates Language*, Cambridge MA: MIT Press

Pinker, S. and R. Jackendoff 2005, "The faculty of language: what's special about it?" *Cognition* 95, 201-236

Pinker, S. and R. Jackendoff 2009, "The reality of a universal language faculty", *Behavioral and Brain Sciences*, 32

Pinker, S. and P. Bloom 1990, "Natural selection and natural language", *Behavioral and Brain Sciences* 13, 707-784

Pinto-Correia, C. 1997, *The Ovary of Eve: Egg and Sperm and Preformation,* Chicago: University of Chicago Press

Plaut, D.C., J.L. McClelland, M.S. Seidenberg and K. Patterson 1996, "Understanding normal and impaired word reading: Computational principles in quasi-regular domains", *Psychological Review* 103, 56-115

Ploog, D. 2002, "Is the neural basis of vocalization different in non-human primates and Homo sapiens?", T.J. Crow (ed.) *The speciation of Homo sapiens*, London: British Academy, 121-135

Poelmans, G, *et al.* 2009, "Identification of novel dyslexia candidate genes through the analysis of a chromosomal deletion", *Am. J. Med. Genet. B Neuropsychiatr. Genet.* 150B, 140-147

Poeppel, D. 2005, "The interdisciplinary study of language and its challenges", D. Grimm (ed.) *Jahrbuch des Wissenschaftskollegs zu Berlin*

Poeppel, D. and D. Embick 2005, "Defining the relation between linguistics and neuroscience" A. Cutler (ed.) *Twenty-first century psycholinguistics: Four cornerstones*, Hillsdale NJ: Erlbaum, 173-189

Poeppel, D. and G. Hickok 2004, "Towards a new functional anatomy of language", *Cognition* 92, 1-12

Poizner, H., U. Bellugi and R.D. Tweney 1981, "Processing of formational, semantic, and iconic information in American Sign Language", *Journal of Experimental Psychology. Human Perception and Performance* 7 (5), 1146-1159

Poliak, S. *et al.* 1999, "Caspr2, a new member of the neurexin superfamily, is localized at the juxtaparanodes of myelinated axons and associates with K+ channels", *Neuron* 24: 1037-1047

Popper, K.R. 1963, *Conjectures and Refutations*, London: Routledge

Posner, M.J. 2005, "Timing the Brain: Mental Chronometry as a Tool in Neuroscience", *PLOS Biol 3.2*

Post, E.L. 1944, "Recursively enumerable sets of positive integers and their decision problems", *Bulletin of the American Mathematical Society*, 50 284-316

Postal, P.M. 2009, "The incoherence of Chomsky's 'biolinguistic' ontology" *Biolinguistics* 31, 104-123

Potkin, S. *et al.* 2009, "Genome-wide strategies for discovering genetic influences on cognition and cognitive disorders: methodological considerations", *Cogn. Neuropsychiatry* 14, 391-418

Poulisse, N. and T. Bongaerts 1994, "1st language use in 2nd-language production", *Applied Linguistics* 15, 36-57

Price, C.J. and K.J. Friston 2001, "Functional neuroimaging of neuropsychologically impaired patients", R. Cabeza and A. Kingstone (eds.) *Handbook of functional neuroimaging of cognition*, Cambridge MA: MIT Press, 379-399

Price, C.J., D.W.Green and R. Von Studnitz 1999, "A functional imaging study of translation and language switching", *Brain* 122, 2221-2235

Price, C., P. Indefrey and M. van Turennout 1999, "The neurocognitive architecture of language", C.M. Brown and P. Hagoot (eds.) *The neurocognition of language*,Oxford, Oxford University Press, 211- 240

Ptak, S., W. Enard, V. Wiebe, I. Hellmann, J, Krause, M. Lachmann and S. Pääbo 2009, "Linkage desequilibrium extends across putative selected sites in FOXP2", *Molecular Biology and Evolution*, doi:10.1093/molbev/msp143

Pullum, G.K. and J. Rogers 2006, "Animal patterning-learning experiments. Some mathematical background", http://www.lel.ed.ac.uk./~gpullum/MonkeyMath.pdf.

Pullum, G.K. and B.C. Scholz 2002, "Empirical assessment of stimulus poverty arguments", *The Linguistic Review* 19, 9-50

Pullum, G.K. and B.C. Scholz 2009, "For universals (but not finite-state learning) visit the zoo", *Behavioral and Brain Sciences*, 32

Pulvermüller, F. 1999, "Words in the brain's language", *Behavioral and Brain Sciences* 22, 253-336

Pullvermüller, F. 2002, *The Neuroscience of Language,* Cambridge: Cambridge University Press

Putnam, H. 1961, "Brains and behavior", *History and Philosophy of Science*, Section L, American Association for the Advancement of Science [reprinted in N. Block (ed.), Readings in Philosophy of Psychology. Volume One, Cambridge MA: Harvard University Press, 24-36]

Rakic, P. and D.R. Kornack 2001, "Neocortical expansion and elaboration during primate evolution: a view from neurobiology", D. Falk and

K.R. Gibson (eds.) *Evolutionary Anatomy of the Primate Cerebral Cortex*, Cambridge: Cambridge University Press, 30-56

Raff, R.A. 1996, *The Shape of Life: Genes, Development and the Evolution of Animal Form,* Chicago: University of Chicago Press

—. 2000, "Evo-devo: The evolution of a new discipline", *Nature Rev. Genet.* 1, 74-79

Raimy, E. 2000, *The phonology and morphology of reduplication*, Berlin: Walter de Gruyter

Ramsay, M. 2000, "Communication genes clustered on 7q31", *Mol. Med. Today* 6, 380-381

Ramus, F., M.D. Hauser, C. Miller, D. Morris and J. Mehler 2000, "Language discrimination by human newborns and by cotton-top tamarin monkeys", *Science* 288, 349-351

Ramus, F. 2006, "Genes, brain, and cognition: a roadmap for the cognitive scientist", *Cognition* 101, 247-269

Rapin, I. and D.A. Allen 1983, "Developmental language disorders: Nosologic considerations", U. Kirk (ed) *Neuropsychology of Language, Reading and Spelling,* New York: Academic Press, 155-184

Rapp, B., and M. Goldrick 2000, "Discreteness and interactivity in spoken word production", *Psychological Review* 107, 460-499

Redford, M.A. 1999, *An articulatory basis for the syllable*. Austin TX: University of Texas dissertation

Redford, M.A., C.C. Chen and R. Miikkulainen 2001, "Constrained emergence of universals and variation in syllable systems", *Language and Speech* 44, 27-56

Reiner, A., S.E. Brauth and H.J. Karten 1984, "Evolution of the amniote basal ganglia", *Trends in Neurosciences* 7, 320-325

Reuland, E. 2011, *Anaphora and language design*, Cambridge MA: MIT Press

Rezac, M. 2011, *Phi-features and the modular architecture of language*, Dordrecht: Spriger

Richards, M. 2009, "Stabilizing syntax: On instability, optionality, and other indeterminacies", Presented at the Syntax workshop, Universität Stuttgart, March 2010

Rieppel, O. 1994, "Homology, topology, and typology: The history of modern debates", B.K. Hall (ed.) *Homology: The hierarchical basis of comparative biology*, San Diego: Academic Press, 63-100

Risch, N. and K.R. Merikangas 1996, "The future of genetic studies of complex human diseases", *Science* 273, 1516-1517

Rizzi, L. 1978, "Violations of the Wh-island Constraint in Italian and the Subjacency Condition", C. Dubuisson, D. Lighfoot and Y.C. Morin (eds.) *Montreal Working Papers in Linguistics* 11, 155-190

—. 2005, "On the grammatical basis of language development: A case study", G. Cinque and R. Kayne (eds.) *The Oxford Handbook of Comparative Syntax*, Oxford: Oxford University Press, 70-109

—. 2009, "The discovery of language invariance and variation, and its relevance for the cognitive sciences", *Behavioral and Brain Sciences*, 32

Robert, J.S. 2003, "Developmental Systems Theory". En B.K. Hall and W.M. Olson (eds.) *Keywords and Concepts in Evolutionary Developmental Biology*, Cambridge MA: Harvard University Press, 94-97

—. 2004, *Embriology, epigenesis, and evolution. Taking development seriously*, Cambridge: Cambridge University Press

—. 2008, "Taking old ideas seriously: Evolution, development, and human behavior", *New Ideas Psychol.* 26, 387-404

Robert, J.S., B.K. Hall and W.M. Olson 2001, "Bridging the gap between developmental systems theory and evolutionary developmental biology", *BioEssays* 23, 954-962

Rochefort, C., X. He, S. Scotto-Lomassese and C. Scharff 2007, "Recruitment of FoxP2-expressing neurons to area X varies during song development", *Developmental Neurobiology* 67, 809-17

Roe, S.A. 1981, *Matter, Life, and Generation*, Cambridge: Cambridge University Press

Rooryck, J., N.V. Smith, A. Liptak and D. Blakemore 2010, "Editorial introduction to the special issue of *Lingua* on Evans and Levinson's 'The myth of language universals'", *Lingua* 120: 2651-2656

Rosen, G.D. *et al.* 2007, "Disruption of neuronal migration by RNAi of *Dyx1c1* results in neocortical and hippocampal malformations", *Cereb. Cortex.* 17, 2562-2572

Rosenberg, A. 1997, "Reductionism redux: Computing the embryo", *Biol. Philos.* 12, 445-470

Rossi, S., M. Gugler, A.D. Friederici and A. Hahne 2006, "The impact of proficiency on second-language processing of German and Italian: evidence from Event-Related Potentials", *Journal of Cognitive Neuroscience* 18 (2), 2030-2048

Rosinski, R.R., R.M. Golinkoff and K.S. Kukish 1975, "Automatic semantic processing in the picture-word interference task", *Child development* 46, 247-253

Rothstein, S. and A. Treves 2010, "Computational constraints on compositional interpretation: Refocusing the debate on language universals", *Lingua* 120, 2717-2722

Russell, E.S. 1916, *Form and function. A contribution to the history of animal morphology*, London: J. Murray

Sabourin, L. and L.A. Stowe 2008, "Second language processing: When are first and second languages processed similarly?", *Second Language Research* 24, 397-430

Sabourin, L., L.A. Stowe and G.J. de Haan 2006, "Transfer effects in learning a second language grammatical gender system", *Second Language Research* 22, 1-29

Saffran, E.M. 1997, "Aphasia: Cognitive neuropsychological aspects", T.E. Feinberg and M.J. Farah (eds.) *Behavioural neurology and neuropsychology,* New York: Mc Graw-Hill, 151-166

Safir, K. 2010, "How not to find linguistic universals", *Lingua* 120, 2723-2726

Saltzman, I.J. and W.R. Garner 1948, "Reaction time as a measure of span of attention", *Journal of Psychology* 25, 227-241

Sampedro, J. 2002, *Deconstruyendo a Darwin. Los enigmas de la evolución a la luz de la nueva genética*, Crítica: Barcelona

Sampson, G. 2002, "Exploring the richness of the stimulus", *The Linguistic Review* 19, 73-104

Samuels, B. 2009a, "The third factor in phonology", *Biolinguistics* 3, 355-382

—. 2009b, *The structure of phonological theory*, Doctoral Dissertation, Harvard University

—. 2009c, Structure and specification in harmony, *Proceedings of NELS 38 vol 2,* 283-296

—. 2010, "Consequences of Phases for morphophonology". Ms., University of Maryland [Under review for Á. Gallego (ed.) *Phase theory: developing the framework*. Berlin: Walter de Gruyter]

—. 2011, *Phonological architecture: a biolinguistic perspective,* Oxford: Oxford University Press

Samuels, B. and C. Boeckx 2009, "What emerges from Merge in phonology?", Paper presented at OCP 6, Edinburgh

Samuels, B., M.D. Hauser and C. Boeckx *To appear*, "Do animals have Universal Grammar? A case study in phonology", I.G. Roberts (ed.) *The Oxford Handbook of Universal Grammar*, Oxford: Oxford University Press

Sandler, W. 1986, "The spreading hand autosegment of American Sign Language", *Sign Language Studies* 50, 1-28

Santesteban, M. and A. Costa 2006, "Does L1 syntax affect L2 processing? A study with highly proficient early bilinguals", B. Fernández and I. Laka (eds.) *Andolin gogoan. Essays in honour of Professor Eguzkitza*, Bilbao: EHUko argitalpen zerbitzua, 817-834

Sapir, E. 1921, *Language. An introduction to the study of speech*, Harcourt, Brace and Company

Sapp, J. 2003, "Inheritance: Extragenomic", B.K. Hall and W.M. Olson (eds.) *Keywords and Concepts in Evolutionary Developmental Biology*, Cambridge MA: Harvard University Press, 201-209

Saussure, F. de 1916, *Cours de linguistique générale*, Paris: Payot

Schachner, A., T.F. Brady, I.M. Pepperberg and M.D. Hauser 2009, "Spontaneous motor entrainment to music in multiple vocal mimicking species", *Current Biology* 19, 831-836

Schaffner, K.F. 1998, "Genes, behavior, and developmental emergentism: One process, indivisible?", *Philos. Sci.* 65, 209-252

Scharff C. and S. Haesler 2004, "An evolutionary perspective on FoxP2: strictly for the birds?", *Curr. Opin. Neurobiol.* 15 (6), 694-703

Schiller, N.O. 1997, *The role of the syllable in speech production. Evidence from lexical statistics, metalinguistics, masked priming, and electromagnetic midsagittal articulography*, PhD dissertation, Nijmegen University

—. 1998, "The effect of visually masked primes on the naming latencies of words and pictures", *Journal of Memory and Language* 39, 484-507

—. 1999, "Masked syllable priming of English nouns", *Brain and Language* 68, 300-305

—. 2000, "Single word production in English: The role of subsyllabic units during speech production", *Journal of Experimental Psychology: Learning, Memory and Cognition*, 26, 512-528

Schiller, N.O. and A. Costa 2006, "Activation of segments, not syllables, during phonological encoding in speech production", *The Mental Lexicon* 1, 231-250

Schiller, N.O., A. Costa and A. Colomé 2002, "Phonological encoding of single words: In search of the lost syllable", C. Gussenhoven and N. Warner (eds.) *Papers in Laboratory Phonology 7*, Berlin: Mouton de Gruyter, 35-59

Schleicher, A. 1863, *Die Darwinsche Theorie und die Sprachwissenschaft* H. Böhlau: Weimar [Quoted from the 1869 English version *Darwinism Tested by the Science of Language* by J.C. Hotten, included in Koerner, ed., 1-71]

Schroer, R. J. *et al.* 1998, "Autism and maternally derived aberrations of chromosome 15q", *Am. J. Med. Genet.* 76, 327-336

Schusterman, R.J. and D. Kastak 1993, "A California sea lion (Zalophuscalifornianus) is capable of forming equivalence relations", *The Psychological Record* 43, 823-839

Schwartz, A.I., J.F. Kroll and M.Diaz 2007, "Reading words in Spanish and English: mapping orthography to phonology in two languages", *Language and Cognitive Processes* 22, 106-129

Scott, S.K. and R.J. Wise 2004, "The functional neuroanatomy of prelexical processing in speech perception", *Cognition* 92, 13-45

Scoville, W.B. and B. Milner 1957, "Loss of recent memory after bilateral hippocampal lesions", *Journal of Neurology, Neurosurgery and Psychiatry* 20, 11-21

Searle, J.R. 1995, *The construction of social reality*, New York: The Free Press

Sears, C.R., Y. Hino and S.J. Lupker 1995, "Neighborhood frequency and neighborhood size effects in visual word recognition", *Journal of Experimental Psychology: Human Perception and Performance* 21, 876-900

Sebastián-Gallés, N. and L. Bosch 2002, "The building of phonotactic knowledge in bilinguals: The role of early exposure", *Journal of Experimental Psychology: Human, Perception and Performance* 28, 974-989

Sebastián-Gallés, N., S. Echeverría and L. Bosch 2005, "The influence of initial exposure on lexical representation: Comparing early and simultaneous bilinguals", *Journal of Memory and Language* 52, 240-255

Sebastián-Gallés, N. and J.F.K. Kroll 2003, "Phonology in bilingual language processing: Acquisition, perception and production", N.O. Schiller and A.S. Meyer (eds.) *Phonetics and phonology in language comprehension and production*, Berlin: Mouton de Gruyter, 279-317

Sebastián-Gallés, N. and S. Soto-Faraco 1999, "On-line processing of native and non-native phonemic contrasts in early bilinguals", *Cognition* 72, 112-123

Seilacher, A. 1970, "Arbeitskonzept zur Konstruktions-morphologie" *Lethaia* 3, 393-396

Seyfarth, R.M., D.L. Cheney and T.J. Bergman 2005, "Primate social cognition and the origins of language", *Trends in Cognitive Sciences* 9, 264- 266

Shallice, T. 1988, *From neuropsychology to mental structure*, Cambridge: Cambridge University Press

Shao, Y. *et al.* 2003, "Fine mapping of autistic disorder to chromosome 15q11-q13 by use of phenotypic subtypes", *Am. J. Hum. Genet.* 72: 539-548

Shaywitz S., R. Morris and B. Shaywitz 2008, "The education of dyslexic children from childhood to young adulthood", *Annu. Rev. Psychol.* 59: 451-475

Shaywitz, S.E. *et al.* 1998, "Functional disruption in the organization of the brain for reading in dyslexia", *Proc. Nat. Acad. Sci. U.S.A.* 95, 2636-2641

Shettleworth, S.J. 1998, *Cognition, evolution, and behavior*, Oxford: Oxford University Press

Sholtis, S. and K.M. Weiss 2005, "Phenogenetics: genotypes, phenotypes, and variation", B. Hallgrímsson and B.K. Hall (eds.) *Variation,* Amsterdam: Elsevier, 499-523

Shriberg, L.D., K.J. Ballard, J.B. Tomblin, J.R. Duffy, K.H. Odell and C.A. Williams 2006, "Speech, prosody, and voice characteristics of a mother and daughter with a 7,13 translocation affecting *FOXP2*", *J. Speech Lang. Hear. Res.* 49, 500-525

Shriberg, L.D., J.B. Tomblin and J.L. McSweeny 1999, "Prevalence of speech delay in 6-year-old children and comorbidity with language impairment", *J. Speech Lang. Hear. Res.* 42, 1461-1481

Shu, W. *et al.* 2005, "Altered ultrasonic vocalization in mice with a disruption in the *Foxp2* gene", *Proc. Natl. Acad. Sci. U.S.A.* 102, 9643-9648

Shu, W., H. Yang, L. Zhang, M.M. Lu and E.E. Morrisey 2001, "Characterization of a new subfamily of wingedhelix/forkhead (Fox) genes that are expressed in the lung and act as transcriptional repressors", *J. Biol. Chem.* 276, 27488-27497

Shubin, N., C. Tabinand S. Carroll 2009, "Deep homology and the origins of evolutionary novelty", *Nature* 457, 818-823

Sigurðsson, H.Á. 2009, "Language Quarks", *Iberia* 1 (1), 169-183

Simon, H.A. 1962, "The architecture of complexity", *Proceedings of the American Philosophical Society* 106, 467-482 [reprinted in: H.A Simon 1969, *The Sciences of the Artificial.* Cambridge MA: MIT Press, 183-216]

Skinner, B.F. 1957, *Verbal Behavior.* East Norwalk CT: Appleton-Century-Crofts

Slater, P.J.B. 1999, *Essentials of animal behavior*, Cambridge: Cambridge University Press

—. 2000, "Birdsong repertoires: their origins and use", N. Wallin, B. Merker and S. Brown (eds.) *The origins of music*, Cambridge MA: MIT Press, 49-64

SLI Consortium 2002, "A genomewide scan identifies two novel *loci* involved in specific language impairment", *Am. J. Hum. Genet.* 70, 384-398

—. 2004, "Highly significant linkage to the SLI1 *locus* in an expanded sample of individuals affected by specific language impairment", *Am. J. Hum. Genet.* 74, 1225-1238

Smith, J.W. 1977, *The behavior of communicating: An ethological approach*, Cambridge MA: Harvard University Press

Smith, K., and S. Kirby 2008, "Natural selection for communication favors the cultural evolution of linguistic structure", A.D.M. Smith, K. Smith and R. Ferrer i Cancho (eds.) *The evolution of language*, Singapore: World Scientific, 283-290.

Smith, L.B. and E. Thelen 2003, "Development as a dynamic system", *Trends in Cognitive Science* 7, 343-348

Smith, N. 1999, *Chomsky. Ideas and Ideals*, Cambridge: Cambridge University Press

Smith, N. V. and I.M. Tsimpli 1995, *The Mind of a Savant: Language-Learning and Modularity*, Oxford: Blackwell

Smith, S.D. 2007, "Genes, language development, and language disorders", *Ment. Retard. Dev. Disabil. Res. Rev.* 13, 96-105

Smith, S.D., J.W. Gilger and B.F. Pennington 1996, "Dyslexia and other specific learning disorders" D.L. Rimoin, J.M. Connor and R.E. Pyeritz (eds.) *Principles and practice of medical genetics,* New York: Churchill Livingstone, 1767-1789

Sorger, B., R. Goebel, C. Schiltz and B. Rossion 2007, "Understanding the functional neuroanatomy of acquired prosopagnosia", *Neuroimage* 35, 836-852

Spelke, E.S. 1998, "Nativism, empiricism, and the origins of knowledge", *Infant. Behav. Dev.* 21, 181- 200

Spelke, E. 2003, "What makes us smart? Core knowledge and natural language", D. Gentner and S. Goldin-Meadow (eds.) *Language and Mind: Advances in the study of language and thought*, Cambridge MA: MIT Press, 277-311

Spencer, J.P., M.S. Blumberg, B. McMurray, S. Robinson, R. Scott, L.K. Samuelson and J.B. Tomblin 2009, "Short arms and talking eggs: Why we should no longer abide the nativist-empiricist debate", *Child Development Perspectives* 3, 79-87

Sperber, D. 2001, "In defense of massive modularity", D. Emmanuel (ed.) *Language, Brain, and Cognitive Development: Essays in Honor of Jacques Mehler*, Cambridge: MIT Press, 47-57

Spiteri, E. *et al.* 2007, "Identification of the transcriptional targets of *FOXP2*, a gene linked to speech and language, in developing human brain", *American Journal of Human Genetics* 81, 1144-1157

Stein, C.M. *et al.* 2006, "Speech sound disorder influenced by a *locus* in 15q14 region", *Behav. Genet.* 36, 858-868

Stein, J. and V. Walsh 1997, "To see but not to read: the magnocellular theory of dyslexia", *Trends Neurosci.* 20, 147-152

Steinhauer, K., E.J. White and J.E. Drury 2009, "Temporal dynamics of late second language acquisition: evidence from event-related brain potentials", *Second Language Research* 25 (1), 13-41

Stemberger, J.P. 2004, "Neighbourhood effects on error rates in speech production", *Brain and Language* 90, 413-422

Stemberger, J.P. and B. MacWhinney 1986, "Frequency and the lexical storage of regularly inflected forms", *Memory and Cognition* 14, 17-26

Sterelny, K. and P.E. Griffiths 1999, *Sex and Death: An Introduction to the Philosophy of Biology*, Chicago: University of Chicago Press

Stevens, P.F. 1984, "Homology and phylogeny: Morphology and systematics", *Systematic Botany* 9, 395-409

Stich, S. 1975, "The idea of innateness", S. Stich (ed.) *Innate Ideas*, Berkeley: University of California Press, 1-22

Stirling, J. and R. Elliott 2008, *Introducing Neuropsychology,* Hove: Psychology Press

Stokoe, W. 1960, "Sign language structure: an outline of the visual communication systems of the american deaf", *Studies in linguistics. Occasional papers 8,* Silver Spring MD: Linstok Press

Stokoe, W., D. Casterline and C. Croneberg 1965, *A dictionary of American Sign Language on linguistic principles*. Wahsington DC: Gallaudet College Press

Stokoe, W. 1991, "Semantic phonology", *Sign Language Studies* 71, 107-114

Striedter, G.F. 2005, *Principles of Brain Evolution*. Sunderland MA: Sinauer

Stromswold, K. 2001, "The heritability of language: A review and metaanalysis of twin, adoption and linkage studies", *Language* 77: 647-723

Stroop, J. R. 1935, "Studies of interference in serial verbal reactions", *Journal of Experimental Psychology: General* 18, 643-662

Swanson, H.L., C.B. Howard and L. Saez 2006, "Do different components of working memory underlie different subgroups of reading disabilities?", *J. Learn. Disabil.* 39, 252-269

Taft, M. 1991, "Reading and the mental lexicon", Hove: Erlbaum

Taft, M. and M. Radeau 1995, "The influence of the phonological characteristics of a language on the functional units in reading: A study in French", *Canadian Journal of Experimental Psycholgy* 49, 330-346

Tager-Flusberg, H. 2006, "Defining language phenotypes in autism", *Clin. Neurosci. Res.* 6, 219-224

Taipale, M. *et al.* 2003, "A candidate gene for developmental dyslexia encodes a nuclear tetratricopeptide repeat domain protein dynamically regulated in brain", *Proc. Natl. Acad. Sci. U.S.A.* 100, 11553-11558

Takahashi, K., F.C. Liu, K. Hirokawa and H. Takahashi 2003, "Expression of *Foxp2*, a gene involved in speech and language in the developing and adult striatum", *Journal of Neuroscience Research* 73, 61-72

Tallal, P. and M. Piercy 1973, "Defects of nonverbal auditory perception in children with developmental dysphasia", *Nature* 241, 468-469

Tallal, P., R. Sainburg and T. Jernigan 1991, "The neuropathology of developmental dysphasia: behavioral, morphological, and physiological evidence for a pervasive temporal processing disorder", *Read. Writ.* 3, 363-377

Tanebe, H. *et al.* 1987, "Conduction aphasia and arcuate fasciculus", *Acta Neurologica Scandinavica* 76, 422-427

Tattersall, I. 1998, *Becoming Human: Evolution and Human Uniqueness*, Oxford: Oxford University Press

Taub, S. 2001, *Language from the body: iconicity and metaphor in American Sign Language*, Cambridge: Cambridge University Press

Taylor, R.C., B.A. Klein, J. Stein and M.J. Ryan 2008, "Faux frogs: Multimodal signaling and the value of robotics in animal behavior", *Animal Behaviour* 76, 1089-1097

Temple, E. *et al.* 2000, "Disruption of the neural response to rapid acoustic stimuli in dyslexia: evidence from functional MRI", *Proc. Nat. Acad. Sci. U.S.A.* 97, 13907-13912

ten Cate, C, C.A.A. van Heijningen and W. Zuidema 2010, "Reply to Gentner et al.: As Simple As Possible, but Not Simpler", *PNAS* 107, E66-E67

Teramitsu, I. *et al.* 2004, "Parallel FoxP1 and FoxP2 expression in songbird and human brain predicts functional interaction", *Journal of Neuroscience* 24, 3152-3163

Teramitsu I. and S.A. White 2006, "FoxP2 regulation during undirected singing in adult songbirds", *Journal of Neuroscience* 26, 7390-4

Tettamanti, M. *et al.* 2005, "Basal ganglia and language: phonology modulates dopaminergic release", *Brain Imaging* 16, 397-401

Thomas, M. 2002, "Development of the concept of 'the poverty of the stimulus'", *The Linguistic Review* 19, 51-71

Thompson, R., K. Emmorey and T.H. Gollan 2005, "'Tip of fingers' experiences by deaf signers. Insights into the organization of a sign-based lexicon", *Psychological Science,* 16 (11), 856-860

Thompson, R.L., D.P. Vinson and G. Vigliocco 2009, "The link between form and meaning in american sign language: Lexical processing effects", *Journal of Experimental Psychology. Learning, Memory, and Cognition* 352, 550-557

Thompson, R.L., D.P. Vinson and G. Vigliocco *In press*, "The link between form and meaning in British Sign Language: Lexical processing effects in a phonological decision task", *Journal of Experimental Psychology. Learning, Memory, and Cognition.*

Thornton, R. and K. Wexler 1999, *Principle B, Ellipsis, and Interpretation in Child Grammar*, Cambridge MA: MIT Press

Threlkeld, S.W. *et al.* 2007, "Developmental disruptions and behavioral impairments in rats following in utero RNAi of *Dyx1c1*", *Brain Res. Bull.* 71, 508-514

Tinbergen, N. 1951, *The Study of Instinct*, New York: Clarendon Press

—. 1963, "On aims and methods of ethology", *Zeitschrift für Tierpsychologie* 20, 410-433

Todt, D. and H. Hultsch 1998, "How songbirds deal with large amounts of serial information: retrieval rules suggest a hierarchical song memory", *Biological Cybernetics* 79, 487-500

Tomasello, M. 1999, *The Cultural Origins of Human Cognition.* Cambridge MA: Harvard University Press

Tomasello, M. 2009a, *Why we cooperate*, Cambridge MA: MIT Press

—. 2009b, "Universal Grammar is dead", *Behavioral and Brain Sciences*, 32

Toro, J.M., J.B. Trobalón and N. Sebastián-Galles 2005, "The effects of backward speech and speaker variability in language discrimination by rats", *Journal of Experimental Psychology: Animal Behavior Processes* 31, 95-100

Tournell, C.E., R.A. Bergstrom and A. Ferreira 2006, "Progesterone-induced agrin expression in astrocytes modulates glia-neuron interactions leading to synapse formation", *Neuroscience* 141, 1327-1338

Truckenbrodt, H. 1995, "Phonological phrases: Their relation to syntax, focus, and prominence", Cambridge MA: MIT Dissertation

True, J.R. and Carroll, S.B. 2002, "Gene co-option in physiological and morphological evolution", *Annu. Rev. Cell Dev. Biol.* 18, 53-80

Turing, A.M. 1952, "The chemical basis of morphogenesis", *Philosophical Transactions of the Royal Society of London Series B* 237, 37-72

Újhelyi, M. 1996, "Is there any intermediate stage between animal communication and language?", *Journal of Theoretical Biology* 180, 71-76

Ullman, M.T. 2001a, "The declarative/procedural model of lexicon and grammar", *J. Psycholinguist Res.* 30, 37-69

—. 2001b, "The neural basis of lexicon and grammar in first and second language: The declarative/procedural model", *Bilingualism: Language and Cognition* 41, 105-122

—. 2004, "Contributions of memory circuits to language: The declarative/procedural model", *Cognition* 921 (2), 231-270

Underwood, G. 1976, "Semantic interference from unattended printed words", *British Journal of Psychology* 67, 327-338

Uriagereka, J. 1998, *Rhyme and reason: An introduction to minimalist syntax.* Cambridge MA: MIT Press

—. 1999, "Multiple Spell-Out", S.D. Epstein and N. Hornstein (eds.) *Working minimalism*, Cambridge MA: MIT Press, 251-282

—. 2008a, *Syntactic Anchors: On Semantic Structuring*, Cambridge: Cambridge University Press

—. 2008b, "Desperately evolving syntax", A.D.M. Smith, K. Smith and R. Ferrer i Cancho (eds.) *The evolution of language*. Singapore: World Scientific, 331-337

Van der Lely, H.K.J. 2005, "Domain-specific cognitive systems: Insight from Grammatical-SLI", *Trends in Cognitive Sciences* 9, 53-59

van Heijningen, C.A., A. Jos de Visser, W. Zuidema and C. ten Cate 2009, "Simple Rules Can Explain Discrimination of Putative Recursive Syntactic Structures by a Songbird Species", *PNAS* 106 (48), 20538-43

Van Hell, J.G., and T. Dijkstra 2002, "Foreign language knowledge can influence native language performance in exclusively native contexts", *Psychonomic Bulletin and Review* 9, 780-9

Van Heuven, W.J.B., A. Dijkstra and J. Grainger 1998, "Orthographic neighborhood effects in bilingual word recognition", *Journal of Memory and Language* 39, 458-83

Vanhaeren, M. *et al.* 2006, "Middle paleolithic shell beds in Israel and Algeria", *Science* 312, 1785-1788

Vargha-Khadem, F., D.G. Gadian, A. Copp and M. Mishkin 2005, "*FOXP2* and the neuroanatomy of speech and language", *Nat. Rev. Neurosci.* 6, 131-138

Vargha-Khadem, F., K.E. Watkins, K.J. Alcock, P. Fletcher, and R.E. Passingham 1995, "Praxic and nonverbal cognitive deficits in a large family with a genetically transmitted speech and language disorder" *Proc. Natl. Acad. Sci. U.S.A.* 92, 930-933

Vargha-Khadem, F. *et al.* 1998 "Neural basis of an inherited speech and language disorder" *Proc. Nat. Acad. Sci. U.S.A.* 95, 12695-12700

Velayos-Baeza, A., C. Toma, S. Paracchini and A.P. Monaco 2008, "The dyslexia-associated gene KIAA0319 encodes highly N- and O-glycosylated plasma membrane and secreted isoforms", *Hum. Mol. Genet.* 17, 859-871

Verhage, M. *et al.* 2000, "Synaptic assembly of the brain in the absence of neurotransmitter secretion", *Science* 287, 864-869

Vernes, S.C. *et al.* 2006 "Functional genetic analysis of mutations implicated in a human speech and language disorder", *Hum. Mol. Genet.* 15, 3154-3167

—. *et al.* 2008, "A functional genetic link between distinct developmental language disorders", *N. Engl. J. Med.* 359: 2337-45

—. *et al.* 2007, "High-throughput analysis of promoter occupancy reveals direct neural targets of *FOXP2*, a gene mutated in speech and language disorders", *Am. J. Hum. Genet.* 81, 1232-1250

Vigliocco, G., T. Antonini and M.F. Garrett 1997, "Grammatical gender is on the tip of Italian tongues", *Psychological Science* 8, 314-317

Vitevitch, M.S 1997. The neighborhood characteristics of malapropisms. *Language and Speech, 40 (3,* 211-228.

—. 2002, "The influence of phonological similarity neighborhoods on speech production", *Journal of Experimental Psychology: Learning, Memory, and Cognition* 28, 735-747

Vitevitch, M.S. and P. Luce 1998, "When words compete: Levels of processing inspoken word perception", *Psychological Science* 9, 325-329

Vitevitch, M.S. and P. Luce 1999, "Probabilistic phonotactics and neighborhood activation in spoken word recognition", *Journal of Memory and Language* 40, 374-408

Vitevitch, M.S. and E. Rodríguez 2005, "Neighborhood density effect in spokenword recognition in Spanish", *Journal of Multilingual Disorders* 3, 64-73

Vitevitch, M.S. and M.S. Sommers 2003, "The facilitative influence of phonological similarity and neighbourhood frequency in speech production in younger and older adults", *Memory and Cognition* 31, 491-504

Vitevitch, M.S., and M.K. Stamer 2006, "The curious case of competition in Spanish speech production", *Language and Cognitive Processes* 21, 760-770

Volpi, E.V. and L.M. Bridger 2008, "FISH glossary: an overview of the fluorescence in situ hybridization technique", *Biotechniques* 45, 385-386, 388, 390 *passim.*

Wada, J.A. 1949, "A new method for the determination of the side of cerebral speech dominance", *Igaku to Seibutsugaku* 14, 221-222

Waddington, C.H. 1957, *The Strategy of the Genes*, London: Allen and Unwin

Wagner, G.P. 1996, "Homologues, natural kinds and the evolution of modularity", *Amer. Zool.* 36, 36-43

Wagner, G.P. and L. Altenberg 1996, "Complex adaptations and the evolution of evolvability", *Evolution* 50, 967-976

Wagner, M. 2010, "Prosody and recursion in coordinate structures and beyond", *Natural Language and Linguistic Theory* 28, 183-237

Wahlsten, D. 1999, "Single-gene influences on brain and behavior", *Annu. Rev. Psychol.* 50, 599-624

Walker, E. (ed.) 1978, *Explorations in the Biology of Language*, Montgomery VT: Bradford Books

Walsh, D. 2007, "Development: three grades of ontogenetic involvement", M. Matthen and C. Stephens (eds.) *Handbook to the Philosophy of Science: Vol 3, Philosophy of Biology*, Amsterdam: North-Holland, 179-199

Walter, M.A. 2007, *Repetition avoidance in human language*. Doctoral Dissertation, MIT

Wartenburger I., H.R. Heekeren, J. Abutalebi, S.F. Cappa, A. Villringer and D. Perani 2003, "Early Setting of Grammatical Processing in the Bilingual Brain", *Neuron* 37, 159-170

Watkins, K.E. *et al.* 2002a, "Behavioural analysis of an inherited speech and language disorder: comparison with acquired aphasia", *Brain* 125, 452-64

Watkins, K.E. *et al.* 2002b, "MRI analysis of an inherited speech and language disorder: structural brain abnormalities" *Brain* 125, 465- 478

Webb D.M. and J. Zhang 2005, "FoxP2 in song-learning birds and vocal-learning mammals", *J Hered.* 96 (3), 212-6

Weber, B. and D.J. Depew 2001, "Developmental systems, Darwinian evolution, and the unity of science", S. Oyama, P.E. Griffiths and R.D.

Gray (eds.) *Cycles of Contingencies. Developmental Systems and Evolution*, Cambridge MA: MIT Press, 239-253

Weber-Fox, C. and H. Neville 1996, "Maturational Constraints on Functional Specializations for Language Processing: ERP and Behavioural Evidence in Bilingual Speakers", *Journal of Cognitive Neuroscience* 8 (3), 231-256

Webster G. and B. Goodwin 1996, *Form and Transformation:* Generative *and Relational Principles in Biology,* Cambridge: Cambridge University Press

Wechsler, D. 1958, *The measurement and appraisal of adult intelligence,* Baltimore MD: Williams and Wilkins

Werker, J.F. and A. Voloumanos 2000, "Language: Who's got rhythm?", *Science* 288, 280-281

West-Eberhard, M.J. 2003, *Developmental plasticity and evolution,* Oxford: Oxford University Press

—. 2005, "Developmental plasticity and the origin of species differences", *Proc. Natl. Acad. Sci. U.S.A* Sup. 102, 6543-6549

Wexler, K. 1990, "Innateness and maturation in linguistic development", *Developmental Psychobiology* 23, 645-660

—. 1999, "Maturation and growth of grammar", W.C. Ritchie and T.K. Bhatia (eds.) *Handbook of Language Acquisition*, San Diego: Academic Press, 55-109

White, L., 2003, "Fossilization in steady state L2 grammars: persistent problems with inflectional morphology", *Bilingualism: Language and Cognition* 6, 129- 141

White, S.A. 2010, "Genes and vocal learning", *Brain Lang.* 115: 21-28

White, S. and I. Teramitsu 2006, "FoxP2 regulation during unidirected singing in adult songbirds", *Journal of Neuroscience* 26, 7390-7394

Wilbur, R. 1993, "Syllables and segments: Hold the movement and move the holds", G. Coulter (ed.) *Phonetics and phonology, Vol. 3: Current issues in ASL phonology*, San Diego CA: Academic Press. 135-166

Williams, G.C. 1992, *Natural Selection: Domains, Levels and Challenges,* Oxford University Press: Oxford

Williams, H. and K. Staples 1992, "Syllable chunking in zebra finch (Taeniopygiaguttata) song", *Journal of Comparative Psychology* 106, 278-286

Williams, J. and M.C. O'Donovan 2006, "The genetics of developmental dyslexia", *Eur. J. Hum. Genet.* 4, 681-689

Willmes, K. and K. Poeck 1993, "To what extent can aphasic syndromes be localized?", *Brain* 116, 1527-1540

Wilshire, C.E., and J.L. Nespoulous 2003, "Syllables as units in speech production: Data from aphasia", *Brain and Language* 84, 424-447

Wimsatt, W.C. 1999, "Generativity, entrenchment, evolution, and innateness: Philosophy, evolutionary biology, and conceptual foundations of science", V. Hardcastle (ed.) *Where Biology Meets Psychology. Philosophical Essays*. Cambridge MA: MIT Press, 139-179

Winterer, G. and D. Goldman 2003, "Genetics of human prefrontal function", *Brain Res. Rev.* 43, 134-163

Wise, R.J. *et al.* 2001, "Identifying separate neural sub-systems within Wernicke's area", *Brain* 124, 83-95

Wright, A.A., J. Rivera, S.H. Hulse, M. Shyan and J. Neiworth 2000, "Music perception and octave generalization in rhesus monkeys", *Journal of Experimental Psychology: General* 129, 291-307

Yang, C.D. 2002, *Knowledge and Learning in Natural Language* Oxford: Oxford University Press

—. 2004a, "Universal Grammar, statistics or both?" *Trends in Cognitive Sciences* 8, 451-456

—. 2004b, "Toward a theory of language growth", L. Jenkins (ed.) *Variation and Universals in Biolinguistics*, Amsterdam: Elsevier, 37-56

—. 2005, "On productivity", *Linguistic Variation Yearbook* 5, 265-302

—. 2006, *The Infinite Gift. How Children Learn and Unlearn the Languages of the World*, New York: Scribner

—. 2009, "Who's Afraid of George Kingsley Zipf?", ms., University of Pennsylvania

—. 2010, "Three factors in language variation", *Lingua* 120, 1160-1177.

Yin, H.H., M.I. Davis, J.A. Ronesi and D.M. Lovinger 2006, "The role of protein synthesis in striatal long-term depression", *J. Neurosci.* 26, 11811-11820

Yip, M. 2006a, "Is there such a thing as animal phonology?", E. Baković, J. Ito and J. McCarthy (eds.) *Wondering at the natural fecundity of things: studies in honor of Alan Prince*, Santa Cruz CA: Linguistics Research Center

—. 2006b, "The search for phonology in other species", *Trends in Cognitive Sciences* 10, 442-445

Yu, G., T. Zerucha, M. Ekker and J.L.R. Rubenstein 2001, "Evidence that GRIP, a PDZ-domain protein which is expressed in the embryonic forebrain, coactivates transcription with DLX homeodomain proteins", *Brain Res. Dev. Brain Res.* 130, 217-230

Zawiszewski A., E. Gutierrez, B. Fernández and I. Laka 2011, "Language distance and non-native syntactic processing: evidence from event-related potentials", *Bilingualism: Language and Cognition* 14 (3), 400–411

Ziegler J., M. Muneaux and J. Grainger 2003, "Neighborhood effects in auditory word recognition: Phonological competition and orthographic facilitation", *Journal of Memory and Language* 48, 779-793

Zilhão, J. *et al.* 2010, "Symbolic use of marine shells and mineral pigments by Iberian Neandertals", *Proceedings of the National Academy of Sciences USA*, 107(3), 1023-8

Zondervan, K.T. and L.R. Cardon 2007, "Designing candidate gene and genome-wide case-control association studies", *Nature Protocols* 2, 2492-2501

CONTRIBUTORS

Sergio Balari is Professor of General Linguistics at the Universitat Autònoma de Barcelona since 1993 and a fellow of the Centre de Lingüística Teòrica (UAB). He previously visited the Departments of Linguistics of the City University of New York and of Stanford University, and the Department of Computational Linguistics of the University of the Saarland in Saarbrücken (Germany), where he carried out his research for his PhD under the supervision of Prof. Hans Uszkoreit. During the early 1990s he was mostly involved in computational linguistics projects, but he steadily moved towards the study of the biological foundations of language, a topic on which he has published articles in such journals as *Biolinguistics*, *Biological Theory* or the *International Journal of Evolutionary Biology,* and in the *Cambridge Handbook of Biolinguistics* to be published by Cambridge University Press. He is also the coauthor (with Anna Gavarró) of a brief introductory handbook on biolinguistics.

Cristina Baus is a post-doctoral researcher at the *Speech Production and Bilingualism* research group at the Universitat Pompeu Fabra (Barcelona). Her main research interest is the neural correlates of lexical access during bilingual language production.

Antonio Benítez-Burraco, PhD in Biology and in Linguistics, is Assistant Professor at the Universidad de Huelva. His main research interest is the genetic and molecular basis of language and its evolution. He has published on these issues in edited books and reviews (*International Journal of Evolutionary Biology*, *The Cambridge Handbook of Biolinguistics*, eds. C. Boeckx & K. Grohmann, *The Biolinguistic Turn*, eds. J. Rosselló & J. Martín), and is the author of *Genes y Lenguaje: aspectos ontogenéticos, filogenéticos y cognitivos* (Barcelona, Reverté, 2009).

Cedric Boeckx is Research Professor at the Catalan Institute for Advanced Studies (ICREA), and a member of the Linguistics Department at the Universitat de Barcelona. He is the author of numerous books, including *Islands and Chains* (2003), *Linguistic Minimalism* (2006), *Understanding Minimalist Syntax* (2007), *Bare Syntax* (2008), and *Language in Cognition*

(2009); and the founding co-editor, with Kleanthes K. Grohmann, of the Open Access journal *Biolinguistics*.

Manuel Carreiras is an Ikerbasque Research Professor and Scientific Director of the *Basque Center on Cognition, Brain and Language*. His main research topics are related to reading, bilingualism, visual word recognition, sentence processing, sign language and neurodegeneration of language.

Fernando Cuetos is Full Professor of Psychology at the University of Oviedo and Head of the Cognitive Neuroscience Research Group. He has published extensively in scientific journals (*Journal of Cognitive Neuroscience*, *Cortex*, *Brain and Language*, *Cognition*, *Neuropsychologia*), and is the author of several books, including *Neurociencia del Lenguaje* (2011), *Psicología de la Lectura* (2008), *Psicolingüística del Español* (1999), and *Evaluación y Rehabilitación de las Afasias* (1998).

María del Carmen Horno-Chéliz is Professor of General Linguistics at the University of Zaragoza. Her research has focused on the syntax/semantics interface. In 2001 she presented her PhD dissertation about the syntactic and semantic foundations of prepositional arguments (partially published one year after as *Lo que la preposición esconde*). At present she is interested in the nature of the lexicon and its relationships with syntactic computation, and with memory and cognition.

Itziar Laka is Full Professor at the Department of Linguistics and Basque Studies and director of *The Bilingual Mind* research group at the University of the Basque Country. She graduated in 1990 from MIT with a dissertation entitled *Negation in Syntax: On the Nature of Functional Categories and Projections*, published by Garland in 1994. Her current research combines theoretical linguistics and experimental methods to study the representation and processing of variable/invariable aspects of linguistic structure in bilinguals of Basque and Spanish. She is the author of *A Brief Grammar of Euskara, the Basque Language* (1996), freely available on the web, and of a number of papers on theoretical and experimental linguistics on ergativity, argument structure, and syntactic processing.

Víctor M. Longa is Professor of Linguistics at the University of Santiago de Compostela. His main research interests focus on the Minimalist Program as a biologically feasible theory of language evolution and development. As regards language development, he is also interested in the application of theoretical models from developmental biology like Development Systems Theory. He has published on these issues in the journals *Biolinguistics*, *Folia Linguistica*, *Journal of Anthropological Sciences*, *Lingua*, *Linguistics*, or *Theoretical Linguistics*. He is co-author of chapters in *The Oxford Handbook of Linguistic Minimalism* (Oxford University Press, 2011) and *The Handbook of Biolinguistics* (Cambridge University Press, to appear), and co-author of *Homo Loquens. Biología y evolución del lenguaje* (Lugo, Tris Tram, 2003).

Guillermo Lorenzo is Professor at the University of Oviedo, where he teaches Linguistics. He is the author of several articles on the development and evolution of language in specialized journals like *Biolinguistics*, *Biological Theory*, *Linguistics* or *Lingua*. He has also published two books on Biolinguistics in Spanish: *Homo Loquens. Biología y evolución del lenguaje* (2003), with Víctor M. Longa, and *El vacío sexual, la tautología natural y la promesa minimalista* (2006). Since 2007 he coordinates an interuniversity group of researchers in the field of Biolinguistics funded by the Spanish Government and the European Union.

Txuss Martín is a post-doctoral researcher at the Biolinguistic Initiative at the University of Barcelona. He got his PhD in Linguistics at New York University, with a dissertation entitled "Deconstructing Catalan Object Clitics", written under the supervision of Professor Richard Kayne. His research interests involve the syntax-semantics interface, and the nature and evolution of the human faculty of language. He has worked on these topics, where he authored several published papers, among them "The internal structure of clitics", and "Another linguistics is possible". He's also the co-editor, with Joana Rosselló, of a volume entitled *The Biolinguistic Turn*.

José-Luis Mendívil-Giró is Professor of General Linguistics at the University of Zaragoza. His research has focused on the theory of grammar and the nature and extent of linguistic change and diversity. He was the main researcher of the project *Typology and internal variation of the case & agreement systems in the world's languages*, and is the author of *Gramática natural. La gramática generativa y la tercera cultura* (Machado Libros, Madrid, 2003), and *Origen, evolución y diversidad de las lenguas. Una aproximación biolingüística* (Peter Lang, Frankfurt, 2009).

Joana Rosselló is Professor of Linguistics at the Universitat de Barcelona. She is mainly interested in the biocognitive foundations of human language. She first worked on the theory of syntax *tout court* (*pro-drop* parameter specially) and, transitorily, on first language acquisition. More recently, keeping on with syntactic research (*Journal of Portuguese Linguistics*, 7-2, 109-130, 2008), she enlarged her focus by exploring the components that enter the faculty of language from an evolutionary point of view ("Combinatorial properties at the roots of language") and the structure of language in the gestural-visual modality ("On sloppy readings, ellipsis and pronouns: Missing arguments in Catalan Sign Language (LSC) and other argument-drop languages").

Bridget Samuels is currently Scientific Editor & Coordinator in the Division of Chemistry & Chemical Engineering at the California Institute of Technology. She is the author of *Phonological Architecture: A Biolinguistic Perspective* (Oxford University Press, 2011) and numerous other publications on phonology, the evolution of language, and their relationship.

INDEX